STELLAR ROTATION

STELLAR ROTATION

PROCEEDINGS OF THE IAU COLLOQUIUM
HELD AT THE OHIO STATE UNIVERSITY, COLUMBUS, O.,
U.S.A., SEPTEMBER 8–11, 1969

Edited by

ARNE SLETTEBAK
The Ohio State University, Columbus, O., U.S.A.

D. REIDEL PUBLISHING COMPANY

DORDRECHT-HOLLAND

Library of Congress Catalog Card Number 76–118131

ISBN-13:978-94-010-3301-5   e-ISBN-13:978-94-010-3299-5
DOI: 10.1007/978-94-010-3299-5

Dedicated to the memory of

Armin J. Deutsch

ARMIN J. DEUTSCH

1918–1969

DEDICATION

It is sadly appropriate that Armin Deutsch's last public remarks should be those that brought the Columbus Symposium on Stellar Rotation to a close. Sad, because, at the time of his premature death one month later, he was in the midst of fruitful scientific pursuits. Appropriate, because, in his role as final speaker – a role he filled so admirably on many occasions – his colleagues from all parts of the world had one last opportunity to express their esteem for him and for his contributions to astronomy. Though his scientific interests were catholic, the problems of stellar rotation, especially those related to the peculiar A-type stars, constituted a persistent theme to which Armin Deutsch returned throughout his career.

In 1945 Deutsch became intrigued by the enigmatic spectrum variations of certain peculiar A-type stars. In his thesis, *A Study of the Spectrum Variables of Type A*, he described a number of new variables that he had discovered with the 40-inch refractor at Yerkes Observatory, determined periods for several of them, and compiled a first catalogue of such objects. Simultaneous with this work Horace Babcock at Mt. Wilson had begun to discover variable magnetic fields in these same stars. Within a few years Babcock had found magnetic periodicities for several stars; the periods of the magnetic variations and spectrum variations were identical. It was Babcock who first considered that the magnetic variations might be caused by the rotation of an inclined magnetic dipole, and shortly thereafter Stibbs worked out the observational consequences of this model. However, it was Armin Deutsch who saw in this idea an explanation for *all* of the observed characteristics of a magnetic-spectrum variable. Deutsch's convictions were bolstered by his discovery of an inverse relation between the periods of Ap stars and the widths of their absorption lines – a relation that could be understood quantitatively if the periods were rotational periods. In a report to Commissions 27 and 29 of the Eighth General Assembly of the IAU in September 1952, Deutsch concluded that "in the spectrum variables we observe the rotation of A stars that exhibit intensely magnetic areas within which the peculiar line strengths are produced... On this model it is possible to give at least a qualitative explanation of the radial velocities..., of the doubling of certain lines..., of the non-zero mean magnetic field..., and of the 'cross-over' effect observed by Babcock...".

The spectroscopic patches advocated by Deutsch provoked debate that persisted for nearly two decades. Were the patches due to surface nuclear reactions, to paramagnetic migration, to selective radiation pressure, or to what? Do they really exist? Putting aside questions of the physical processes that were involved, Deutsch went on to devise a method for mapping the magnetic fields and abundances of the elements on his rotating stars by use of the observed variations in equivalent width, radial velocity, and effective magnetic field. He presented a preliminary report on his method as applied to HD 125 248 in 1956. The method has since been employed in analyses of α^2CVn by Diane Pyper and of HD 173 650 by John Rice; a detailed description of the method appeared posthumously in a recent issue of *The Astrophysical Journal*. Although Deutsch felt that the evidence in favor of the rotator model for Ap stars was overwhelming, he was the first to admit that it was circumstantial; to this day a crucial test of the model has not been devised. That so much circumstantial evidence now exists is due in large measure to the adrenalin, pro and con, that flowed in others because of Deutsch's ideas, and this is perhaps the best measure of his stature as an astronomer.

At the time of his death Armin Deutsch was busily engaged in other studies of stellar rotation: the bimodal distribution of rotational velocities for upper main sequence stars, discussed first in 1965 and most recently in this volume; the large rotational velocities of the blue stragglers in old open clusters and their implications for rapid rotation of the interiors of solar-type stars; and, finally, the variable Ca II K_2 and K_3 components of M-type giants that he had come to believe were due to transient chromospheric activity modulated by stellar rotation. Armin Deutsch has left these problems for others to carry to completion, and he has left us. We will miss him.

GEORGE W. PRESTON

PREFACE

The International Astronomical Union Colloquium on Stellar Rotation was held at the Ohio State University in Columbus, Ohio, U.S.A. from September 8th through 11th, 1969. Forty-four scientists from Argentina, Belgium, Canada, England, Finland, East and West Germany, Italy, Israel, Japan, The Netherlands, and the United States attended and participated in the Colloquium. The present volume, which parallels the actual program closely, contains the papers presented at the Colloquium plus most of the discussion following those papers.

The Colloquium was sponsored by the International Astronomical Union, the Ohio State University, and the National Science Foundation. It is a pleasure to record my thanks to these organizations and especially to Dr. Geoffrey Keller, Dean of the College of Mathematics and Physical Sciences of the Ohio State University, and to Prof. C. de Jager, Assistant General Secretary of the International Astronomical Union, for their kind cooperation.

I am also grateful to H. A. Abt, J. Hardorp, R. P. Kraft, Mrs. A. Massevitch, M. Plavec, I. W. Roxburgh, and E. Schatzman of the Organizing Committee, as well as A. J. Deutsch and G. W. Collins, II, for their help in planning the Colloquium – all of them offered valuable suggestions toward organizing the program.

The program was opened by Dr. Novice G. Fawcett, President of the Ohio State University, who welcomed the participants. Chairmen of the individual sessions were D. N. Limber, R. P. Kraft, J. Hardorp, C. Jaschek, G. W. Preston, J. P. Ostriker, and R. Steinitz. A. J. Deutsch delivered the concluding remarks.

I should also like to acknowledge the cooperation and assistance of R. Bottemiller, F. Damm, R. Wright, and J. Baumert, who helped with many of the arrangements, as did Mrs. Delores Chambers, Mrs. Patricia Mouser, and especially, my wife.

The photograph of Armin J. Deutsch which appears in this volume was chosen by his wife and contributed by Miss Henrietta Swope.

ARNE SLETTEBAK

TABLE OF CONTENTS

PART III / STELLAR ROTATION IN BINARIES, CLUSTERS, AND SPECIAL OBJECTS. STATISTICS OF STELLAR ROTATION

PART IV / THE ROTATION OF THE SUN

ORGANIZING COMMITTEE

H. A. Abt, Kitt Peak National Observatory, U.S.A.

J. Hardorp, State University of New York at Stony Brook, U.S.A.

R. P. Kraft, University of California at Santa Cruz, U.S.A.

A. Massevitch, U.S.S.R. Academy of Sciences, U.S.S.R.

M. Plavec, Ondřejov Observatory, Czechoslovakia

I. W. Roxburgh, Queen Mary College, England

E. Schatzman, Institut d'Astrophysique, France

A. Slettebak, Ohio State University, U.S.A.

LIST OF PARTICIPANTS

H. A. Abt, Kitt Peak National Observatory, U.S.A.

P. L. Bernacca, Osservatorio Astrofisico, Asiago, Italy

W. Buscombe, Northwestern University, U.S.A.

R. C. Cameron, Goddard, NASA, U.S.A.

M. J. Clement, University of Toronto, Canada

G. W. Collins, II, Ohio State University, U.S.A.

A. Cowley, University of Michigan, U.S.A.

D. L. Crawford, Kitt Peak National Observatory, U.S.A.

A. J. Deutsch, Hale Observatories, U.S.A.

R. H. Dicke, Princeton University, U.S.A.

B. Durney, High Altitude Observatory, U.S.A.

S. M. Faber, Harvard College Observatory, U.S.A.

G. G. Fahlman, University of Toronto, Canada

K. J. Fricke, Göttingen University, Germany

J. Hardorp, State University of New York at Stony Brook, U.S.A.

S. Jackson, Joint Institute for Laboratory Astrophysics, U.S.A.

C. Jaschek, University of La Plata, Argentina

P. R. Jordahl, University of Texas, U.S.A.

D. A. Klinglesmith, Goddard, NASA, U.S.A.

R. P. Kraft, University of California at Santa Cruz, U.S.A.

D. N. Limber, University of Virginia, U.S.A.

W. C. Livingston, Kitt Peak National Observatory, U.S.A.

J. W.-K. Mark, Massachusetts Institute of Technology, U.S.A.

D. Marks, University of Michigan, U.S.A.

K. Nariai, Goddard, NASA, U.S.A.

M. D. T. Naylor, University of Toronto, Canada

E. C. Olson, University of Illinois, U.S.A.

J. P. Ostriker, Princeton University Observatory, U.S.A.

G. W. Preston, Hale Observatories, U.S.A.

J. D. Rosendhal, University of Wisconsin, U.S.A.

I. W. Roxburgh, Queen Mary College, England

G. Ruben, Potsdam Astrophysical Observatory, D.D.R.

T. Sakurai, Kyoto University, Japan

C. S. Selley, University of Manchester, England

R. Simon, Université de Liège, Belgium

A. Slettebak, Ohio State University, U.S.A.

S. Sobieski, Goddard, NASA, U.S.A.

R. Steinitz, Tel-Aviv University, Israel
T. R. Stoeckley, Michigan State University, U.S.A.
H.-C. Thomas, Max Planck Institute, Munich, Germany
I. V. Tuominen, University of Helsinki, Finland
E. P. J. van den Heuvel, Sonnenborgh Observatory, Utrecht,
   The Netherlands
L. R. Wackerling, Northwestern University, U.S.A.
G. A. E. Wright, University of Manchester, England

*The following were not able to attend the Colloquium but were authors
or co-authors of papers which were read by colleagues:*

S. P. S. Anand, University of Toronto, Canada
I. J. Danziger, Harvard College Observatory, U.S.A.
R. E. Gershberg, Crimean Astrophysical Observatory, U.S.S.R.
J. B. Hutchings, Dominion Astrophysical Observatory, Canada
R. Kippenhahn, Universitäts-Sternwarte Göttingen, Germany
L. Mestel, University of Manchester, England
E. Meyer-Hofmeister, Max Planck Institute, Munich, Germany
I. Okamoto, International Latitude Observatory, Iwate, Japan
M. Plavec, Ondřejov Observatory, Czechoslovakia
D. M. Pyper, San Francisco, California, U.S.A.
I. J. Sackmann, Göttingen Observatory, Germany
P. A. Strittmatter, Institute of Theoretical Astronomy,
   Cambridge, England
O. Vilhu, University of Helsinki, Finland

PART I

THE EFFECTS OF ROTATION ON STELLAR INTERIORS AND EVOLUTION

OBSERVED ROTATIONAL VELOCITIES OF SINGLE STARS

(Review Paper)

ARNE SLETTEBAK

Perkins Observatory, The Ohio State and Ohio Wesleyan Universities, U.S.A.

Abstract. Mean observed rotational velocities for single, normal, main-sequence stars are reviewed and compared with the mean observed $v \sin i$'s for giant and supergiant stars, Be stars, peculiar A-type and metallic-line stars, and Population II objects.

One advantage of helping to organize a conference such as this one is that one can do some fairly outrageous things; such as scheduling oneself to give the very first paper on the program. I hope that you will excuse my presumption on the grounds that a brief overview of the observations of rotations of single stars would be both appropriate and useful before we become involved in the many facets of stellar rotation on our program for the next $3\frac{1}{2}$ days. My remarks will, in any case, be brief and can, in fact, be summarized in the two figures which accompany this text.

Figure 1 illustrates some recent determinations of mean observed rotational velocities for stars on the main sequence. The means given by the different investigators

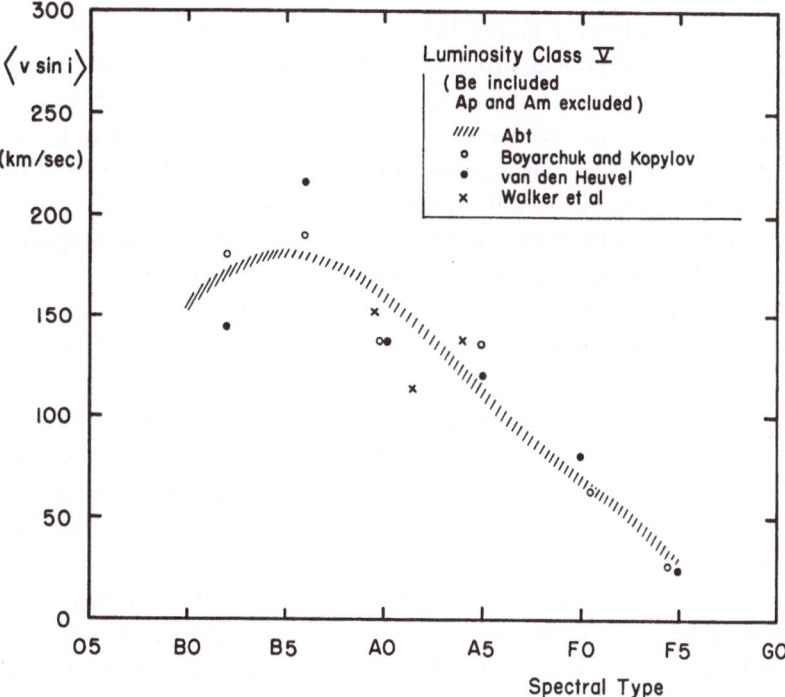

Fig. 1.   The distribution of mean observed rotational velocities with spectral type for single, normal, main-sequence stars.

A. Slettebak (ed.), Stellar Rotation, 3–8. All Rights Reserved

listed on the figure have observed rotational velocities in common and are therefore not necessarily independent, as will be pointed out. In all cases Be stars have been included in the statistics under the assumption that they may be regarded as the most rapidly rotating B-type stars, while Ap and Am stars have been excluded on the grounds that their abnormally low observed rotational velocities may represent later modifications.

The open circles in Figure 1 indicate mean rotational velocities for a total of 485 normal B1-G0 main-sequence stars from the catalogue of Boyarchuk and Kopylov (1964). The means were obtained from Table 2 of their statistical paper (Boyarchuk and Kopylov, 1958) by grouping their data as follows: B1-3, B5-7, B8-A2, A3-7, A9-F2, F3-6, F7-G0. The filled circles are from the curve of mean observed rotational velocities vs. spectral types of Van den Heuvel (1968), which was derived using the values in the Boyarchuk-Kopylov (1964) catalogue plus Kraft's (1965) measurements of stars in the Hyades and Coma clusters and Slettebak's (1966a) measurements of B6-B9e stars. The crosses represent means for 359 normal B9-A5 main-sequence stars from the catalogue of Palmer *et al.* (1968), as listed in the statistical paper by Walker (1965). Finally, the cross-hatched continuous curve is due to Abt (Abt and Hunter, 1962) – you have seen it many times, in his papers on stellar rotation in clusters, as representing the field stars. The curve is derived from the observed rotational velocities of Herbig and Spalding (1955), Slettebak (1954, 1955) and Slettebak and Howard (1955).

An inspection of Figure 1 shows that the various symbols are in fair agreement with Abt's curve. The high mean found by Van den Heuvel at B6 probably represents a disproportionate weighting in favor of B6-7e stars relative to Be stars of earlier type. The low mean at A1.5 from Walker's (1965) paper may be due partly to the admixture of Ursa Major stream stars, as was suggested by Abt and Hunter (1962) to explain the dip at A2 in the averages of Boyarchuk and Kopylov (1958).

The data in Figure 1 have been grouped in order to smooth out irregularities and show the broad features of rotation of single, normal, main-sequence stars. It is interesting to compare this illustration with Figure 3 in the paper by Abt (this volume, p. 193) in which the maxima and minima in the curve have been retained. Whether one believes in the reality of these irregularities or not, certain general features do stand out: rotation increases from very low values in the F-type stars to some kind of maximum in the B-type stars. The exact behavior of the curve as it moves on into the region of the early B-type and O-type stars is still quite uncertain since we do not know the relative importance of rotation and macroturbulence as line-broadening agents in these objects.

Several attempts to explain the observed distribution of rotational velocities for main-sequence stars of different spectral types have been made. The work of Schatzmann (1962), Roxburgh (1964a, b), Dicke (1964), Huang (1965), and Kraft (1968) should especially be mentioned in this respect.

Figure 2 shows Abt's (Abt and Hunter, 1962) curve for main-sequence stars again, plus mean rotational velocities for a number of other types of objects. The region of the O-type stars is shown, with values of the mean rotational velocities (Slettebak, 1956) marked with crosses. None of the O5-O8 stars measured in that sample show really sharp lines, indicating that macroturbulence must play an important role in the line broadening. The relative importance of rotation in these stars is therefore quite uncertain, as has been mentioned. A further uncertainty lies in the difficulty of luminosity classification for the O-type stars.

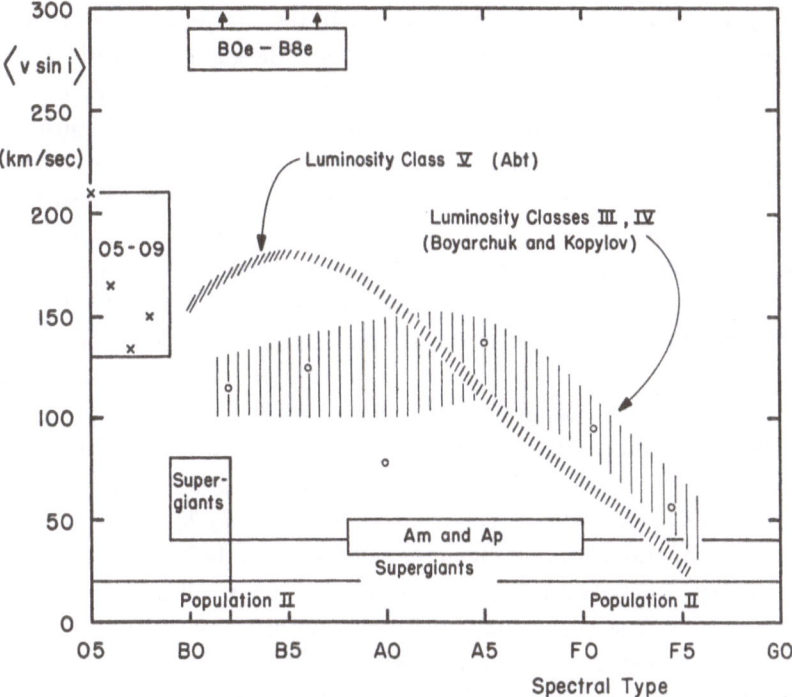

Fig. 2.   Mean observed rotational velocities for a number of different classes of stars as compared with normal main-sequence stars.

The open circles in Figure 2 represent mean rotational velocities for 246 normal luminosity class III and IV stars from the paper of Boyarchuk and Kopylov (1958), arranged in the following spectral-type groups: B1-3, B5-7, B8-A2, A3-7, A9-F2, F3-6. The circles (except for the point at A0) are connected by a broad cross-hatched band, suggesting the uncertainties in the means. The curve indicates that early-type giants appear to rotate more slowly than main-sequence stars of corresponding type, that giants and main-sequence stars rotate with comparable velocities in the middle A's, and that giants rotate more rapidly than their main-sequence counterparts in the late A- and F-types. This behavior of the giant relative to the main-sequence rotation curve can be interpreted as an evolutionary effect: F-type giants, which evolved from rapidly rotating B- and A-type main-sequence stars, still show more rotational broadening than the intrinsically slowly rotating main-sequence stars of corresponding type. Several investigators, including Oke and Greenstein (1954), Sandage (1955), Abt

(1957, 1958), and Kraft (1966), have discussed stellar rotation along evolutionary tracks, and you will hear several papers on this subject at this Colloquium.

The very low point at A0 in the giant curve can probably be interpreted in terms of selection effects. The classification of stars into luminosity classes near A0 is very difficult. In addition to differences in the strength of the Balmer lines (which is not a very sensitive criterion in the luminosity range III–V), one looks for enhanced lines of Fe II, Si II, and other ions in the giants. The latter lines, which appear weak on low-dispersion spectrograms, are more easily visible in the sharp-lined stars than in rotating stars and therefore one is more likely to associate the giant stars with low rotation.

The supergiant stars are shown schematically near the bottom of Figure 2. Super-giants of all spectral types have been shown by a number of investigators never to show conspicuous line broadening. On the other hand Huang and Struve (1954) first pointed out that while supergiant stars have relatively small values of $v \sin i$, none has a zero velocity, suggesting the presence of a broadening agent with spherical symmetry, such as macroturbulence. Attempts to distinguish between rotation and macrotur-bulence in these stars have been made by Huang and Struve (1953), Slettebak (1956), Abt (1958), Kraft (1966), Rosendhal (this volume, p. 122), and others, but the problem is a difficult one.

The Be stars are shown separately in Figure 2, with arrows indicating that their mean rotational velocities are in reality larger than shown. The position of the Be box derives from measures of 40 B6-8e stars (Slettebak, 1966a) and 47 B0-5e stars (Boyarchuk and Kopylov, 1964), averaging 278 km/sec and 284 km/sec, respectively, but the inclusion of gravity darkening in the calibration of the $v \sin i$'s (Slettebak, 1949, 1966b; Friedjung, 1968; Hardorp and Strittmatter, 1968) will tend to increase these means significantly. The corresponding 'true' equatorial rotational velocities, com-puted under the assumption that the axes of rotation are randomly distributed in space, are very close to the calculated velocities such stars should have for equatorial 'break-up'. Thus, the original picture of Be stars as rotationally-distorted B-type stars with equatorial gas rings, proposed by Struve in 1931, is still supported by the present data.

Mean rotational velocities for the peculiar A-type stars and metallic-line stars are shown in the box near the bottom of Figure 2, centered at about 40 km/sec (Slettebak, 1955). It may be seen that the mean rotational velocities for these objects are consider-ably smaller than the means for normal stars of corresponding spectral types. The reasons for this have been discussed by Abt (1961, 1965a, b), Deutsch (1958, 1965), Babcock (1958), Steinitz (1964), Kraft (1968), and others, and you will hear more about these matters at this Colloquium.

The box at the very bottom of Figure 2 indicates schematically that Population II stars have in general small rotation, a result due originally to Greenstein (1960). I have examined slit spectra at moderate dispersion of early-type halo population objects at both the north and south galactic poles, including O- and B-type subdwarfs, horizon-tal-branch stars, and F- and G-type subdwarfs, and have found no evidence of ro-tational line broadening (i.e., no values of $v \sin i$ higher than about 40 km/sec) for these stars. Greenstein (1969) has examined coudé spectrograms of several horizontal-branch stars and finds their lines to appear sharp even at high dispersion.

Many interesting classes of objects (i.e., pre-main sequence stars, pulsating stars, etc.) have been omitted in Figure 2, and of course only single stars have been considered; observations of axial rotation in double stars and in stars in clusters will be discussed later at this conference.

In conclusion, I would like to point out that, with the exception of the supergiants and the O-type stars, in which the precise role of rotation vs. macroturbulence is not yet clear, rotational velocities can be estimated from spectrograms with an accuracy of 10 to 15%. Obtaining accurate spectral types and luminosity classes so that meaningful correlations can be made continues to be a problem, however, especially in view of the fact that rapidly rotating stars are more difficult to classify than sharp-lined stars. The use of photometric indices should help here, but rotation will also affect the colors of the stars, as several investigators have shown. We will hear more about these matters in the papers to follow.

References

Abt, H. A.: 1957, *Astrophys. J.* **126**, 503.
Abt, H. A.: 1958, *Astrophys. J.* **127**, 658.
Abt, H. A.: 1961 *Astrophys. J. Suppl.* **6**, 37.
Abt, H. A.: 1965a, *Astrophys. J. Suppl.* **11**, 429.
Abt, H. A.: 1965b, in *AAS-NASA Symp. on Magnetic and Related Stars* (ed. by R. C. Cameron), Mono Book Corp., Baltimore, p. 173.
Abt, H. A. and Hunter, J. H., Jr.: 1962, *Astrophys. J.* **136**, 381.
Babcock, H. W.: 1958, *Astrophys. J.* **128**, 228.
Boyarchuk, A. A. and Kopylov, I. M.: 1958, *Soviet Astron. J.* **35**, 804.
Boyarchuk, A. A. and Kopylov, I. M.: 1964, *Pub. Crimean Astrophys. Obs.* **31**, 44.
Deutsch, A. J.: 1958, in *Handbuch der Physik* **51** (ed. by S. Flügge), Springer-Verlag, Berlin, p. 689.
Deutsch, A. J.: 1965, *AAS-NASA Symp. on Magnetic and Related Stars* (ed. by R. C. Cameron), Mono Book Corp., Baltimore, p. 181.
Dicke, R. H.: 1964, *Nature* **202**, 432.
Friedjung, M.: 1968, *Astrophys. J.* **151**, 781.
Greenstein, J. L.: 1960, in *Stellar Atmospheres* (ed. by J. L. Greenstein), University of Chicago Press, Chicago, Ill., chap. 19.
Greenstein, J. L.: 1969, private communication.
Hardorp, J. and Strittmatter, P. A.: 1968, *Astrophys. J.* **153**, 465.
Herbig, G. H. and Spalding, J. F., Jr.: 1955, *Astrophys. J.* **121**, 118.
Huang, S. S.: 1965, *Astrophys. J.* **141**, 985.
Huang, S. S. and Struve, O.: 1953, *Astrophys. J.* **118**, 463.
Huang, S. S. and Struve, O.: 1954, *Ann. Astrophys.* **17**, 85.
Kraft, R. P.: 1965, *Astrophys. J.* **142**, 681.
Kraft, R. P.: 1966, *Astrophys. J.* **144**, 1008.
Kraft, R. P.: 1968, in *Otto Struve Memorial Volume* (ed. by G. H. Herbig), in preparation.
Oke, J. B. and Greenstein, J. L.: 1954, *Astrophys. J.* **120**, 384.
Palmer, D. R., Walker, E. N., Jones, D. H. P., and Wallis, R. E.: 1968, *Royal Obs. Bull* **135**, E385.
Roxburgh, I.: 1964a, *Monthly Notices Roy. Astron. Soc.* **128**, 157.
Roxburgh, I.: 1964b, *Monthly Notices Roy. Astron. Soc.* **128**, 237.
Sandage, A. R.: 1955, *Astrophys. J.* **122**, 263.
Schatzman, E.: 1962, *Ann. Astrophys.* **25**, 18.
Slettebak, A.: 1949, *Astrophys. J.* **110**, 498.

Slettebak, A.: 1954, *Astrophys. J.* **119**, 146.
Slettebak, A.: 1955, *Astrophys. J.* **121**, 653.
Slettebak, A.: 1956, *Astrophys. J.* **124**, 173.
Slettebak, A.: 1966a, *Astrophys. J.* **145**, 121.
Slettebak, A.: 1966b, *Astrophys. J.* **145**, 126.
Slettebak, A. and Howard, R. F.: 1955, *Astrophys. J.* **121**, 102.
Steinitz, R.: 1964, *Bull. Astron. Inst. Netherlands* **17**, 504.
Struve, O.: 1931, *Astrophys. J.* **73**, 94.
Van den Heuvel, E. P. J.: 1968, *Bull. Astron. Inst. Neth.* **19**, 309.
Walker, E. N.: 1965, *Observatory* **85**, 162.

Discussion

Abt: The speaker is overly generous in attaching my name to a mean curve, because the curve is based entirely on Slettebak's data. The curve ignored some of the maxima and minima in those data, but in view of some of the recent data, such as Walker's *et al.*, we begin to take the maxima and minima more seriously.

Jaschek: According to recent work by the Jascheks and the Cowleys, the average $v \sin i$ velocities for giants and subgiants do not drop at all between B9 and F0. It is therefore questionable how much weight should be attached to Kopylov and Boyarchuk's single point at F5 which determines the decrease of the mean $v \sin i$.

ROTATION AND STELLAR INTERIORS

(Review Paper)

IAN W. ROXBURGH

Queen Mary College, University of London, England

1. Introduction

Research into the effect of rotation on the internal structure of stars has a long and detailed history. The names of McLaurin, Jacobi, Kelvin and Jeans are associated with detailed work on the structure and stability of rotating liquid masses assuming uniform rotation. Much of this work is summarized in Jeans' (1929) *Astronomy and Cosmogony* and more recent summaries are given in Lyttleton (1953) and Lebovitz (1967).

As more was discovered about the physical conditions inside stars it became clear that stars were gaseous and compressible and Jeans (1929) also considered the effect of uniform rotation on polytropic gases, that is a gas whose pressure and density are related by a power law $P = K\varrho^{1+(1/n)}$. In particular Jeans showed, in an analysis remarkable for its inaccuracies, and even more remarkable for the accuracies of its results, that a uniformly rotating polytrope has a point of bifurcation, that is, admits a non-axially symmetric configuration if $n < 0.8$. For n larger than this the equatorial regions reach Keplerian velocities and the equilibrium series terminates.

During the same period, it became apparent due to the pioneering work of Eddington that in large portions of a star energy was transported by radiation not by convection. This led Milne (1923) and von Zeipel (1924) to consider uniformly rotating radiative stars and led through von Zeipel's theorem to the discovery of internal circulation in stars by Vogt (1925) and Eddington (1925, 1929). The origin of such circulation is easy to see. The hydrodynamic energy and mass conservation equations for a rotating gaseous star are

$$\varrho \frac{\partial v^i}{\partial t} + \varrho v^k \frac{\partial v^i}{\partial x^k} = -\frac{\partial p}{\partial x^i} + \varrho \mathbf{g} + \frac{\partial}{\partial x^j}\left[\eta\left\{\left(\frac{\partial v^i}{\partial x^j} + \frac{\partial v^j}{\partial x^i}\right) - \frac{2}{3}\frac{\partial v^n}{\partial x^k}\right\}\right]$$

$$\varrho T \frac{\partial S}{\partial t} + \mathbf{v}\cdot\nabla S = \nabla\cdot\mathbf{F} - \varepsilon\varrho$$

$$\frac{\partial \varrho}{\partial t} + \nabla\varrho\mathbf{v} = 0$$

where **v** is the velocity, S the entropy, η the coefficient of viscosity and the other symbols have their usual meaning. The radiative flux F is given by

$$F = -C\nabla T, \quad C = \frac{4ac\,T^3}{3\,\kappa\varrho}$$

A. Slettebak (ed.), Stellar Rotation, 9–19. All Rights Reserved
Copyright © 1970 by D. Reidel Publishing Company, Dordrecht-Holland

where κ is a known function of pressure and temperature. Von Zeipel assumed that the only motion was uniform rotation and hence deduced that the energy generation must be

$$\varepsilon = \text{const.} (1 - \Omega^2/2\pi G\varrho),$$

a clearly unphysical result. Vogt and Eddington pointed out that this conclusion followed from assuming the only motion was rotation and that it disappeared if meridional circulation was invoked. The reason for such circulation is simple. If we assume no circulation, then due to the departure from spherical form – and the latitude dependence of pressure, temperature, etc., radiative equilibrium cannot be achieved everywhere. This breakdown forces circulation currents to carry excess energy. The speed of such circulation is also simple to estimate. The breakdown in thermal equilibrium in a star is re-adjusted in the Kelvin-Helmholtz time

$$t_{KH} = GM^2/RL$$

which is essentially the ratio of the gravitational energy to the rate at which energy is radiated. If the breakdown in thermal equilibrium is due to the departure from spherical form, this time-scale is increased by a factor

$$GM/\Omega^2 R^3$$

giving

$$t_{circ} = G^2 M^3/\Omega^2 L R^4$$

and a velocity R/t of

$$v_{circ} = \Omega^2 L R^5/G^2 M^3.$$

For the sun the circulation time-scale is of the order of 10^{12} years – a lot longer than its evolutionary time-scale. More generally stars on the main sequence have $L \propto M^4$ and as the evolutionary time-scale

$$t_{evol} \propto M/L \propto 1/M^3$$

we have

$$\frac{t_{circ}}{t_{evol}} = \left(\frac{\Omega^2 R^3}{GM}\right) \frac{M}{R}$$
$$= \lambda 10^3$$

where λ is the ratio of centrifugal force to gravity and M/R is a slowly varying function of mass. For rapidly rotating stars $\lambda \simeq 1$ and the meridional circulation completes several circulations and hence radically alters the angular velocity field.

The original work of Vogt and Eddington has been revised and improved by Sweet (1950) and Baker and Kippenhahn (1959) and a review is given by Mestel (1965).

This discovery of the circulation immediately gave rise to the assumption that rapidly rotating stars were kept chemically homogeneous by the circulation. This is not necessarily true and Mestel (1953) showed that the chemical composition changes prevented such a mixing by showing that the circulation produced a non-spherically symmetric distribution of chemical composition which 'choked' back the circulation.

2. Steady State Solutions

For those stars where the rotation is sufficiently strong to drive circulation currents many times round the star while the star is still on the main sequence, some redistribution of angular momentum will take place and we may expect some steady state distribution to be achieved. Such solutions have been investigated by Schwarzschild (1947) and Roxburgh (1964). The difficulty is that the viscosity in stellar interiors is so small that unless there are enormous gradients of angular velocity the viscous forces are negligibly small. The balance of forces in the azimuthal direction then reduces to

$$\mathbf{v} \cdot \nabla \Omega \omega^2 = R_\phi(\Omega, \eta) = 0$$

where R_ϕ is the azimuthal component of the viscous force. The only solutions of this equation that have been found are $\mathbf{v} = 0$, no meridian circulation, although the present author did find solutions for $v \neq 0$ in the case of polytropes. With no circulation the equations determining the steady state are

$$\nabla P/\varrho = -\mathbf{g} + \Omega^2 \omega$$
$$\nabla \cdot c \nabla T = 0$$

where the nuclear energy generation is negligible except in the very central regions.

The form of the solutions of this equation are readily seen by taken $C = $ constant, writing $P = P_0 + P_1$, etc., and then solving the perturbation equations

$$\nabla P_1 = -\varrho_1 \mathbf{g} + \varrho_0 \Omega^2 \omega$$
$$\nabla^2 T_1 = 0$$
$$P_1/P_0 = \varrho_1/\varrho_0 + T_1/T_0 .$$

Elimination of ϱ_1 gives

$$\nabla \left(\frac{P_1}{P_0} \right) = + \frac{\mathbf{T}_1}{T_0} \frac{\varrho_0}{P_0} \mathbf{g} + \frac{\varrho_0}{P_0} \Omega^2 \omega$$

on taking the curl we have

$$\nabla T_1 \times \frac{\varrho_0 \mathbf{g}}{T_0 P_0} + \mathrm{curl} \left(\frac{\varrho_0}{P_0} \Omega^2 \omega \right) = 0 .$$

Now ϱ_0, T_0, P_0, etc., are functions only of radial distance r, and if we expand T_1 in spherical harmonics

$$T_1 = \sum T_n P_n, \quad T_n = A_n r^n + B_n r^{-(n+1)}$$

we have

$$\sum T_n \frac{\partial P_n}{\partial \theta} = \frac{T_0^2}{g} \mathrm{curl} \left(\frac{\Omega^2 \omega}{T_0} \right) .$$

A solution of this equation with $\Omega = \Omega(r)$ is readily obtained by taking $n = 2$. In this case

$$\mathrm{curl} \left(\frac{\Omega^2}{T} \omega \right) = r \frac{\mathrm{d}}{\mathrm{d}r} \left(\frac{\Omega^2}{T} \right) \sin \theta \cos \theta$$

and so Ω is given by

$$\frac{T^2}{g}\frac{d}{dr}\left(\frac{\Omega^2}{T}\right) = Ar + \frac{B}{r^3}$$

where the integration constants are determined by the boundary conditions. This is in essence the solution obtained by Roxburgh (1964) where he included the dependence of c on ϱ and T. This solution is given in Table I where $\beta = \Omega^2/\Omega_c^2$ is tabulated against fractional radius $x = r/R$.

It should be emphasized that this is only the first step in a procedure to determine

TABLE I

Variation of $\beta = \Omega^2/\Omega_c^2$ for a steady state solution with no circulation currents

x	β	x	β	x	β
0.00	1.000	0.34	0.612	0.68	0.469
0.02	1.000	0.36	0.592	0.70	0.467
0.04	1.000	0.38	0.574	0.72	0.465
0.06	1.000	0.40	0.558	0.74	0.463
0.08	1.000	0.42	0.545	0.76	0.462
0.10	1.000	0.44	0.534	0.78	0.460
0.1217	1.000	0.46	0.524	0.80	0.459
0.14	0.984	0.48	0.515	0.82	0.458
0.16	0.950	0.50	0.507	0.84	0.457
0.18	0.905	0.52	0.501	0.86	0.456
0.20	0.858	0.54	0.495	0.88	0.455
0.22	0.812	0.56	0.490	0.90	0.454
0.24	0.769	0.58	0.485	0.92	0.453
0.26	0.730	0.60	0.481	0.94	0.452
0.28	0.695	0.62	0.478	0.96	0.452
0.30	0.663	0.64	0.475	0.98	0.451
0.32	0.636	0.66	0.472	1.00	0.451

the steady state solution. With the Ω so determined the viscous force is non-zero so that there must be a non-zero circulation velocity driven essentially by the viscous force. In the complete steady state solution there will be differential rotation and meridian circulation. Such refinements, while of marginal interest for stellar interiors, could be of importance for the structure of planetary atmospheres.

Other steady state solutions have been determined when magnetic fields are introduced (Roxburgh, 1967), the most realistic case being one where a magnetic field maintains uniform rotation in spite of the meridian circulation. The transport of angular momentum by the circulation is just balanced by a magnetic torque. More recently Maheswaran (1968) has looked at the effect of slight departures from non-uniform rotation but still constant on field lines.

3. Stability of Steady State Solutions

In the last few years there has been considerable progress in examining the stability of differential rotation, notably by Fricke (1968) and by Goldreich and Schubert (1967). For a rotating liquid the condition of equilibrium requires

$$\partial \Omega / \partial z = 0 \quad \text{Taylor-Proudman Theorem.}$$

While the condition of dynamical stability requires

$$\partial (\Omega^2 \omega^4) / \partial \omega > 0 \quad \text{Rayleigh condition.}$$

If either of these two conditions is violated then the system changes in time.

For a compressible fluid the situation is different. The Taylor-Proudman Theorem is no longer valid and equilibrium solutions with $\Omega = \Omega(\omega, z)$ are possible. Again, Rayleigh's criterion is no longer true for adiabatic displacements, the stability effect of the stratification being dominant. The analogue of the Rayleigh condition is (Randers, 1942)

$$\frac{1}{\omega^3} \frac{\partial}{\partial \omega} (\Omega^2 \omega^4) + \mathbf{g} \cdot \left(\frac{\nabla \varrho}{\varrho} - \frac{1}{\gamma} \frac{\nabla P}{P} \right) < 0 .$$

However in a gas there are other possibly unstable modes where the radiative diffusion is important and such modes have been analysed by Fricke (1968) and Goldreich and Schubert (1967) assuming axisymmetric disturbances. For a perturbation that is small compared to a scale height they find instability if

$$\frac{k_z}{\Omega^3 \gamma} \left(g_\omega + \Omega^2 \omega - \frac{k_\omega}{k_z} g_z \right) \left(k_\omega \frac{\partial}{\partial z} - k_z \frac{\partial}{\partial \omega} \right) \ln \left(\frac{T}{\varrho^{\gamma - 1}} \right)$$

$$+ \frac{2}{\gamma \omega \Omega} k_z \frac{\sigma}{\Omega} \left(k_\omega \frac{\partial}{\partial z} - k_z \frac{\partial}{\partial \omega} \right) (\Omega \omega^2) + \frac{\sigma v^2 k^4}{\gamma \Omega^3} < 0 \quad \text{for any} (k_\omega, k_z)$$

where (k_ω, k_z) is the wave number of the disturbance, $k^2 = k_z^2 + k_\omega^2$, and σ is the Prandtl number, the ratio of the radiative conductivity to kinematic viscosity. In the conditions that prevail inside stars σ is large $\simeq 10^5 - 10^6$. For infinite σ the stability criterion reduces to

$$\partial \Omega / \partial z = 0 \quad \partial / \partial \omega (\Omega^2 \omega^4) > 0 ,$$

which is just the Taylor-Proudman Theorem and the Rayleigh condition. For sufficiently large thermal diffusivity a disturbance can radiate away its buoyancy without diffusing its momentum so that the fluid behaves like a classical liquid, and we recover the stability conditions applicable to a liquid.

For the real case the differential rotation that can be stabilized is

$$\Delta (\Omega \omega^2) / \Omega \omega \lesssim g / \sigma \Omega^2 \omega$$

where Δ is either $\partial / \partial \omega$ or $\partial / \partial z$. For rapidly rotating stars $g / \Omega^2 \omega \simeq 1$ and we have just the Rayleigh and Taylor-Proudman conditions, since $\sigma \gg 1$. For slowly rotating stars the possibility arises that $g / \sigma \Omega^2 \omega > 1$ and a substantial degree of differential rotation can be tolerated. The value of the Prandtl number is therefore important. The viscosity has two contributions, molecular and radiative; these give

$$v = \frac{2m^{1/2}(kT)^{5/2}}{\varrho e^4 \ln \Lambda} + \frac{16\sigma T^4}{15c^2\kappa\varrho^2}$$

where

$$\Lambda = \frac{3kT}{e^2}\left(\frac{m}{\varrho}\right)^{1/3},$$

whereas the radiative diffusivity is

$$k = 16\sigma T^3/3\kappa\varrho^2 c_v.$$

For stellar conditions this gives

$$\sigma = (k/v) \sim 10^6.$$

Only in very slowly rotating stars can Ω vary with z or $\Omega\omega^2$ decrease outwards. In such stars the meridian circulation would be so slow that a steady state could not be achieved in the lifetime of the star. In any case where rotation is important stability conditions require

$$\Omega = \Omega(\omega); \quad \frac{\delta}{\delta\omega}(\Omega\omega^2) > 0.$$

On the other hand there are no steady state solutions with $\partial\Omega/\partial z = 0$ (Roxburgh, 1966), and we therefore reach the important conclusion that there are *no stable steady state* solutions for rotating stars. Inside a rapidly rotating star we expect a battle between the meridian circulation and the turbulence driven by the differential rotation the circulation produces. It seems probable that such turbulence produces uniform rotation.

4. Rotation and Convective Zones

In recent years the problem of the differential rotation of the sun has been attacked with great vigour, from many different viewpoints – some of which will be discussed subsequently during this meeting.

The two types of approach can be described as turbulent and liquid. The difficulty about all these attempts is that we know virtually nothing about turbulence on the one hand, and the sun is not an incompressible Boussinesq liquid on the other hand. The most striking result and probably the reason why almost all attempts to explain the equatorial acceleration meet with success is that meridian circulation in a viscous spherical shell can produce equatorial acceleration.

To show this consider a steady state situation in a rotating spherical shell. Let the rotation be Ω and the meridian circulation be given by a stream function $S(r, \theta)$.

$$\varrho v = (\mathbf{k} \times \nabla S)/(r \sin \theta).$$

For simplicity we assume constant velocity and density (although this is not necessary).

The azimuthal component of the equation of motion is then

$$k \times \nabla S \cdot \nabla (\Omega r^2 \sin^2 \theta) = r \left(\sin \theta \frac{\partial^2 (\Omega r^2)}{\partial r^2} + \frac{\partial}{\partial \theta} \frac{\partial \Omega}{\sin \theta \, \partial \theta} \sin^2 \theta \right).$$

For large viscosity this equation can be solved by successive approximation. The first approximation is $\Omega = \Omega_0 =$ constant and the departure from Ω_0, say Ω_1, is then given by

$$\frac{\partial S}{\partial z} \Omega_0 = v \sin \theta \frac{\partial^2 \Omega_1}{\partial r^2} + \frac{\partial}{\partial \theta} \frac{1}{\sin \theta} \frac{\partial}{\partial \theta} \Omega_1 \sin^2 \theta.$$

Obviously by a suitable choice of $S(\omega, z)$, Ω_1 can be made to give equatorial acceleration. In particular a stream function

$$S = \Lambda(r) \sin^2 \theta \cos \theta$$

gives an angular velocity distribution

$$\Omega = \Omega_0 + \omega_1(r) + \omega_2(1) \sin^2 \theta$$

in agreement with the observed equatorial acceleration of the sun provided $d\Lambda/dr > 0$ at the surface.

The turbulent theories have their origin in work of Wasiutinski (1946) who developed the viscous stress tensor for turbulence. This derivation has been improved by Elsasser (1966) and Roxburgh (1969). The essential point is that if the turbulence is anisotropic differential rotation can result (cf. Wasiutinski, p. 86). A similar point was made, Biermann (1951), and subsequently developed by Kippenhahn (1963), Cocke (1966), Roxburgh (1963). Assuming various forms for the viscous stress tensor these authors arrived at various results – all explaining the equatorial acceleration. The point should be emphasized that almost any departures from spherical symmetry will give equatorial acceleration.

A slightly different approach was adopted by Durney and Roxburgh (1969). Arising out of the observations by Dicke (1967) of the oblateness of the sun Roxburgh (1967) was led to deduce a latitude dependence in the convective efficiency of heat transport, an inference supported by the work of Chandrasekhar (1961) and Durney (1968a, b; 1970). Thus although the turbulence may be locally isotropic the effective 'eddy conductivity' varies with latitude due to the latitude effect of rotation. Global thermal equilibrium then demands a circulation which maintains the equatorial acceleration.

The other method of approach has been to test the stability and steady state convection of a Boussinesq (incompressible) liquid in a spherical shell. This approach has been used by Durney (1968a, b, 1970) and Busse (1969). They find that when rotation is included the most unstable mode is a 'banana-shaped mode' proportional to a surface harmonic T_l^l where l is determined by the thickness of the shell ($l \simeq 10$). This convective mode produces a transport of angular momentum towards the equator which maintains the differential rotation against viscous braking. Durney (1970) has followed the evolution of the system until an asymptotic state is reached using the

Herring approximation for developed convection. He finds equatorial acceleration and a preference for energy to go through the equatorial regions. This may be linked to Dicke's oblateness measurements (see Durney and Roxburgh, 1969).

Other work on the equatorial acceleration of the sun has invoked Rossby waves travelling around the sun as the momentum transport mechanism. The asymptotic periodic time dependence found by Durney is very similar to such waves and not surprisingly both give similar results.

5. Effect of Rotation on the Main Sequence

Considerable effort has been expended on this problem in recent years. Following earlier work, Sweet and Roy (1953) determined the effect of uniform rotation on a simple Kramer's opacity Cowling Model star for slow rotation.

Other models have since been constructed by Roxburgh *et al.* (1965) and a set of

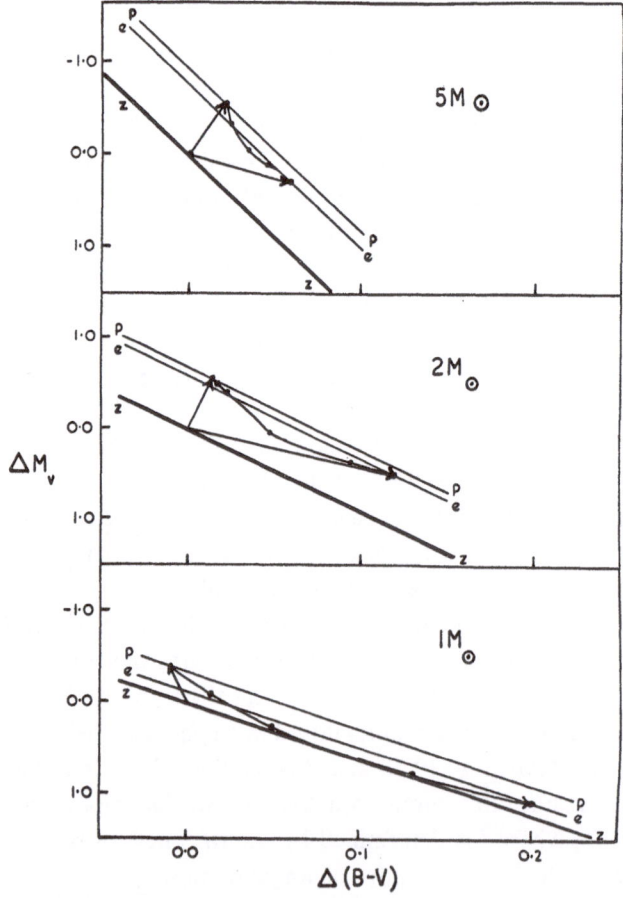

Fig. 1.   Variation of ΔM_V and $\Delta (B-V)$ with aspect angle $i = 0$, to $\pi/2$, and maximum rotational velocity (reprinted courtesy of the *Astrophysical Journal*).

accurate models were calculated by Faulkner *et al.* (1968). The effects of gravity darkening and distortion on the observed properties of stars was first determined by Collins (1963) which was followed by a determination including the effect of rotation on the interior carried out by Roxburgh and Strittmatter (1965). Recent work by Faulkner *et al.* (1968) has provided a set of detailed main-sequence models for a range of mass and angular velocity, and a simple prescription for determining any other models. The results of this latter work are given in Figures 1 and 2.

In the last few years interest has been growing in the effect of non-uniform rotation for many reasons. Firstly the effect of rotation is enhanced if the central regions spin faster than the surface and the work of Mark and Ostriker illustrates this effect. Secondly, it has been suggested by Dicke that the interior of the sun may be more rapidly rotating than the outside. Thirdly, substantial differential rotation could lead to fissional instability and the formation of binary stars as I suggested in 1966. I am sure we will have many other reasons for looking at non-uniform rotation before this colloquium is over.

I have not attempted in this talk to be comprehensive and I have left out several important topics. I make no apology for this since I am sure they will all be covered during the course of the meeting, and a recent review article by Strittmatter (1969) also covers much of this material.

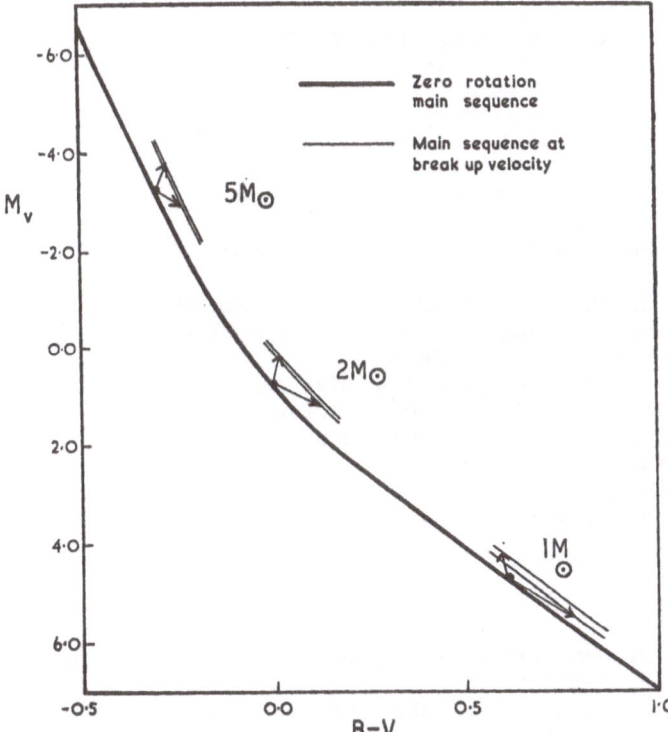

Fig. 2. Effect of rapid rotation on the main sequence (reprinted courtesy of the *Astrophysical Journal*).

References

Baker, N. and Kippenhahn, R.: 1959, *Z. Astrophys.* **48**, 140.
Biermann, L.: 1951, *Z. Astrophys.* **28**, 304.
Busse, S.: 1969, to be published.
Chandrasekhar, S.: 1961, *Hydrodynamic and Hydromagnetic Stability,* Oxford.
Cocke, W. J.: 1966, *Astrophys. J.* **150**, 1044.
Collins, G. W.: 1963, *Astrophys. J.* **138**, 1134.
Dicke, R. and Goldenberg, M.: 1967, *Phys. Rev. Letters* **18**, 313.
Durney, B.: 1968a, *J. Atmos. Sci.* **25**, 372.
Durney, B.: 1968b, *J. Atmos. Sci.* **25**, 771.
Durney, B.: 1970, this volume, p. 30.
Durney, B. and Roxburgh, I. W.: 1969, to be published.
Eddington, A. S.: 1925, *Observatory* **48**, 73.
Eddington, A. S.: 1929, *Monthly Notices Roy. Astron. Soc.* **90**, 54.
Elsasser, K.: 1966, *Z. Astrophys.* **63**, 65.
Faulkner, J., Roxburgh, I. W., and Strittmatter, P. A.: 1968, *Astrophys. J.* **151**, 203.
Fricke, K.: 1968, *Z. Astrophys.* **68**, 317.
Goldreich, P. and Schubert, G.: 1967, *Astrophys. J.* **150**, 571.
Jeans, J.: 1929, *Astronomy and Cosmogony*, Cambridge.
Kippenhahn, R.: 1963, *Astrophys. J.* **137**, 664.
Lebovitz, N.: 1967, *Adv. Astron.* Vol. 5, p. 465.
Lyttleton, R. A.: 1953, *Stability of Rotating Liquid Masses.*
Maheswaran, M.: 1968, *Monthly Notices Roy. Astron. Soc.* **140**, 93.
Mestel, L.: 1953, *Monthly Notices Roy. Astron. Soc.* **113**, 716.
Mestel, L.: 1965, 'Meridian Circulation in Stars', in *Stellar Structure* (ed. by Allen and McLaughlin), Chicago.
Milne, E. A.: 1923, *Monthly Notices Roy. Astron. Soc.* **83**, 118.
Ostriker, J. P. and Mark, J. W.: 1968, *Astrophys. J.* **151**, 1075.
Randers, S.: 1942, *Astrophys. J.* **95**, 454.
Roxburgh, I. W.: 1963a, Fellowship Thesis, Churchill College, Cambridge.
Roxburgh, I. W.: 1963b, *Monthly Notices Roy. Astron. Soc.* **126**, 67.
Roxburgh, I. W.: 1964, *Monthly Notices Roy. Astron. Soc.* **128**, 157.
Roxburgh, I. W.: 1966, *Astrophys. J.* **143**, 111.
Roxburgh, I. W.: 1967, *Nature* **213**, 1077.
Roxburgh, I. W.: 1969, *Solar Hydrodynamics and Hydromagnetics,* National Center for Atmospheric Research.
Roxburgh, I. W., Griffith, J. S., and Sweet, P. A.: 1965, *Z. Astrophys.* **61**, 203.
Roxburgh, I. W. and Strittmatter, P. A.: 1965, *Z. Astrophys.* **63**, 15.
Schwarzschild, M.: 1947, *Astrophys. J.* **106**, 407.
Sweet, P. A.: 1950, *Monthly Notices Roy. Astron. Soc.* **110**, 548.
Sweet, P. A. and Roy, A. E.: 1953, *Monthly Notices Roy. Astron. Soc.* **113**, 701.
Strittmatter, P. A.: 1969, *Adv. Astron.*, Vol. 6, p. 665.
Wasiutinski, J.: 1946, *Astron. Norv.* **4**.
Vogt, H.: 1925, *Astron. Nachr.* **223**, 229.
Von Zeipel, H.: 1924, *Monthly Notices Roy. Astron. Soc.* **84**, 665.

Discussion

Clement: Is an equilibrium distribution of angular velocity possible for which departures from cylindrical symmetry are small?

Roxburgh: If there is no magnetic field, then the answer is no. Cylindrical rotation with circulation would require velocities parallel to the rotation axis, i.e., matter leaving the star. Equilibrium solutions with no circulation are approximately spherically symmetric.

Clement: It is not correct to say that equilibrium solutions with no circulation are spherically symmetric. Spherical symmetry is only a special case (cf. Clement: 1969, *Astrophys. J.* **156**, 1051).

Mark: Concerning the statement that 'large differential rotation will cause large changes in stellar interior parameters': in fact, only $\Omega_p/\Omega_e \simeq 2\text{-}3$ can give very large changes. What seems to matter is the kind of differential rotation that does not give an equilibrium model with an equatorial cusp which terminates the sequence.

Roxburgh: I call an inward increase of 2–3 in Ω large; this increases the effect of rotation in the central regions by a factor of 4–9 and therefore the change in, say, luminosity is increased by a similar factor. This could give a luminosity change of 30 to 70% on an extrapolation of the uniform rotation results.

Fricke: You mentioned in connection with Randers' work, that each Ω-distribution should be dynamically stable in a gaseous star due to the stable stratification of the pressure-temperature-density field. This is not the case. You can well construct unstable angular momentum distributions for any given stratification against adiabatic motions.

Concerning the Rayleigh criterion, I may point out that it is generally not a sufficient criterion for the stability of a rotating liquid. Shear instabilities may occur even if Rayleigh's criterion is satisfied.

Roxburgh: What you say is true. If the angular momentum gradient is sufficiently large then an adiabatic instability can exist. For this to be the case, the ratio of the scale of variation of the square of the angular momentum to the radius of the star must exceed the ratio of gravity to centrifugal force. If the scale of variation of the angular momentum is of the same order as the radius of the star then the distribution will be, in general, stable. Of course, shearing instabilities could occur but this is usually a more severe criterion than the Rayleigh criterion. Near the axis of a star it may be possible to have shearing instabilities and be stable in the Rayleigh case.

Ostriker: I would like to make a small objection to an argument which was forwarded by Roxburgh as only schematic. It appears that a contracting rotating star in which viscous and magnetic forces may be neglected will never shed mass at the equator. Rotational 'mass loss' does not occur – at least not under the stated conditions.

Roxburgh: I agree that there are difficulties in understanding how the mass loss occurs but this may just be our ignorance. My own conjecture is that when the gravity and temperature in the equatorial regions is the same as that in pulsating stars a pulsation instability sets in which ejects material – this may be the cause of Be stars.

Dicke: Perhaps it should be remarked that the Goldreich-Schubert instability is model-dependent and that it would be dangerous to conclude that a rapidly rotating stellar core is *impossible*. The stabilizing effects of compositional gradient and meridional flow or oscillation are omitted. Also the complexity and possibly stabilizing effects of a magnetic field (if it should occur) has been omitted.

Roxburgh: Of course this is true and composition gradients could indeed stabilize differential rotation although they are only to be expected in the center of a star. It is more difficult to see how meridional velocities would produce a stabilization since they are so very slow. If they were fast enough they would be able to do so but I doubt if this is the case for a real star.

A magnetic field is more difficult, but I think Klaus Fricke will talk about this later in the meeting.

A SIMPLE METHOD FOR THE SOLUTION OF THE STELLAR STRUCTURE EQUATIONS INCLUDING ROTATION AND TIDAL FORCES

R. KIPPENHAHN and H.-C. THOMAS

Universitätssternwarte Göttingen and Max-Planck-Institut für Physik und Astrophysik, München, Germany

Abstract. In the following a method is presented for computing the internal structure of nonspherical stars assuming that the force per gram causing the deviation from spherical symmetry is conservative. The method has the advantage that in a normal (spherical) stellar structure code only slight changes have to be made in order to obtain nonspherical stellar models. The method can be applied as well to rotating stars as to stars distorted by tidal effects. Although it is similar to that of Faulkner *et al.* (1968) in the case of purely rotating stars, it is not necessary to use the division into two zones, where either slow rotation or negligible contribution to the gravitional potential is assumed.

1. Mathematical Definitions

We consider a potential function $\Psi(x, y, z)$. The equipotential surfaces then are defined by $\Psi = \text{const.}$ We assume them to be topologically equivalent to spheres. The volume enclosed by a surface $\Psi = \text{const.}$ may be V_Ψ, the surface area S_Ψ.

For any function $f(x, y, z)$ we define the mean value over an equipotential surface $\Psi = \text{const.}$ by

$$\bar{f} = \frac{1}{S_\Psi} \int_{\Psi = \text{const.}} f \, d\sigma, \tag{1}$$

where $d\sigma$ is the surface element of the surface $\Psi = \text{const.}$ By definition

$$S_\Psi = \int_{\Psi = \text{const.}} d\sigma. \tag{2}$$

The mean value $\bar{f}$ can be determined for each equipotential; it therefore is a function of Ψ only, $\bar{f} = \bar{f}(\Psi)$. The distance dn between two neighboring equipotential surfaces $\Psi = \text{const.}$ and $\Psi + d\Psi = \text{const.}$ is in general not constant. We define a function $g(x, y, z)$ by

$$g = d\Psi/dn. \tag{3}$$

From the function g the mean values $\bar{g}$ and $\overline{g^{-1}}$, both being functions of Ψ, can be determined. One then gets for the volume dV_Ψ between the surfaces Ψ and $\Psi + d\Psi$

$$dV_\Psi = \int_{\Psi = \text{const.}} dn \, d\sigma = d\Psi \int_{\Psi = \text{const.}} \frac{dn}{d\Psi} \, d\sigma = \overline{g^{-1}} S_\Psi \, d\Psi. \tag{4}$$

In analogy to the sphere we define a 'radius' r_Ψ by

$$V_\Psi = \frac{4\pi}{3} r_\Psi^3. \tag{5}$$

The new quantity r_Ψ is a function of Ψ. The surface area S_Ψ in general is not equal to $4\pi r_\Psi^2$; we therefore define a new quantity u by

$$u = S_\Psi/4\pi r_\Psi^2, \tag{6}$$

where again u is a function of Ψ only, $u=u(\Psi)$. If the surfaces $\Psi=$ const. are spheres, u is equal to 1. We define by M_Ψ the mass enclosed by the surface $\Psi=$ const. If Ψ defines the mechanical potential of a vector field $\mathbf{g}$:

$$\mathbf{g} = -\nabla\Psi, \tag{7}$$

we can define the two functions

$$v = \bar{g}\,\frac{r_\Psi^2}{GM_\Psi}, \quad w = \overline{g^{-1}}\,\frac{GM_\Psi}{r_\Psi^2}, \tag{8}$$

where G is the gravitational constant. If the function Ψ is the gravitational potential of a selfgravitating sphere the surfaces $\Psi=$ const. are spheres $(u=1)$, and

$$g = GM_\Psi/r_\Psi^2 \tag{9}$$

is constant on these spheres and therefore v and w are constant and equal to 1.

The Equations (1)–(8) are pure mathematical definitions. We will apply them to gravitational potential fields distorted by perturbing forces. It is worth mentioning that the Equations (1)–(8) do not contain any simplifying assumptions.

2. Stellar Structure Equations

A star may be distorted by centrifugal or tidal forces. The perturbing acceleration may be conservative. The potential of the total (i.e. gravitational plus perturbing) acceleration may be Ψ. In the following we assume the function $\Psi(x, y, z)$ to be known. We then derive the stellar structure equations for nonspherical stars, the nonsphericity being defined by the nonspherical equipotential surfaces. We start with spherical stars and use the well-known fact that P, ϱ in hydrostatic equilibrium are constant on equipotentials. The mass dM_Ψ between two equipotentials Ψ and $\Psi+d\Psi$ is according to (5)

$$dM_\Psi = \varrho(\Psi)\,dV_\Psi = 4\pi r_\Psi^2\varrho(\Psi)\,dr_\Psi. \tag{10}$$

We therefore get

$$dr_\Psi/dM_\Psi = 1/4\pi r_\Psi^2\varrho. \tag{11}$$

From (4), (10) we get

$$d\Psi = \left(\frac{dV_\Psi}{d\Psi}\right)^{-1} dV_\Psi = \left(\frac{dV_\Psi}{d\Psi}\right)^{-1}\frac{dM_\Psi}{\varrho} = \frac{dM_\Psi}{g^{-1}S_\Psi\varrho} \tag{12}$$

and with (6), (8)

$$d\Psi = \frac{GM_\Psi\,dM_\Psi}{\varrho u w 4\pi r_\Psi^4}. \tag{13}$$

The condition for hydrostatic equilibrium, $dP/d\Psi = -\varrho$, can now be written with (13) in the form

$$\frac{dP}{dM_\Psi} = -\frac{GM_\Psi}{4\pi r_\Psi^4} f_P, \tag{14}$$

where

$$f_P = \frac{1}{uw} = \frac{4\pi r_\Psi^4}{GM_\Psi S_\Psi} \frac{1}{\overline{g^{-1}}}. \tag{15}$$

The factor f_P is a function of Ψ. If Ψ is known the equipotential surfaces can be determined, and with them S_Ψ, r_Ψ, $\bar{g}$, and $\overline{g^{-1}}$ for each surface simply from the geometry of the equipotentials. The mass M_Ψ depends on the density distribution $\varrho(\Psi)$ and in principle can be determined from (10) by integration.

If L_Ψ is the energy which passes per second through the equipotential $\Psi = $ const. in the outward direction, then the increase dL_Ψ of L_Ψ between the equipotential surfaces Ψ and $\Psi + d\Psi$ is given by

$$dL_\Psi = \varepsilon \, dM_\Psi \tag{16}$$

where ε is the nuclear energy generation per gram per sec. For chemically homogeneous stars ε depends only on ϱ and T and is therefore constant on equipotential surfaces

$$dL_\Psi/dM_\Psi = \varepsilon. \tag{17}$$

In these energy equations we neglected an additional term which contains the time derivative of the entropy. This may be permitted as long as we consider variations in time which are slow compared to the Kelvin-Helmholtz time scale.

For the case of energy transport by radiation we have

$$F = -\frac{4acT^3}{3\kappa\varrho} \frac{dT}{dn} = -\frac{4acT^3}{3\kappa\varrho} g \frac{dT}{d\Psi}. \tag{18}$$

This equation contains the well-known fact that the radiative flux F varies on an equipotential surface proportionally to g. With Equation (13) we obtain

$$F = -\frac{4acT^3}{3\kappa} g \frac{dT}{dM_\Psi} \frac{uw \cdot 4\pi r_\Psi^4}{GM_\Psi}. \tag{19}$$

In order to get L_Ψ we integrate Equation (19) over the equipotential $\Psi = $ const. and obtain

$$L_\Psi = \int\limits_{\Psi = \text{const.}} F d\sigma = -\frac{16\pi acT^3}{3\kappa} \bar{g} S_\Psi \frac{dT}{dM_\Psi} \frac{uw r_\Psi^4}{GM_\Psi}, \tag{20}$$

and

$$L_\Psi = -\frac{64\pi^2 acT^3 r_\Psi^4}{3\kappa} u^2 vw \frac{dT}{dM_\Psi}, \tag{21}$$

or

$$\frac{dT}{dM_\Psi} = -\frac{3\kappa L_\Psi}{64\pi^2 ac T^3 r_\Psi^4} f_T,$$ (22)

with

$$f_T = \frac{1}{u^2 vw} = \left(\frac{4\pi r_\Psi^2}{S_\Psi}\right)^2 \frac{1}{\bar{g}\,\overline{g^{-1}}}.$$ (23)

Equations (11), (14), (17), (22) become the four well-known stellar structure equations for the case of spherical configurations, for which $f_P = f_T = 1$. In the nonspherical case we assume that f_P, f_T are known as functions of M_Ψ. We can then solve the stellar structure equations. But in order to get a consistent solution one has to check, whether the equipotential surfaces and the functions f_P, f_T are consistent with this model. For instance, that part ϕ of Ψ, which corresponds to the self-gravitation, must satisfy the Poisson equation $\Delta\phi = 4\pi G\varrho$.

3. The Uniformly Rotating Star

In order to solve the stellar structure equations one has to know the correction factors f_P and f_T as functions of r_Ψ. They are pure geometrical quantities, depending only on the form of the equipotential surfaces. In principle it is necessary to solve the Poisson equation simultaneously with the stellar structure equations. In our method we approximate the potential simply by the dimensionless expression (the 'geometry function')

$$\tilde{\Psi} = \frac{1}{x} + \frac{1}{2} x^2 (1 - \mu^2)$$ (24)

with

$$x = (\omega^2/GM)^{1/3}\, r, \quad \mu = \cos\vartheta.$$ (25)

We also redefine v and w as

$$v = \overline{(d\tilde{\Psi}/dn)}/(1/y^2), \quad w = \overline{(d\tilde{\Psi}/dn)^{-1}}/y^2$$ (26)

with

$$y = (\omega^2/GM)^{1/3}\, r_\Psi$$ (27)

instead of using Equations (8). Although in Equations (8) M_Ψ and r_Ψ appear simultaneously, the mass M_Ψ does not appear any more in the definition of v and w. The reason is that the geometry function is only used to describe the deviations from sphericity of the equipotentials.

Although our $\tilde{\Psi}$ is identical with the normalized Roche potential the models computed by the method described are realistic stellar models with finite central density and not Roche models. This can immediately be seen if one applies the method to the limiting case of a nonrotating star. The equipotential surfaces are spheres; therefore $f_P = f_T = 1$ and one gets the normal stellar structure equations and *not* spherical Roche models. The fact that stellar models are regular in the center although a geometry function is used which is singular there is a major advantage of our method.

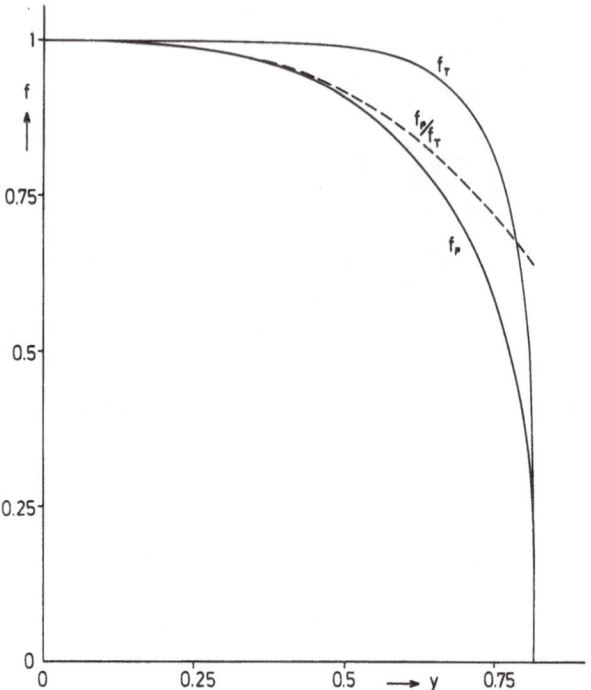

Fig. 1.   The correction factors f_P and f_T as functions of $y = (\omega^2/GM)^{1/3} r_\Psi$ for the case of pure solid body rotation. The ratio of the two functions is plotted in order to show the behavior near the surface.

Once the function Ψ is specified and

$$V_\Psi = \int_0^{\widetilde{\Psi}=\text{const.}} d\tau,$$ (28)

$$S_\Psi = \int_{\widetilde{\Psi}=\text{const.}} d\sigma,$$ (29)

$$\frac{\overline{d\widetilde{\Psi}}}{dn} = \frac{1}{S_\Psi} \int_{\widetilde{\Psi}=\text{const.}} \frac{d\widetilde{\Psi}}{dn} \, d\sigma,$$ (30)

$$\overline{\left(\frac{d\widetilde{\Psi}}{dn}\right)^{-1}} = \frac{1}{S_\Psi} \int_{\Psi=\text{const.}} \left(\frac{d\widetilde{\Psi}}{dn}\right)^{-1} d\sigma$$ (31)

are computed, f_P and f_T can be determined (Figure 1).

In the case of a corotating star in a close binary system

$$\widetilde{\Psi} = \frac{1}{u_1} + \frac{q}{u_2} + \tfrac{1}{2}x^2(1-\mu^2)(1+q) - \tfrac{1}{2}q\left(2 + \frac{q}{1+q}\right)$$ (32)

can be used as a geometry function, with

$$u_1^2 = x^2 + \frac{2q}{1+q} x \cos\varphi \sin\vartheta + \frac{q^2}{(1+q)^2}, \tag{33}$$

$$u_2^2 = x^2 - \frac{2}{1+q} x \cos\varphi \sin\vartheta + \frac{1}{(1+q)^2}, \tag{34}$$

$$q = M_2/M_1, \quad x = r/d, \tag{35}$$

where for the angular velocity

$$\omega^2 = G(M_1 + M_2)/d^3 \tag{36}$$

is taken according to Kepler's law (d=distance between the two components). The origin of the system of polar coordinates r, ϑ, φ is in the center of mass of the binary system.

In principle the geometry function can be adjusted to the problem to be solved. If, for instance, the angular velocity is much higher near the axis of rotation, the equipotentials will have a stronger oblateness for x in the neighborhood of zero, an effect which in principle can be taken into account by adding to the geometry function a term which represents the potential of a quadrupole-like symmetry.

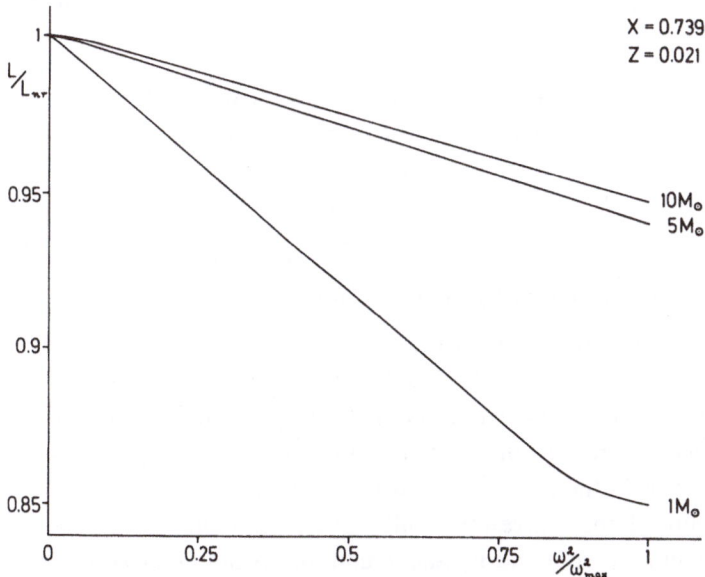

Fig. 2. Luminosity as a function of ω^2 for uniformly rotating zero age main sequence stars of 1, 5, and 10 $M_\odot$ with a chemical composition of $X=0.739$, $Z=0.021$. The units used are: maximum angular velocities ω_{max} for the abscissa, and luminosities of the nonrotating models L_{nr} for the ordinate. The values of these units are:

	1 $M_\odot$	5 $M_\odot$	10 $M_\odot$
$\omega_{max}/\text{sec}^{-1}$	5.124×10^{-4}	2.635×10^{-4}	1.493×10^{-4}
$\log L_{nr}/L_\odot$	-0.1204	2.6558	3.6724

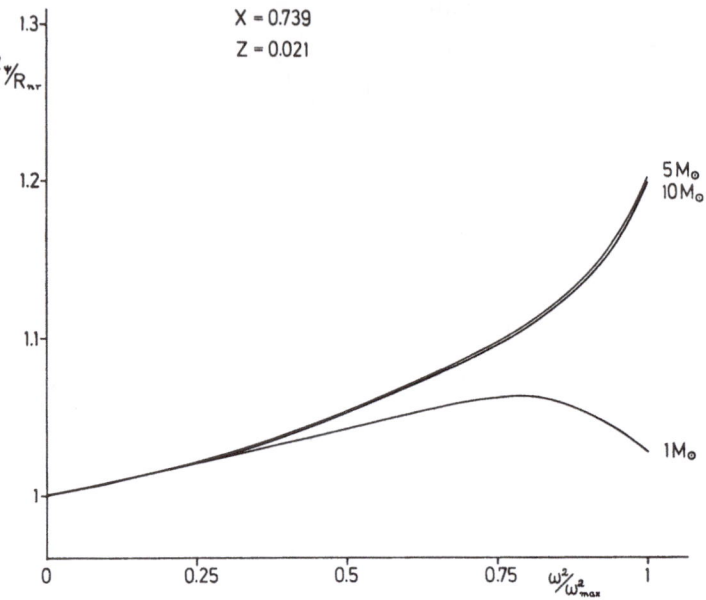

Fig. 3.   Mean radius R_Ψ as a function of ω^2 for uniformly rotating zero age main sequence stars of 1, 5, and 10 $M_\odot$. Units for the abscissa are the same as in Figure 2; for the ordinate the radii R_{nr} of the nonrotating models were used. The values of these radii are

	1 $M_\odot$	5 $M_\odot$	10 $M_\odot$
R_{nr}/cm	6.320×10^{10}	1.767×10^{11}	2.655×10^{11}

One can imagine an iterative procedure, starting with a given geometry function, in which one solves the stellar structure equations and uses the density distribution obtained to improve the geometry function via the Poisson equation.

4. Applications

A. UNIFORMLY ROTATING MAIN SEQUENCE MODELS

Sets of models of different angular velocities have been computed for 1, 5, and 10 $M_\odot$ using the geometry function (24). The tabulated correction factors were built into a normal stellar structure code based on the Henyey method. Starting with nonrotating models only a few iterations in the Henyey method were necessary to obtain the rotating models, even in the case of models rotating with the maximum possible angular velocity. Figures 2 and 3 give the results. The results agree well with those obtained by Faulkner et al. (1968) and by Sackmann and Anand (1969); our values are only slightly smaller. The only remarkable difference is the change in the mean radius for the 1 $M_\odot$-model.

B. UNIFORMLY ROTATING HELIUM MAIN SEQUENCE STARS

For 1, 5, and 10 $M_\odot$, helium stars, on the helium main sequence, have been computed for different angular velocities. The results are given in Figures 4 and 5.

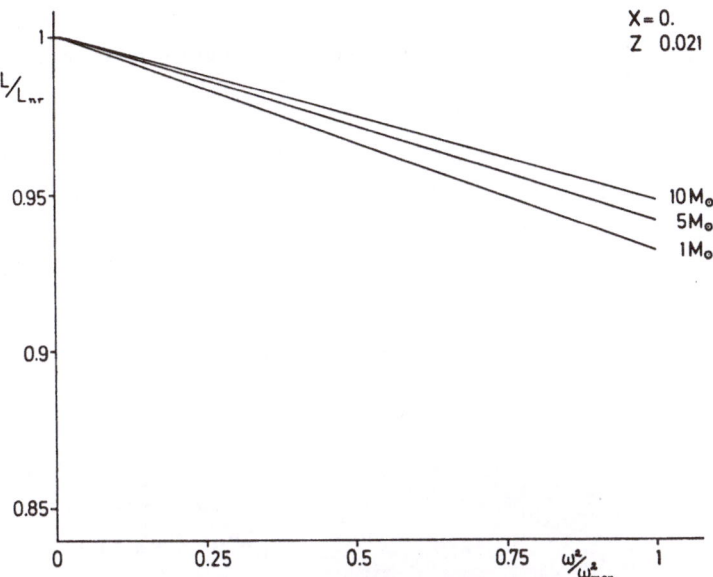

Fig. 4. Luminosity as a function of ω^2 for uniformly rotating helium stars on the helium main sequence with masses 1, 5, and 10 $M_\odot$ and a chemical composition of $Y = 0.979$, $Z = 0.021$. Units are defined as in Figure 2, their values being:

	1 $M_\odot$	5 $M_\odot$	10 $M_\odot$
$\omega_{max}/\text{sec}^{-1}$	4.106×10^{-2}	1.699×10^{-2}	1.268×10^{-2}
$\log L_{nr}/L_\odot$	2.3727	4.4647	5.1374

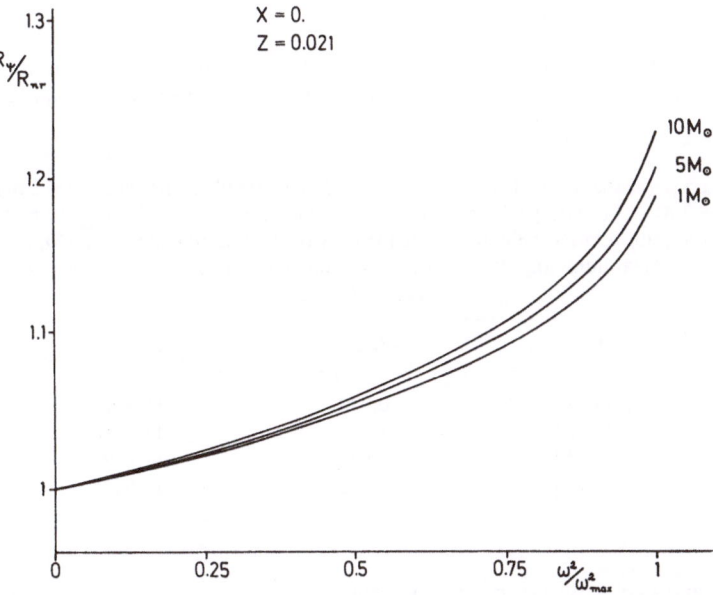

Fig. 5. Mean radius R_Ψ as a function of ω^2 for uniformly rotating helium stars on the helium main sequence with masses 1, 5, and 10 $M_\odot$. Units are defined as above, their values being:

	1 $M_\odot$	5 $M_\odot$	10 $M_\odot$
R_{nr}/cm	1.363×10^{10}	4.136×10^{10}	6.219×10^{10}

TABLE I

Comparison of a nonrotating and a fully rotating cooling white dwarf of the same central temperature

$\log L/L_\odot$	L/L_{nr}	$\log R_\Psi$	R_Ψ/R_{nr}	$\log P_c$	$\log T_c$
-2.4285	1	9.2099	1	21.6534	7.0371
-2.6110	0.657	9.2794	1.174	21.5761	7.0371

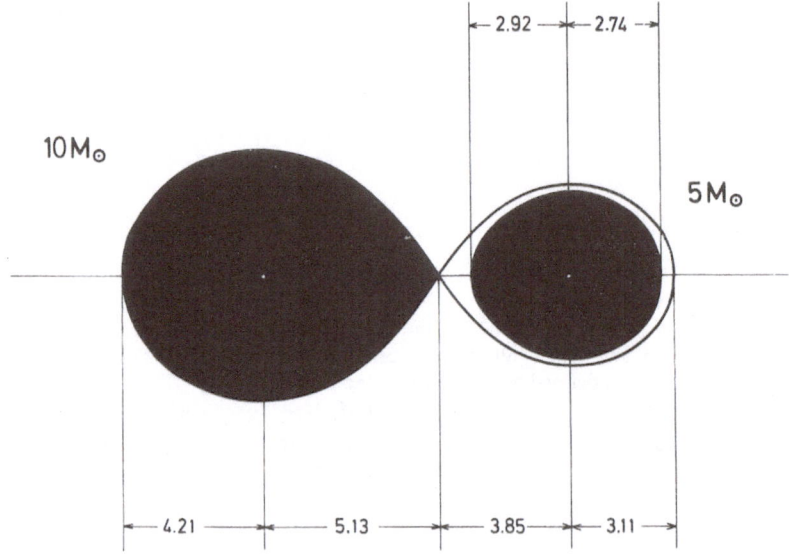

Fig. 6. The close binary system of $10 + 5\ M_\odot$ in the plane of symmetry vertical to the rotation axis. Both stars are in the zero age stage. The numbers give distances in solar radii.

TABLE II

A zero age main sequence star of $10M_\odot$ ($X = 0.739$, $Z = 0.021$) filling its critical lobe in a close binary system of $10 + 5\ M_\odot$. For comparison, results for a nonrotating star of $10\ M_\odot$ (line 2), a star rigidly rotating with the same angular velocity as the primary of the binary system (line 3), and a star rigidly rotating with the maximum angular velocity possible (line 4) are given. The value for $\omega_{max} = 1.493 \times 10^{-4}\ \text{sec}^{-1}$.

ω^2/ω^2_{max}	$\log L/L_\odot$	L/L_{nr}	$\log R_\varphi$	R_φ/R_{nr}
0.3655	3.6632	0.979	11.4414	1.041
0	3.6724	1	11.4240	1
0.3655	3.6632	0.979	11.4392	1.036
1	3.6463	0.942	11.5028	1.200

C. A FULLY ROTATING WHITE DWARF MODEL

A nonrotating white dwarf model of 0.23 $M_\odot$ consisting of a helium core with an inactive hydrogen rich envelope has been taken from the computations of Kippenhahn *et al.* (1968) and compared with a fully rotating white dwarf model of the same central temperature and the same chemical structure. The latter again was computed with the method described above. Table I gives the comparisons.

D. A COROTATING CLOSE BINARY SYSTEM

The method has been applied to a system of $10+5\,M_\odot$. The distance has been chosen in such a way that in the zero age stage the primary just fills its critical lobe (Figure 6). For both models the geometry function (32) has been used. While the secondary is only slightly distorted the primary shows the maximum effects due to tidal distortion. Table II gives the comparison between a spherical model for the primary, the nonspherical model derived by our method, and the model of a rotating star, rotating with the angular velocity of the binary system. One can see that within an accuracy of 10^{-4} the change in luminosity is caused by the rotation of the corotating binary system while the effect of the tidal forces can only be seen in a somewhat larger increase of the mean radius compared with the purely rotating stellar model. Similar conclusions were reached by Jackson (1970), who extended the method of Faulkner, Roxburgh, and Strittmatter by including tidal effects from a companion in a close binary system.

References

Faulkner, J., Roxburgh, I. W., and Strittmatter, P. A.: 1968, *Astrophys. J.* **151**, 203.
Jackson, S.: 1970, *Astrophys. J.* (in press).
Kippenhahn, R., Thomas, H.-C., and Weigert, A.: 1968, *Z. Astrophys.* **69**, 265.
Sackmann, I. J. and Anand, S. P. S.: 1969, preprint.

Discussion

Jordahl: Please specify more fully the variation in angular velocity with radius and why it was chosen.

Thomas: It was chosen such that $\omega^2 = \omega_s^2 \cdot (M_r/M)$, ω_s being the angular velocity at the surface, in order to simplify the geometry function. ω starts decreasing only when f_p and f_T are already close to 1, so the error will be negligible.

Ostriker: Can the change in sign you found (for the one solar mass model) in the quantity $d\,r_\psi/d\,(\omega^2)$ be due to a change from primarily CNO cycle to PP chain as the central temperature decreases with increasing rotation?

Thomas: No, because in our $1\,M_\odot$ model the contribution of the CNO-cycle to nuclear energy production is small, even in the spherical model.

THE INTERACTION OF ROTATION WITH CONVECTION

BERNARD R. DURNEY

High Altitude Observatory, National Center for Atmospheric Research*, Boulder, Colo., U.S.A.

Abstract. The equations for a rotating convective spherical shell are solved in the Herring approximation as an initial value problem. The main results are

(1) The most unstable modes (those that maximize the heat flux) correspond to convective cells stretching from pole to pole.

(2) The calculations of the Reynolds stresses show transport of angular momentum towards the equator. That is, differential rotation sets in with equatorial acceleration.

(3) The convective heat transport is maximum at the equator. This would give rise to an equator-pole flux difference.

(4) If convection is non-axisymmetric (as in the most unstable modes) there are no time independent solutions. The time dependence is oscillatory and of the form $\omega t + m\phi$.

1. Theory

In a system of coordinates rotating with an angular velocity Ω, the basic equations can be written:

$$\frac{1}{\sigma}\frac{\partial}{\partial t}\mathbf{V}\times\mathbf{U} - \mathbf{V}\times\mathbf{V}^2\mathbf{U} = -\frac{1}{\sigma}\mathbf{V}\times(\mathbf{U}\cdot\mathbf{V})\,\mathbf{U}$$

$$+ \mathscr{R}_1\mathbf{V}\times g(r)\,\mathbf{r}T + \mathscr{T}_1(\hat{\omega}\cdot\mathbf{V})\,\mathbf{U} \tag{1a}$$

$$\text{div}\,\mathbf{U} = 0 \tag{1b}$$

$$\left(\frac{\partial}{\partial t} - \mathbf{V}^2\right)T = -\mathbf{V}\cdot(\mathbf{U}T). \tag{1c}$$

In Equations (1) all quantities are dimensionless, σ is the Prandtl number, $\mathscr{R}_1 = \mathscr{R}/(1-\eta)^3$ where $\mathscr{R}$ is the Rayleigh number and $\eta = 0.8$ is the internal radius of the spherical shell in units of R_0, the external radius;

$$\mathscr{T}_1 = 2\Omega R_0^2/\nu = \mathscr{T}_a^{1/2}/(1-\eta)^2$$

where $\mathscr{T}_a$ is the Taylor number; and $\hat{\omega}$ is a unit vector in the direction of the angular velocity. The unit of distance is R_0, and not, d, the thickness of the spherical shell, this explains the factors $(1-\eta)^3$ and $(1-\eta)^2$ in $\mathscr{R}_1$ and $\mathscr{T}_1$, respectively. The factor $g(r)$ takes into account the variation of gravity with radial distance.

We expand the velocity field in basic poloidal and toroidal vectors:

$$\mathbf{U} = \sum_{L,\,m}\left[\mathbf{P}(p_L^m(r,t)\,Y_L^m(\theta,\varphi)) + \mathbf{T}(t_L^m(r,t)\,Y_L^m(\theta,\varphi))\right]. \tag{2a}$$

---

* The National Center for Atmospheric Research is sponsored by the National Science Foundation.

A. Slettebak (ed.), Stellar Rotation, 30–36. All Rights Reserved
Copyright © 1970 by D. Reidel Publishing Company, Dordrecht-Holland

In spherical coordinates the components of $\mathbf{P}\,(p_L^m\,Y_L^m)$ and $\mathbf{T}(t_L^m\,Y_L^m)$ are given by

$$P^{(r)} = \frac{(L+1)\,L}{r^2}\,p_L^m\,Y_L^m$$

$$P^{(\theta)} = \frac{1}{r}\frac{\partial p_L^m}{\partial r}\frac{\partial Y_L^m}{\partial \theta}, \tag{3a}$$

$$P^{(\varphi)} = \frac{1}{r\sin\theta}\frac{\partial p_L^m}{\partial r}\frac{\partial Y_L^m}{\partial \varphi}$$

and

$$T^{(r)} = 0$$

$$T^{(\theta)} = \frac{t_L^m}{r\sin\theta}\frac{\partial Y_L^m}{\partial \varphi} \tag{3b}$$

$$T^{(\varphi)} = -\frac{t_L^m}{r_i}\frac{\partial Y_L^m}{\partial \theta}.$$

The poloidal and toroidal vectors defined by (3) form a complete orthogonal set for solenoidal vector fields.

For the temperature we take

$$T = \frac{1}{1-\eta}\left(\frac{\eta}{r}-1\right) + \psi(r,t) + \Theta(\mathbf{r},t), \tag{2b}$$

with

$$\Theta(\mathbf{r},t) = \sum_{L,\,m}\Theta_L^m(r,t)\,Y_L^m(\theta,\varphi). \tag{2c}$$

We take the spherical harmonics appearing in (2) and (3) as defined by Condon and Shortley (1951).

In Herring's approximation (1963, 1964) the fluctuating self interactions, that is, the terms $(1/\sigma)\,(\mathbf{U}\cdot\nabla\mathbf{U} - \overline{U\cdot\nabla U})$ and $(U\cdot\nabla\Theta - \overline{U\cdot\nabla\Theta})$ are neglected. The main effect of the small scale part of these terms is to give rise to a turbulent viscosity and conductivity; the Rayleigh and Taylor number should then be defined in terms of these last quantities.

The equations for $\Theta_L^m(r,t)$ and $\psi(r,t)$ (the fluctuating and average parts of the distortion in temperature from its purely conductive value) and the equations for the poloidal and toroidal components of the velocity field are found to be:

$$\left(\frac{\partial}{\partial t}-\mathscr{D}_L-\frac{(L+1)\,L}{r^2}\right)\psi = -\frac{1}{4\pi r^2}\sum_{L,\,m}^{|m|\leqslant L}(L+1)\,L\,\frac{\partial}{\partial r}(rP_L^m\Theta_L^{*m}), \tag{4a}$$

$$\left(\frac{\partial}{\partial t}-\mathscr{D}_L\right)\Theta_L^m = \frac{(L+1)\,L}{r}\,P_L^m\left[\frac{\eta}{(1-\eta)\,r^2}-\frac{\partial\psi}{\partial r}\right], \tag{4b}$$

$$\frac{1}{\sigma}\frac{\partial}{\partial t}\mathscr{D}_L P_L^m - \frac{im\mathscr{T}_1}{(L+1)L}\mathscr{D}_L P_L^m - \mathscr{D}_L^2 P_L^m$$

$$= -\mathscr{R}_1 g(r)\Theta_L^m + \mathscr{T}_1\{A(L,m)(L-1)(L-2)T_{L-1}^m/L$$
$$- A(L+1,m)(L+2)(L+3)T_{L+1}^m/(L+1)$$
$$- A(L,m)(L-1)rT_{L-1}'^m/L - A(L+1,m)(L+2)rT_{L+1}'^m/(L+1)\}$$

(4c)

$$\frac{1}{\sigma}\frac{\partial}{\partial t}T_L^m - \frac{im\mathscr{T}_1}{(L+1)L}T_L^m - D_L T_L^m$$

$$= -\mathscr{T}_1\{A(L,m)(L-1)^2 P_{L-1}^m/Lr^2 - A(L+1,m)$$
$$\times (L+2)^2 P_{L+1}^m/(L+1)r^2 - A(L,m)(L-1)P_{L-1}'^m/rL$$
$$- A(L+1,m)(L+2)P_{L+1}'^m/r(L+1)\}$$

(4d)

where

$$\mathscr{D}_L = \frac{d^2}{dr^2} + \frac{2}{r}\frac{d}{dr} - \frac{(L+1)L}{r^2}$$

$$D_L = \frac{d^2}{dr^2} + \frac{4}{r}\frac{d}{dr} + \frac{2-(L+1)L}{r^2}$$

$$P_L^m = p_L^m/r; \quad T_L^m = t_L^m/r^2; \quad T_L'^m = \frac{dT_L^m}{dr}; \quad P_L'^m = \frac{dP_L^m}{dr}$$

$$A(L,m) = [(L+m)(L-m)/(2L+1)(2L-1)]^{1/2}.$$

Equations (4) are the fundamental equations of the problem. We assume free surface boundary conditions at $r=\eta$, 1. It is easily seen that they imply $P_L^m = P_L'^m = T_L'^m = 0$ at $r=\eta$, 1. The boundary conditions for the temperature are as usual $\Theta = \psi = 0$ at $r=\eta$, 1.

We neglect the time derivatives in Equations (4c) and (4d). In the absence of rotation modes with different L's do not interact. The integration in time of Equations (4) shows then that after a sufficiently long time only the mode with $L=10$ remains different from zero (Durney, 1968a). Another value of L for the non-zero mode would have been found for a different thickness of the spherical shell (in the present case, the thickness is 0.2 R_0).

Rotation couples, through the Taylor number, modes with different L's. For $m=0$, there exist time-independent solutions (Durney, 1968b). For $m \neq 0$ no time-independent solutions were found. This was attributed to the fact for $m \neq 0$ the convective modes of polytropes show overstability (Durney and Skumanich, 1968). It was then decided to integrate in time Equations (4) for a given value of m by keeping only the $L=8$, 10, 12 modes for P_L^m and Θ_L^m and the $L=9$, 11 modes for T_L^m. This should be a good approximation for small Taylor numbers.

To perform the time integration P_L^m, Θ_L^m, $\psi(r,t)$ and T_L^m were expanded in terms of a complete set of functions of r satisfying the boundary conditions. The coefficients in these expansions are time-dependent, and from Equations (4) it is possible to obtain ordinary differential equations for these coefficients.

2. Numerical Results

The values of the Rayleigh number and Taylor number were chosen to be equal to 1500 and 4, respectively.

A. THE MOST UNSTABLE MODE

Table I gives the value of the heat flux as function of m.

TABLE I

Heat Flux × 10⁻²	0.999 19	0.999 24	0.999 41
m	0	1	2
0.999 68	1.000 07	1.000 56	
3	4	5	
1.001 16	1.001 88	1.002 7	
6	7	8	
1.003 57	1.004 59		
9	10		

The most unstable mode is thus the one with $m = 10$. This is in agreement with Busse's (1969) results who finds that the Rayleigh number for the onset of convection is a minimum for $m = L$.

B. DIFFERENTIAL ROTATION

The average of the azimuthal velocity U_φ over φ is zero. Differential rotation, as observed in the sun, should, however, be a property of a rotating convective spherical shell. Once the velocities are known in the Herring approximation, it is possible to evaluate the torque $T(\theta) \, d\theta$, exerted by the Reynolds stresses on an annulus of thickness d (the thickness of the spherical shell). The polar angle of the annulus is θ. This torque is given in Figure 1 and shows equatorial acceleration.

C. EQUATOR-POLE FLUX DIFFERENCE

Figure 2 shows the convective flux $\overline{U_r \Theta}$ (the average is over φ) as function of θ for $r = 0.9$ which is the surface halfway between the inner $(r = 0.8)$ and outer $(r = 1)$ surfaces. The convective flux is larger at the equator. This could give rise to an equator-pole flux difference as observed by Dicke and Goldenberg (1967). A possible interpretation of this effect is based on the interaction of rotation and convection (Roxburgh, 1967a, b, 1969; Durney and Roxburgh, 1969).

D. GRAVITATIONAL-GYROSCOPIC WAVES

The velocity field U and the fluctuation in temperature Θ are time-dependent. The

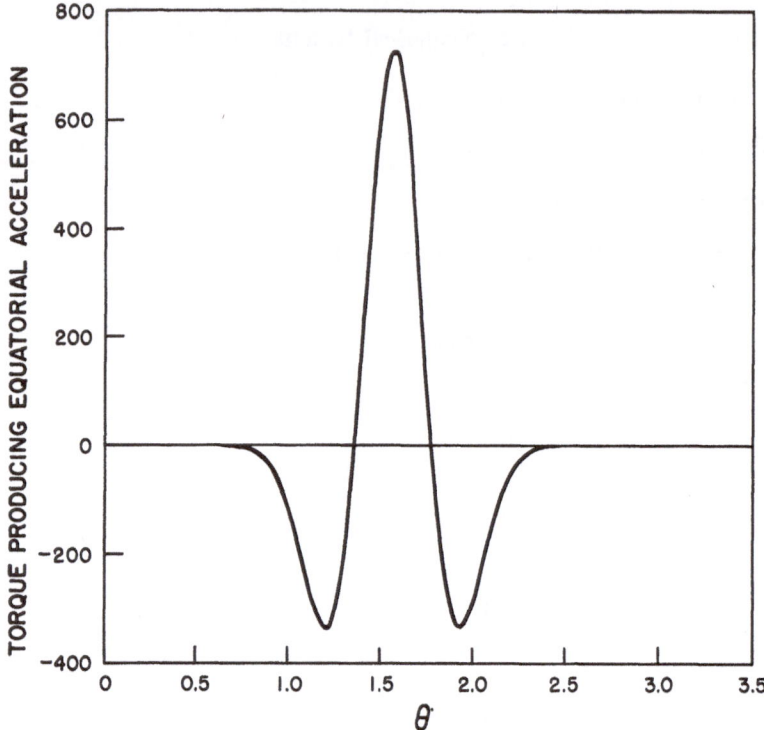

Fig. 1. Torque exerted by the Reynolds stresses on an annulus of polar angle θ. This torque produces differential rotation.

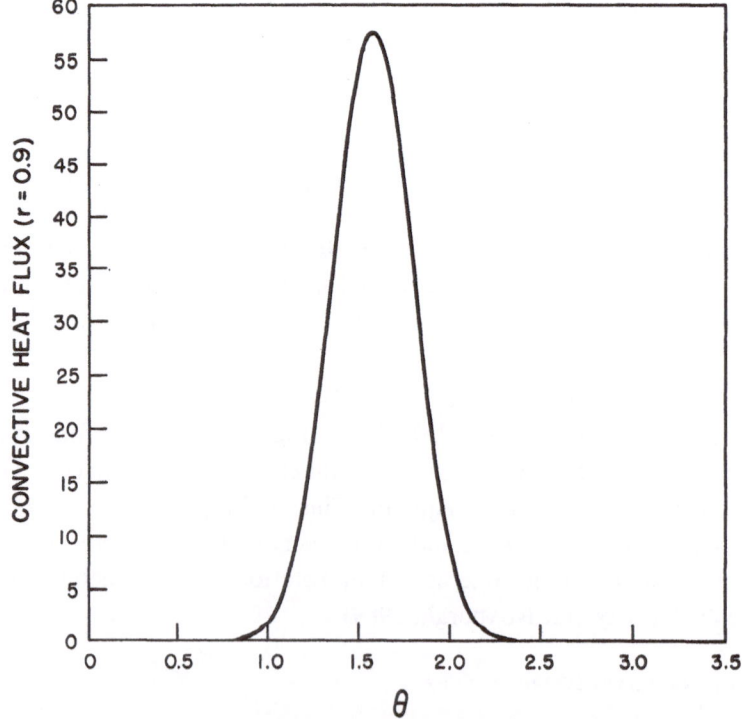

Fig. 2. Convective heat flux at a surface halfway between the inner and outer surfaces.

torque and the convective heat flux as given by Figures 1 and 2 are nevertheless time-independent. More generally if A and B stand for any of the quantities $\mathbf{U}$ and Θ then the average of AB over φ $\overline{(AB)}$ is constant in time for sufficiently large values of the time. As A or B are of the form $A = A_1(r, \theta, t)\cos m\varphi + A_2(r, \theta, t)\sin m\varphi$ the constancy of $\overline{AB}$ suggests the following time-dependence:

$$A = A_1(r, \theta)\cos(\omega t + m\varphi) + A_2(r, \theta)\sin(\omega t + m\varphi).$$

The value of ω was found to be equal to $\omega = 3.153$.

This time dependence is identical to that of the states of marginal stability of the convective modes of a rotating polytrope (Durney and Skumanich, 1968). These 'gravitational-gyroscopic' waves have some similarities with Rossby waves as introduced by Ward (1965, 1966), Starr and Gilman (1965) and Gilman (1967) to explain the differential rotation of the sun. If giant convective cells exist in the sun, the use of the Herring approximation can be justified to explain qualitatively certain features of the convective zone of the sun. On theoretical grounds Simon and Weiss (1969) have suggested the existence of giant convective cells and Howard (1969) has observed upward and downward motions extending from pole to pole.

Acknowledgements

The author is grateful to Professors P. Gilman and I. Roxburgh and to Dr. A. Skumanich for many illuminating discussions.

References

Busse, F. H.: 1969, Max Planck Institute Report MPI-PAE/Astro 15.
Condon, E. U. and Shortley, G. H.: 1951, *The Theory of Atomic Spectra*, Cambridge University Press, p. 52.
Dicke, R. H. and Goldenberg, H. M.: 1967, *Nature* **214**, 1294.
Durney, B. R.: 1968a, *J. Atmos. Sci.* **25**, 372.
Durney, B. R.: 1968b, *J. Atmos. Sci.* **25**, 771.
Durney, B. R. and Roxburgh, I. W.: 1969, *Nature* **221**, 646.
Durney, B. R. and Skumanich, A.: 1968, *Astrophys. J.* **152**, 225.
Gilman, P. A.: 1967, Part I, II, III, *J. Atmos. Sci.* **24**, 101.
Herring, J. R.: 1963, *J. Atmos. Sci.* **20**, 325.
Herring, J. R.: 1964, *J. Atmos. Sci.* **21**, 277.
Howard, R.: 1969, unpublished.
Roxburgh, I. W.: 1967a, *Nature* **213**, 1077.
Roxburgh, I. W.: 1967b, *Nature* **216**, 1286.
Roxburgh, I. W.: 1969, *The Application of Modern Physics to the Earth and Planetary Interiors*. Wiley London.
Simon, G. W. and Weiss, N. O.: 1969, *Z. Astrophys.* **69**, 435.
Starr, V. P. and Gilman, P. A.: 1965, *Astrophys. J.* **141**, 1119.
Ward, F.: 1965, *Astrophys. J.* **141**, 534.
Ward, F.: 1966, *Astrophys. J.* **145**, 416.

Discussion

Roxburgh: It is important to emphasize that giant cells have been observed by Howard and that these have the same banana-mode shape.

Ruben: Generally a rotating star should be hotter at the pole than at the equator. Does your effect of mainly equatorial heat transport compensate this?

Durney: The effect of rotation on convection should be much larger than the von Zeipel effect.

Fricke: I would mention that Fritz Busse in a recent paper concerning the equatorial acceleration of the sun came to similar results with such an analysis.

Durney: Yes. Busse's treatment is based on a double expansion: one, in the convective amplitudes and the other in the Taylor number.

ROTATIONAL PERTURBATION OF A RADIAL OSCILLATION
IN A GASEOUS STAR

R. SIMON

Institut d'Astrophysique de l'Université de Liège, Cointe-Sclessin, Belgium

The perturbation method has been applied to the problem of the oscillations of a gaseous star rotating around a fixed z-axis according to a general law of the type

$$\Omega = \Omega(r, \theta).$$

This rotation was assumed to be small and the analysis included all the effects of order Ω and Ω^2. Particularly, the rotational distortion of the star has been taken into account by means of a mapping between the spheroidal volume of the rotating star and the spherical volume of a neighboring nonrotating configuration. Care has also been taken, in the perturbation method, to account for the existence of the so-called 'trivial modes' of the sphere, i.e. those oscillations of vanishing frequencies in which the displacement is transversal and divergence-free.

A final and practical result has been obtained in the case of the perturbation of a radial mode (Simon, 1969).

In the case of the standard model (polytrope of index 3), for a rigid rotation, the following result was obtained numerically for the perturbation of the fundamental radial mode:

$$\sigma_R^2 = \tfrac{2}{3}\Omega^2,$$

for $\gamma = \tfrac{4}{3}$; and

$$\sigma_R^2 = \sigma^2 - 3.8576\,\Omega^2,$$

for $\gamma = \tfrac{5}{3}$, with

$$\sigma^2 = (0.0569)\,4\pi G\varrho_c,$$

where ϱ_c denotes the central density of the rotating star.

Finally, in the case of the homogeneous model, for a rigid rotation, the following result was obtained analytically for the perturbation of the fundamental radial mode, in the case of a constant γ:

$$\sigma_R^2 = \sigma^2 + (5 - 3\gamma)\,\tfrac{2}{3}\Omega^2,$$

with

$$\sigma^2 = (3\gamma - 4)\,\tfrac{4}{3}\pi G\varrho,$$

where ϱ is the density of the rotating star.

This latter result can easily be generalized to any radial mode; a point which was overlooked in the paper summarized here. We get in this way the following expression:

$$\sigma_R^2 = (1 - \Omega^2/2\pi G\varrho)\,\sigma^2 + \tfrac{2}{3}\Omega^2,$$

A. Slettebak (ed.), Stellar Rotation, 37–38. All Rights Reserved
Copyright © 1970 by D. Reidel Publishing Company, Dordrecht-Holland

in which ϱ is the density of the rotating star and σ the frequency of the radial oscillation of the nonrotating, spherical, homogeneous model which has the same density ϱ as the actual star. In the case of a constant γ:

$$\sigma^2 = [3\gamma - 4 + k(2k + 5)\gamma] \tfrac{4}{3}\pi G\varrho,$$

with $k=0$ for the fundamental mode, $k=1$ for the first harmonic, etc. Consequently:

$$\sigma_R^2 = \sigma^2 + [5 - 3\gamma - k(2k + 5)\gamma] \tfrac{2}{3}\Omega^2.$$

Reference

Simon, R.: 1969, *Astron. Astrophys.* **2**, 390.

ROTATIONAL VELOCITIES OF EVOLVED A AND F STARS*

SANDRA M. FABER† and I. J. DANZIGER

*Harvard College Observatory, Cambridge, Mass. and
Kitt Peak National Observatory,‡ U.S.A.*

Abstract. Mean rotational velocities are presented for stars on and above the main sequence from spectral type A5 to F9. The velocities are shown to be inconsistent with the hypothesis of stellar rotation in completely uncoupled shells for stars that have expanded in radius by less than a factor of four. They do, however, support the hypothesis of solid-body rotation for these stars. The data for the giant stars are discussed, but no firm conclusions about the mode of rotation in these stars can be drawn.

1. Introduction

Rotational velocity is an important parameter in stellar structure and one which, for the interior of stars at least, is largely unknown. Investigations by many workers in the past have produced an extensive collection of $v \sin i$'s, and the behavior of surface equatorial velocity along the main sequence is consequently quite well determined. However, our only possibility for understanding internal rotational structure lies in determination of surface $v \sin i$ along an evolutionary track, together with theoretically predicted stellar radii and moments of inertia. Several investigators, notably Oke and Greenstein (1954), Sandage (1955), Abt (1957, 1958), and Rosendhal (1968), have utilized this technique. Following the approach pioneered by Oke and Greenstein, all these workers have set themselves a modest goal, to determine whether angular momentum is conserved as stars evolve and, if so, which of two hypotheses, solid-body rotation or rotation in completely uncoupled shells, agrees more closely with the observed data. No definite evidence has been found for non-conservation of angular momentum in any of these investigations. However, attempts to discriminate between the two modes of rotation have either been inconclusive, because of too small a sample of available velocities, the possible presence of turbulence, especially in the supergiants, and inaccurate theoretical values for moments of inertia; or have produced conflicting results, apparently because the groups of stars under consideration in these various papers have occupied different regions of the HR diagram. The present investigation has employed the same approach while attempting to circumvent some of these difficulties.

2. The Data

A vertical strip in the HR diagram, from spectral type A5 to F9, was selected. All stars in the *Catalog of Bright Stars* (Hoffleit, 1964) known to lie above the main sequence in this strip, together with all catalog members in this region without a listed

---

* Contributions from the Kitt Peak National Observatory, No. 491.
† National Science Foundation Fellow.
‡ Operated by the Association of Universities for Research in Astronomy, Inc., under contract with the National Science Foundation.

A. Slettebak (ed.), Stellar Rotation, 39–47. All Rights Reserved
Copyright © 1970 by D. Reidel Publishing Company, Dordrecht-Holland

luminosity class, were taken as candidates. In this way we sought to include every possible star in the catalog lying above the main sequence between A5 and F9. The literature was searched for all available rotational velocities for these stars. Velocities for the great majority of the remainder, some 360 stars, were obtained by us, yielding a total population of 579 stars for further analysis.

Since the details of this work will be published elsewhere, it is not appropriate here to discuss at length the procedure used in determining the rotational velocities. In brief, the values of $v \sin i$ were obtained by comparing observed profiles, traced from spectra with a dispersion of 13.6 Å/mm, with a grid of standard profiles. The standards were lines chosen from observed zero-velocity or near zero-velocity stars of a variety of spectral types and luminosity classes, distributed throughout the region of the HR diagram under consideration. These zero-velocity profiles were then mathematically rotated by varying amounts in $v \sin i$. Only one novel aspect of this procedure deserves mention. In addition to three normal profiles of single lines, the 'profiles' referred to above contained two extended sections of spectrum totaling over 200 Å in length. These sections were traced on each star and compared to the grid of model sections, the entire lengths of which had been mathematically rotated. These extended regions proved to be extremely helpful velocity indicators at all ranges of $v \sin i$, in effect providing over 150 lines and blends for comparison with the model profiles.

In an effort to obtain more accurate and up-to-date values for stellar moments of inertia, we have utilized unpublished data on stellar models kindly made available by Professor Iben. These new data alter significantly the conclusions to be drawn for the

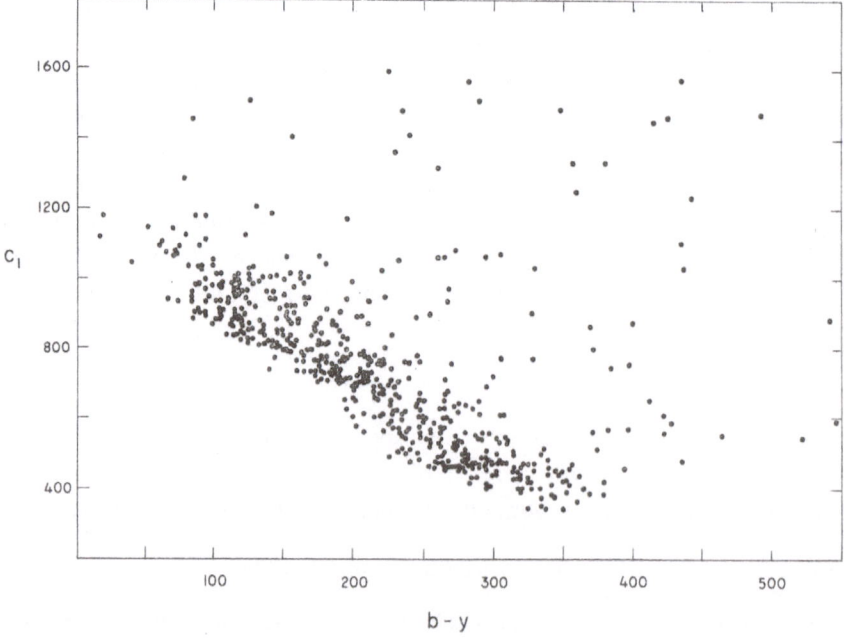

Fig. 1.   The $c_1 - (b - y)$ diagram for all stars.

most massive stars. Iben's models predict that the velocities expected under the hypothesis of solid-body rotation, or case A, are fully 5 times greater for Ia supergiants in this region than the velocities expected under the hypothesis of rotation in shells, hereafter called case B. This value of 5 is to be compared with the factor of 2 generally adopted in previous investigations.

For comparison with theoretical evolutionary tracks, accurate estimates of luminosity and effective temperature for each star are necessary. For this purpose, *uvby* photometry was obtained as an aid to spectral classification for all stars not in the catalog by Strömgren and Perry (1962). The resultant $c_1 - (b-y)$ diagram for all stars is shown in Figure 1.

3. Stars in the Main Sequence Band

The stars were divided into two groups. The first contained all stars lying in the main sequence band. For these stars, the *uvby* photometry was used as an indicator of luminosity and spectral type. Iben's tracks for 1.5 $M_\odot$ and 1.25 $M_\odot$ were transferred to the $c_1 - (b-y)$ diagram and the stars broken down into smaller groups for the determination of mean $v \sin i$'s as shown in Figure 2. The groups were located so that the stars in regions B and C are evolutionary descendents of region A, and similarly for regions D, E, and F. The initial starting points of the 1.5 $M_\odot$ and 1.25 $M_\odot$ tracks are also indicated in the figure.

The final data pertaining to these stars are contained in Table I and Figure 3. The errors quoted for the mean values of $v \sin i$ were determined by considering the distribution of velocities to be a Gaussian and computing the standard error of the mean according to the usual formula

$$\sigma = \sqrt{\frac{\sum\limits_{i}^{N}((v\sin i)_i - \overline{v \sin i})^2}{N(N-1)}},$$

where N is the total number of stars in each group. Groups A and D being assumed to represent the main sequence, the ratio

$$K = \frac{\overline{v \sin i}_{ms}}{\overline{v \sin i}},$$

TABLE I

Stars in the main sequence band

Group	No. of members	$\overline{v \sin i}$	K_{obs}	K_A	K_B
A	102	132 ± 6	1.00	1.00	1.00
B	77	129 ± 8	$1.02 \pm .08$	1.04	1.19
C	61	120 ± 7	$1.10 \pm .07$	1.07	1.50
D	136	48 ± 3	1.00	1.00	1.00
E	45	51 ± 5	$0.94 \pm .12$	0.97	1.13
F	33	50 ± 7	$0.96 \pm .15$	1.00	1.36

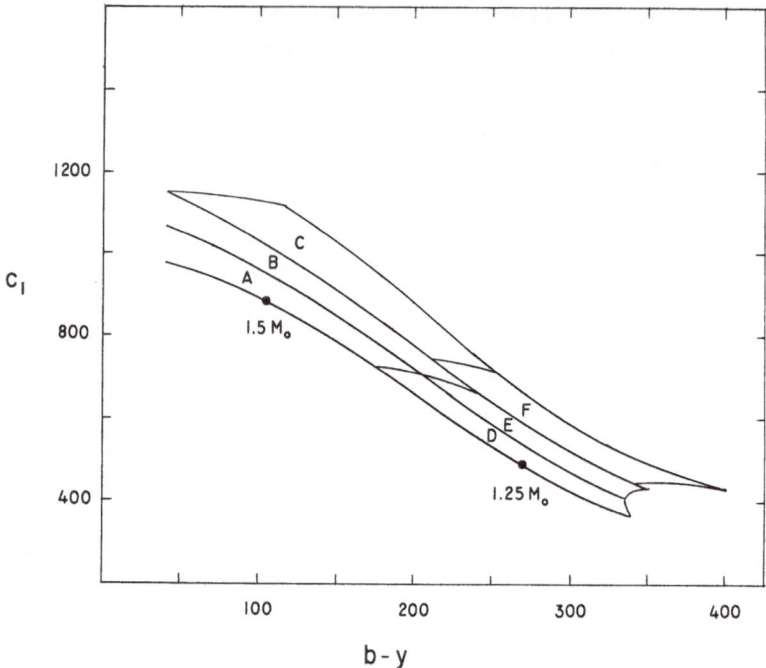

Fig. 2.   The groups into which the stars near the main sequence were divided.

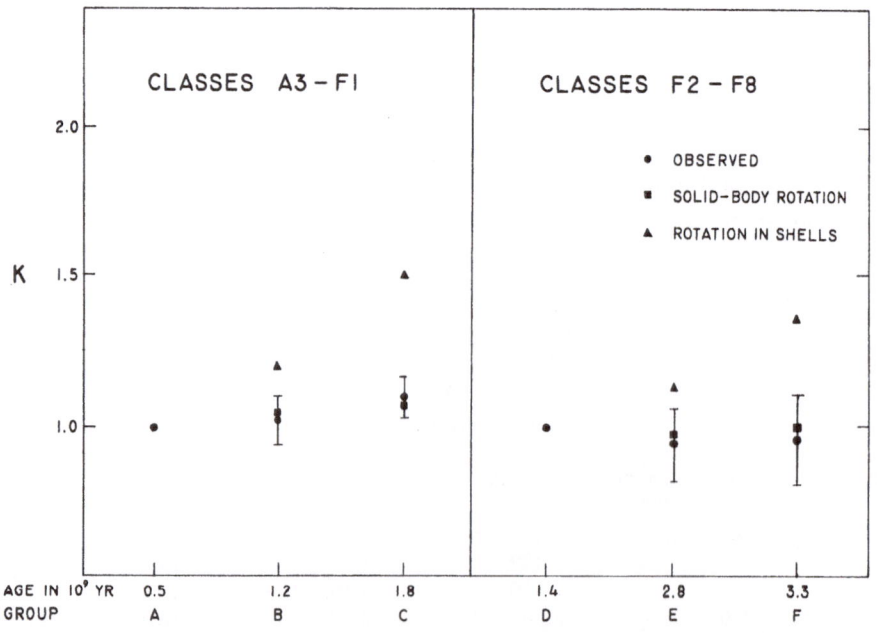

Fig. 3.   The observed and theoretical values of K for stars near the main sequence. Ages on the
x-axis are the mean ages for each group estimated from Iben's models.

where $v \sin i_{ms}$ is the mean value of $v \sin i$ on the main sequence, was determined from Iben's models according to hypotheses A and B. These predicted values, K_A and K_B, are compared with the observed values of K in the table and in the figure. The data appear to exclude the possibility of rotation in uncoupled shells for these stars, whereas they are extremely compatible with the hypothesis of solid-body rotation.

We note, however, that the identical result might be obtained if the photometry should not in fact accurately indicate luminosity and, hence, age. If this were true, we might expect stars of differing ages to be uniformly spread out over the main sequence band, yielding roughly equal $\overline{v \sin i}$'s for all groups in the same sequence. This is exactly what was found. Moreover, Kraft and Wrubel (1965) have calculated the changes in c_1 and $(b-y)$ to be expected for a star on the main sequence in this region having various values of $v \sin i$. They find that stars with high $v \sin i$'s are shifted to a higher position in the band, the maximum effect being just about equal to the width of the band. Unfortunately, this movement is in precisely the direction which would tend to obliterate the decline in $v \sin i$ toward the top of the band predicted by theory B. It is important to note that their results are based on the investigations of Sweet and Roy (1953). Although the later efforts of Roxburgh and Strittmatter (1966) would predict an effect only one-third as large, the uncertainty of the situation makes it desirable to verify that the photometry is indeed a reliable age indicator.

Two procedures were carried out to test the photometry. The first was a simple plot of absolute magnitude above the main sequence predicted by the photometry vs. the magnitude expected from the parallax listed in the *Catalog of Bright Stars* if available. A very strong correlation was found, which had a slope near unity. Secondly, histograms of the velocities in all six groups were compared. If the mechanism of Kraft and Wrubel were operative, we would expect no stars with $v \sin i$ greater than 150 km/sec in group A and none with $v \sin i$ over 100 km/sec in group D. We would also expect relatively larger numbers of fast rotators in groups B, C, E, and F. Neither effect is observed. In fact, within statistical uncertainties, the histograms of A, B, and C are identical, the same being true for groups D, E, and F. This is exactly what we would expect if solid-body rotation is true for these stars.

4. Giants and Supergiants

The analysis of the stars above the main sequence band is not as straightforward. The Strömgren indices have not been calibrated for stars as luminous as these, and even if they were, interstellar reddening would still be a problem for most of these stars. We have therefore used spectral types as indicators of luminosity and temperature in this region. Those stars for which accurate types are not available were classified by a combination of criteria: the listed *Catalog of Bright Stars* type, the reddening-free Strömgren indices $[m_1]$ and $[c_1]$ (Strömgren, 1966), the mean reddening line in the $c_1-(b-y)$ diagram (Strömgren, 1963), and the appearance of the spectra on our plates. Stars that were so classified comprised slightly less than one-half the total number above the main sequence band. Cases for which the various criteria gave conflicting results were eliminated from further consideration.

The stars were divided into five classes according to luminosity, as follows: Ia supergiants, Ib supergiants, class II, class II–III, and class III; and were further subdivided into an early and a late group within each class. $v \sin i$ and values of the ratio K were again calculated for a mean point within each group. K_A and K_B were obtained by interpolating between Iben's tracks. The results are presented in Table II and Figure 4. In the figure, each triad of data points represents the observed and

TABLE II

Stars above the main sequence band

Lum. class	Spec. types	No. of members	$v \sin i$	K_{obs}	K_A	K_B
Ia	A5–F2	4	24 ± 4	6.3 ± 0.8	4.6–7.2	23.0–35.6
Ib	A5–F2	11	18 ± 2	8.3 ± 0.7	3.4–5.5	7.0–11.6
Ib	F5–F8	6	15 ± 2	10.0 ± 0.8	4.5–7.2	9.4–15.6
II	A5–F3	16	37 ± 9	4.6: ± 1.1	2.5–3.6	4.2– 6.2
II	F5–F9	11	42 ± 15	4.0: ± 1.4	3.1–4.6	5.3– 7.8
II–III	F2–F8	4	92 ± 39	2.0 ± 0.8	2.1–2.7	3.4– 4.3
III	A4–A9	7	115 ± 23	1.5 ± 0.3	1.3–1.4	2.0– 2.3
III	F0–F8	21	89 ± 13	1.9 ± 0.3	1.5–1.9	2.4– 2.8

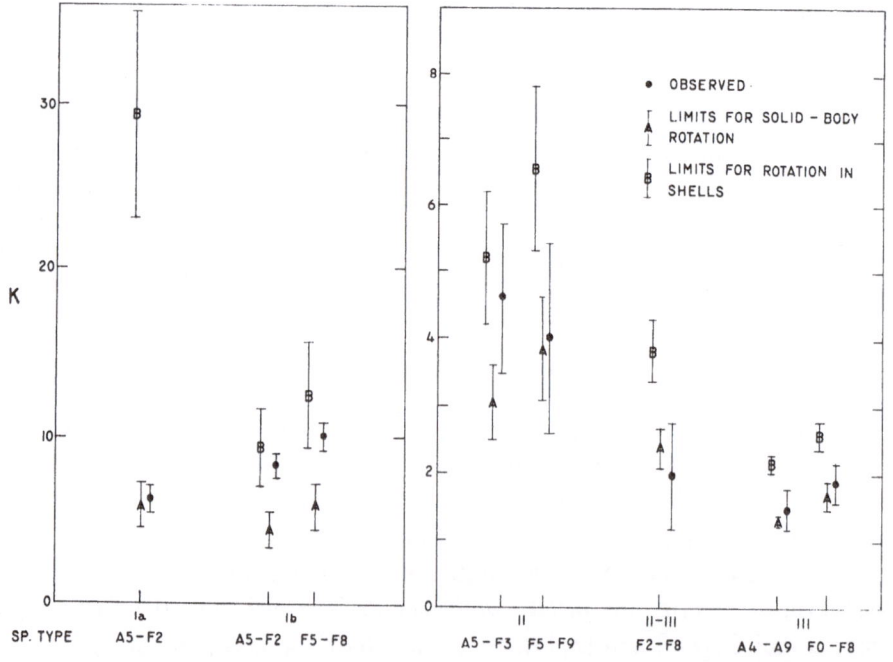

Fig. 4.   The observed and theoretical values of K for giants and supergiants. Each triad of points represents the data for one group of stars as identified below on the x-axis.

predicted values of K for one group of stars. K_A and K_B have been represented in the table and in the figure as a range of possible values. The principal source of uncertainty in these quantities arises from our ignorance of the exact position of the main sequence in the M_{bol}, $\log T_e$ plane and the consequent error in the factor by which a star expands in radius. The upper limits to K_A and K_B have been obtained by using the observed values of the bolometric corrections and temperatures given by Johnson (1966) together with the calibration of M_v for the main sequence by Blaauw (1963). The lower limits are those predicted from observed stellar radii (Harris *et al.*, 1963). These lower limits are also in reasonable agreement with the origin of Iben's tracks in the M_{bol}, $\log T_e$ plane.

The data once again support solid-body rotation for the less luminous stars of classes III and II–III, all of which have expanded by less than a factor of 4. We therefore support the conclusion suggested by Abt (1958) that all stars which have expanded by less than a factor of 4 appear to rotate as solid bodies.

The situation for stars of luminosity class II seems to be intermediate between cases A and B. This is to be expected if these stars have undergone a recent transition from solid-body to some other mode of rotation, such as rotation in uncoupled shells. However, it should be emphasized that the mean values in Table II and Figure 4 for class II are very sensitive to the spectral classification assigned to a few stars with higher-than-average velocities. These are all stars classified primarily on the basis of the photometry. If these stars are omitted, the mean for class II, A5—F3, becomes 19 km/sec instead of 37 km/sec, and for class II, F5–F9, it is 32 km/sec instead of 42 km/sec. These revised numbers would be clear evidence for rotation in shells. A star for which an accurate estimate of spectral type is particularly desirable is HR 6531, which has a $v \sin i$ of 175 km/sec and an estimated type of F5II. If this star is truly a member of class II, it provides clearcut evidence against case B, since the velocity extrapolated back to the main sequence would then considerably exceed break-up velocity.

It is difficult to draw conclusions regarding the supergiants because of the possible effect of turbulence in these stars. The data in Table II and Figure 4 refer to the observed macroscopic broadening uncorrected for turbulence. The true rotational velocities may therefore be considerably lower and the observed data points plotted in the figure higher. Solid-body rotation for the Ib supergiants seems to be fairly definitely excluded, however, except perhaps for the period of time these stars spend near the main sequence. The tantalizing question, of course, is whether the striking agreement between case A and observations for the Ia supergiants is real or merely fortuitous. The observed mean is based on only four stars. However, the value of 24 km/sec obtained adjoins smoothly to Rosendhal's (1968) data (uncorrected for turbulence) for the earlier Ia supergiants. In fact, if we assume that turbulent broadening is unimportant compared to rotational broadening for all Ia supergiants, the mean uncorrected rotational velocities obtained by Rosendhal and by us agree quite well with the velocities predicted under case A at all spectral types earlier than F8.

5. Conclusion

In conclusion, it seems clear that the results of all these investigations rule out the possibility that stars which have expanded in radius by less than a factor of 4 rotate in uncoupled shells. The data are in accordance with the hypothesis of solid-body rotation in these stars, but it is possible that some strange exchange of angular momentum is taking place in their interiors such as to speed up the outer layers at the expense of the central parts. The present technique obviously cannot distinguish between these two processes. For class II giants, the situation is still in doubt, while the velocities for the Ia supergiants, naively interpreted, flatly contradict the results for fainter super-giants. One fact is clear, however. It is no longer lack of velocities for a sufficient number of stars which prevents us from eliminating one of our two hypotheses. It is, rather, a need for more accurate absolute magnitudes together with a method for dealing effectively with the phenomenon of turbulence in these giants. Further research in this area must come to grips with these two problems.

Acknowledgements

We would like to thank the staff of Kitt Peak National Observatory for enabling us to obtain the rotational velocity spectra and the *uvby* photometry. We are also indebted to Professor Iben for his unpublished data and to Mrs. Anne Cowley for spectral types of several stars. Mr. William F. Cahill of the Goddard Space Flight Center, Greenbelt, Md., generously permitted our use of computing facilities at the Center. One of us (S.M.F.) gratefully acknowledges the support of a National Science Foundation Fellowship.

References

Abt, H. A.: 1957, *Astrophys. J.* **126**, 503.
Abt, H. A.: 1958, *Astrophys. J.* **127**, 658.
Blaauw, A.: 1963, in *Basic Astronomical Data* (ed. by K. Aa. Strand), University of Chicago Press, Chicago, Ill., p. 383.
Harris, D. L., Strand, K. Aa., and Worley, C.: 1963, in *Basic Astronomical Data* (ed. by K. Aa. Strand), University of Chicago Press, Chicago, Ill., p. 273.
Hoffleit, D.: 1964, *Catalog of Bright Stars*, Yale University Press, New Haven, Conn.
Johnson, H. L.: 1966, *Ann. Rev. Astron. Astrophys.*, **4**, 193.
Kraft, R. P. and Wrubel, M. H.: 1965, *Astrophys. J.* **142**, 703.
Oke, J. B. and Greenstein, J. L.: 1954, *Astrophys. J.* **120**, 384.
Rosendhal, J. D.: 1968, Thesis, Yale University.
Roxburgh, I. W. and Strittmatter, P. A.: 1966, *Monthly Notices Roy. Astron. Soc.* **133**, 345.
Sandage, A. R.: 1955, *Astrophys. J.* **122**, 263.
Strömgren, B.: 1963, in *Basic Astronomical Data* (ed. by K. Aa. Strand), University of Chicago Press, Chicago, Ill., p. 123.
Strömgren, B.: 1966, *Ann. Rev. Astron. Astrophys.* **4**, 433.
Strömgren, B. and Perry, C. L.: 1962, *Photoelectric uvby Photometry for 1217 Stars Brighter than V = 6.5 mag. mostly of Spectral Classes A, F, and G*, Institute for Advanced Study, Princeton, N.J.
Sweet, P. A. and Roy, A. E.: 1953, *Monthly Notices Roy. Astron. Soc.* **113**, 701.

Discussion

Abt: It seems very likely that among the Ia supergiants, and to a lesser extent the Ib's, the primary contributor to the line broadening is turbulence because at each luminosity the minimum broadening is large and the range in broadening is small, whereas for rotation and random orientation one would expect to observe some narrow-lined spectra and a large range in broadening.

Faber: Our sample of four supergiants is so small that statistical analysis of them is admittedly dangerous. However, the minimum velocity seen, 16 km/sec, and the maximum seen, 40 km/sec, are about what one would expect if the broadening is due to rotation only and the rotation axes were distributed randomly in space. In any case we do not assert that the broadening in these stars is due principally to rotation; only that one can no longer reject rotation solely on the grounds that the observed velocities are higher than can be imagined under any hypothesis.

Jaschek: Did you look into the proportion of spectroscopic binaries in your groups?

Faber: That is something we have planned. I did not show separately the means of $v \sin i$ of the stars we measured and the means of the stars measured in the literature. In almost all cases, the means of this second group of stars are lower than those of the first. We do not think that this could be due to systematic errors in our method because individual $v \sin i$'s measured by us agree very well with those already given in the literature. There is, however, a strikingly higher percentage of multiple stars in this second group, as tabulated in the *Catalogue of Bright Stars*, but whether this high number of binaries is related to the low $v \sin i$'s also found for these stars we do not know at this time.

Buscombe: What is the resolution of the spectrograms, and what interval in the spectrum was studied? Were many of the late A's metallic-line stars?

Faber: The spectra have a dispersion of 13.6 Å/mm. The two extended sections of spectrum I referred to bracketed Hγ and Hδ and included $\lambda\lambda$ 4058–4141 and $\lambda\lambda$ 4263–4382. The other three lines were Fe II, Ti II at λ 4549.6, Fe II at λ 4508.3 and Fe I at λ 4476.0. In general, we avoided metallic-line stars because they are suspected to constitute a peculiar slow-rotating group.

Jaschek: Can you distinguish an Am star from a normal star on the basis of your photometric technique?

Faber: We see no difference in the $c_1 - (b - y)$ diagram used for classifying the main-sequence stars. We have not in general made use of the m_1 index for classification where, presumably, metal abundance effects would be visible.

Roxburgh: Apart from the two cases you considered it would be worthwhile to do other cases; namely, angular momentum conserved in different chemical regions, angular velocity constant in convective zones, and angular momentum conserved in radiative regions or angular velocity constant in radiative regions.

Faber: Definitely.

ROTATION AND EVOLUTION OF Be STARS

JOHANNES HARDORP*

Institute of Theoretical Astronomy, Cambridge, England

and

PETER A. STRITTMATTER

*Mt. Stromlo and Siding Spring Observatories, Canberra, Australia, and
Institute of Theoretical Astronomy, Cambridge, England*

Abstract. The evidence in favor of the hypothesis that Be stars owe their emission properties to material rotationally ejected from the equator is reviewed. The evolutionary state of Be Stars is then discussed with reference to evolutionary sequences of stellar models. It is concluded that (i) Be stars are not confined to the secondary contraction phase as previously proposed (ii) evolution probably proceeds with uniform rotation at least outside the initial convective core. Mechanisms for transporting angular momentum are briefly discussed.

1. Be Stars as Rapid Rotators

The original motivation for this study was to examine how much may be inferred from observational material about the changes in angular momentum distribution inside stars during their evolution from the main sequence. Our discussion here will be concerned in the main with emission-line Be stars and must be considered a preliminary report.

It has long been conjectured that the Be stars owe their emission properties to the presence of a disc distribution of hot gas rotationally ejected from the star. Since such an assumption is crucial to the subsequent discussion we will first of all review the evidence in favor of this hypothesis.

When a star rotates so rapidly that the centrifugal force at the equator balances gravity, the outermost matter loses pressure connection with the underlying layers. This may appear to be a stable configuration. Both radiation pressure and pressure arising from material ejected subsequently will, however, tend to force the matter further outwards and may lead to the formation of a gaseous disc in the equatorial plane which gives rise to the emission. In this paper the velocity at which centrifugal force at the equator balances gravity will, for brevity, also be called 'critical velocity' or even 'break-up velocity', although we don't mean to imply that the star is really breaking apart.

We wish to convince ourselves, by looking at the observed distribution of projected rotational velocities, that Be stars, as a group, are indeed rotating at their critical velocity. We take the values of $v \sin i$ from the work of Slettebak (1954) and Slettebak and Howard (1955), but arrange them according to the MK rather than the Draper spectral type. In Figure Ib the distribution of rotational velocities for B6–B9 stars of

* Now at the Department of Earth and Space Sciences, State University of New York, Stony Brook, N.Y.

luminosity classes III, IV, and V, brighter than $V = 5^m.5$ and north of $-20°$ declination is shown; emission or shell stars are treated separately from the remainder. There are altogether 87 stars, 6 of which, or 7%, show emission or shell characteristics. The observed velocities are here subdivided into intervals of 50 km/sec, with the exception

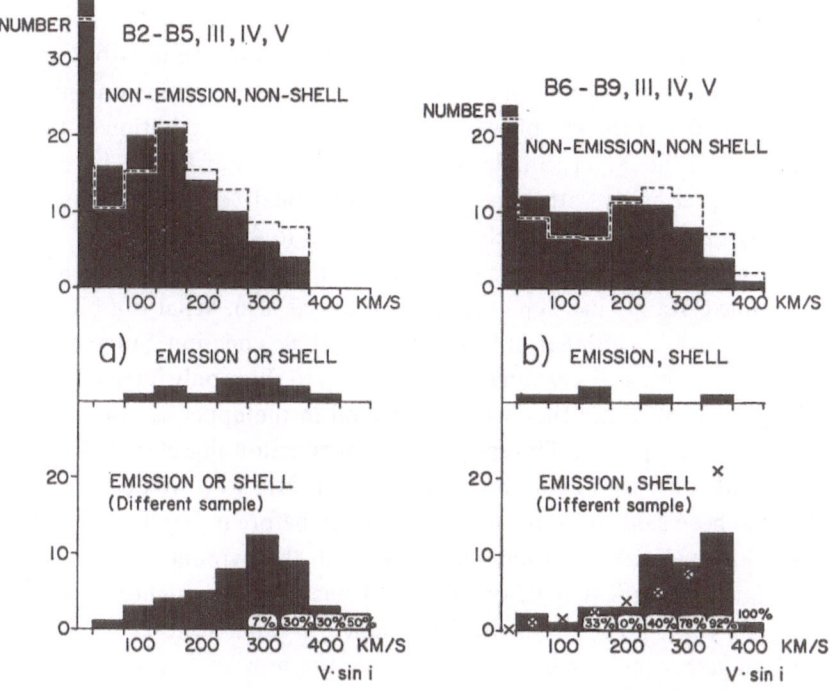

Fig. 1. Distribution of projected rotational velocities of all stars of the indicated spectral types which are brighter than $V = 5^m.5$ and north of $-20°$ declination (top two lines). The bottom histograms correspond to larger samples of emission or shell stars. Top and bottom graphs clearly exhibit different distributions. Bottom graph b is consistent with the view that all emission stars rotate with 375 km/sec, the distribution of $v \sin i$ being an aspect effect (crosses). The percentages of shell stars are indicated and support this view. The broken lines in the upper graphs show the distribution of true rotational velocities.

of the lowest interval which is only 25 km/sec. The number of stars in this last group has been doubled in order to create equal area for each star in the histogram. In this way the overabundance of slow rotators shows up more clearly.

Since it is difficult to discuss the statistics of 6 stars, we have plotted, at the bottom of the diagram, the distribution of a larger sample of emission-line stars, also taken from the work of Slettebak (1949, 1966). The difference between the distributions for the emission and non-emission stars is striking: the emission-line stars have much higher average velocity. The crosses indicate the distribution to be expected if all emission stars in this sample were rotating with $v = 375$ km/sec, and the rotation axes were randomly orientated. We see that the observed data are consistent with the view that all emission-line stars are intrinsically fast rotators, the distribution of $v \sin i$ being an aspect effect.

A further point of evidence is the percentage of shell stars among the Be's: if one looks at the extended disk equator on, the star is hidden behind the disk which causes sharp absorption lines to develop (the shell characteristics). If we interpret the distribution of $v \sin i$ as an aspect effect, we expect a higher percentage of shell stars to show up at higher $v \sin i$; this is indeed observed (see Figure 1b).

The same statements hold for the B2–B5 III, IV, V stars, the distribution of which is displayed in the same manner in Figure 1a. In this case the emission and shell stars comprise about 10% of the sample of 124 stars. Again the distribution of the emission stars, judged from the larger sample shown at the bottom of Figure 1a, can be interpreted as an aspect effect with large intrinsic rotational velocity, although this time we have to assume a mixture of velocities from 325 to 475 km/sec.

Such a mixture is not at all surprising: the breakup-speed decreases during evolution, and our sample certainly has stars of different masses also. What is a bit surprising is the fact that there are a number of fast rotators with *no* emission. Some of them could be at break-up too, because certain stars are known to show only intermittent emission characteristics. At off-times they would be found in the upper sample, but would of course remain breakup-stars. Thus percentages of emission-line stars mentioned above give lower limits only to the number of Be stars at 'break-up' velocity.

All that has been said so far has been mentioned before in the literature, mainly by Slettebak. This evidence is entirely consistent with the hypothesis that Be stars as a class are rotating at critical velocity; this will be assumed throughout the subsequent discussion. In view, however, of its importance in what follows we considered it worthwhile to gather the evidence together here. We now turn to the question of the evolutionary state of the Be stars.

2. Be Stars and Stellar Evolution

Schmidt-Kaler (1964) concluded from the observed location of the Be stars in the HR-diagram (namely $\simeq 1^m$ above the main sequence of non-emission stars) that the Be stars are all in the secondary contraction phase following hydrogen exhaustion in the core (phase 2 to 3 in Iben's tracks as shown in Figure 2). This idea, first mentioned by Crampin and Hoyle (1960), is theoretically attractive: if there is a mechanism that keeps stars in rigid rotation at least during the main-sequence phases (case A), then the rotational velocity should increase rapidly in the secondary contraction phase. The shrinking of the star makes the moment of inertia decrease; the star has to spin faster in order to conserve its total angular momentum. This might bring the rotation to critical speeds, whereupon an extended shell might develop.

Of course this would not work if angular momentum were conserved in independent shells (case B), because then the star would be closest to the critical velocity only at its minimum radius, which is on the zero-age main sequence. We shall therefore simply assume that rigid rotation always holds (with the possible exclusion of the convective

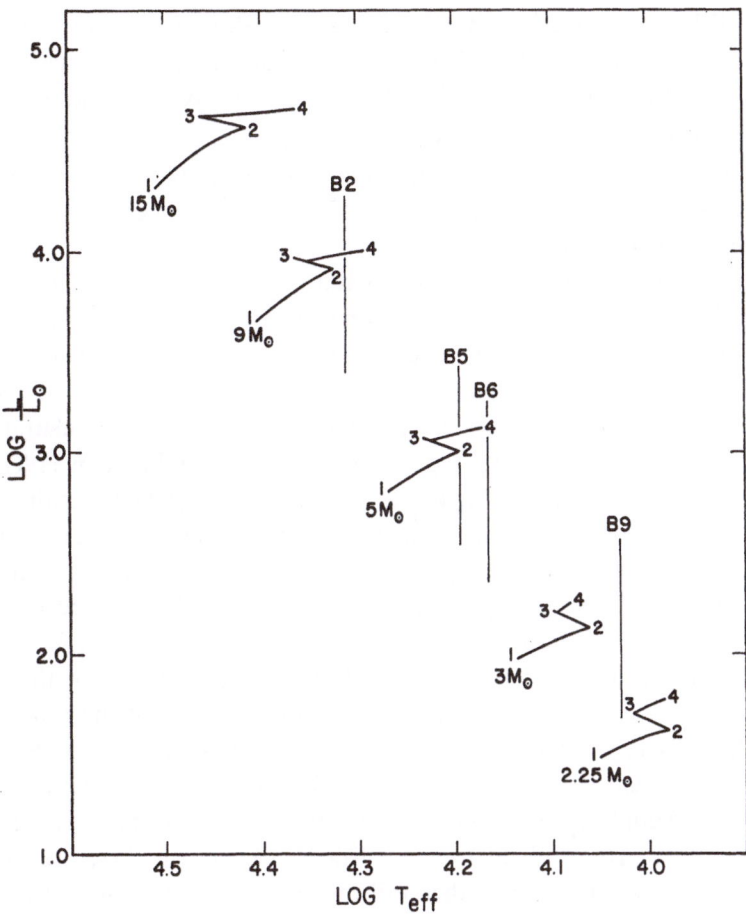

Fig. 2.   The main-sequence part of Iben's evolutionary tracks. For numbered points the moment of inertia was evaluated. The spectral types were taken from Morton and Adams (1968).

core), as was tacitly done by Schmidt-Kaler. If we can show that the above idea does not work even under these more favorable assumptions, then it probably can be excluded.

First we consider some data derived from evolutionary calculations for spherical stars by Iben (1967) for various masses, by Kippenhahn *et al.* (1967) for 2 $M_\odot$, by Hofmeister (1967) for 5 and 9 $M_\odot$, and by Hofmeister *et al.* (1964) for 7 $M_\odot$. Listed in Table I, column 3 is the percentage of its main-sequence lifetime that the star spends in what we call the 'evolved phase'. This includes not only the secondary contraction phase but also the establishment of a thick shell-burning zone; it corresponds to stages 2–4 as indicated in Figure 2. The values are obtained from Iben's calculations. Also shown in Table I are the values of $\lambda = \Omega^2 R^3 / GM$ (arbitrary units), the ratio of centrifugal force to gravity at the equator computed on the basis of uniform rotation and conservation of angular momentum. The values have been derived from computations for spherical stars and will, therefore be inaccurate near the break-up velocity. The accuracy will, however, be adequate for our purpose. We note a considerable difference between values of λ derived by different investigators; these seem much larger than would be ascribed to composition differences. Fortunately the trend is in each case the same. In Table I we also list $1/R$ and $\sqrt{\lambda/R}$ (arbitrary units), because these are the rotational speeds under assumptions B and A, respectively. The critical velocity v_{cr} of Table I was computed with the value 1.45 for the ratio of critical to non-rotating radius, after Faulkner *et al.* (1968).

A further datum required for this discussion is the distribution of true rotational velocity in each of the above samples. This may be obtained, at least approximately, from the distribution of $v \sin i$, and the results are shown by the dashed lines in Figure 1. Finally for the B2–B5 group we adopt a mean mass $\bar{M} \approx 5\,M_\odot$ while for the B6–9 stars $\bar{M} \approx 3\,M_\odot$ is more appropriate. This choice is justified from the relation between effective temperature and spectral type given by Morton and Adams (1968) as illustrated in Figure 2, and is in fair agreement with Popper's (1967) result. (Although masses as high as $9\,M_\odot$ might occur in the B2–B5 sample, most stars will have the lower value simply because of the rapid decrease in luminosity function with increasing mass.) We are now in a position to examine whether the Be stars occur during the 'evolved phase'.

For the B2–B5 group the answer is clearly negative since 10% of the stars have emission lines whereas stars of this mass spend only $\sim 4.5\%$ of their lifetime in the 'evolved phase'. This conclusion is further strengthened when we consider the distribution of intrinsic velocity. Among B2–B5 stars 65% have intrinsic rotational velocities $v < 225$ km/sec. Adopting a mean critical velocity $v_{cr} \approx 350$ km/sec*, we have $\lambda < 0.42$ for this subgroup. From the changes in λ listed in Table I it therefore follows that none of these stars will ever attain their critical velocity and cannot therefore become Be stars. The ratio of Be stars observed to those expected in the evolved phase is thus increased by at least a factor 3.

Among B6–9 stars, 7% are emission line objects while the stars spend $\sim 8\%$ of their main-sequence lifetime in the evolved phase. However, 68% of these stars have intrinsic velocities $v < 275$ km/sec, while the critical velocity v_{cr} is again in excess of 350 km/sec, therefore $\lambda < 0.62$. Using the evolutionary changes in λ for the appropriate mass range as listed in Table I, it is again clear that none of these stars can reach their critical velocity. This reduces the expected number of Be stars in the evolved phase to at most 2.6% compared to the observed proportion of 7%. Again there are too many Be stars and we therefore conclude that the Be phenomenon cannot be confined to the evolved phase and hence most certainly not to the secondary contraction stage.

We now investigate the possibility that the Be stars became rotationally unstable during main-sequence evolution, that is, between stages 1 and 2. According to Iben's calculations this would require that for the B2–B5 group approximately 10% of the stars were within 1% (for $5\,M_\odot$) of their break-up velocity when they reached the main

---

* This value of v_c is a lower limit on both observational and theoretical grounds, and thus maximizes λ. (cf. Faulkner *et al.*, 1968; Hardorp and Strittmatter, 1968).

TABLE I

Evolution of the rotation parameter and of equatorial speeds

Author		Iben						Kippenhahn et al., Hofmeister				Hofmeister			
Composition		$X = 0.71$, $Y = 0.27$						$X = 0.602$, $Y = 0.354$				$X = 0.739$, $Y = 0.240$			
$M/M_\odot$	Phase	Time 2–4	λ	λ^{E}	$1/R$	$\sqrt{\lambda}/R$	v_{cr} km/sec	λ	λ^{E}	$1/R$	$\sqrt{\lambda}/R$	λ	λ^{E}	$1/R$	$\sqrt{\lambda}/R$
1.25	1		1		100	100	389								
	2	30%	1.32		77	101	342								
	3		1.37		78	103	344								
	4		1.61		56	95	291								
2	1							1	1	100	100				
	2							1.13	1.11	67	87				
	3							1.35	1.31	71	98				
	4							1.47	1.42	53	88				
2.25	1		1	1	100	100	449								
	2	10%	1.05	1.03	61	80	352								
	3		1.27	1.23	67	92	366								
	4		1.32	1.28	53	84	326								
3	1		1		100	100	472								
	2	8	1.08		59	80	364								
	3		1.30		64	92	379								
	4		1.34		56	87	354								
5	1		1	1	100	100	520	1	1	100	100	1	1	100	100
	2	4.5	1.02	0.97	58	77	394	1.20	1.15	61	86	1.45	1.37	63	96
	3		1.35	1.27	62	91	407	1.49	1.43	68	101				
	4		1.27	1.18	44	75	344								

(Table I continued)

		Iben						Kippenhahn *et al.*, Hofmeister				Hofmeister			
Author		$X=0.71,\ Y=0.27$						$X=0.602,\ Y=0.354$				$X=0.739,\ Y=0.240$			
Composition															
$M/M_{\odot}$	Phase	Time 2–4	λ	λ^{E}	$1/R$	$\sqrt{\lambda}/R$	v_{cr} km/sec	λ	λ^{E}	$1/R$	$\sqrt{\lambda}/R$	λ	λ^{E}	$1/R$	$\sqrt{\lambda}/R$
7	1								1			1	1		
	2							1.35	1.22			1.41	1.32	100	100
9	1		1	1	100	100	584		1	100	100	1	1	100	100
	2	3.2	1.22	1.11	51	79	418	1.41	1.30	56	89	1.64	1.49	59	98
	3		1.74	1.56	57	100	443	1.18	1.65	62	106				
	4		1.36	1.19	39	73	365								
15	1		1	1	100	100	649								
	2	2.4	1.50	1.27	47	84	442								
	3		2.08	1.71	54	106	476								
	4		1.75	1.39	33	76	373								

sequence. For B6–B9 stars a similar argument requires 7% of the stars to have initial velocities within 4% of the critical value. The precise numbers here clearly depend on details of rotational distortion of the surface, the actual distribution of masses within the spectral group, etc. However, from Iben's calculations it would appear that a considerable peak is required at the break-up velocity in the initial rotational velocity distribution function. This in itself would not be surprising since those stars with excess angular momentum would merely have shed matter during the later stages of pre-main sequence contraction allowing them all to arrive at the ZAMS rotating at their critical speed. Subsequent evolution would, in case A, keep them at break-up velocity and cause further shedding of material to replenish the emission region. (This would not, of course, occur in case B.) More disconcerting, however, is the fact that results derived from similar calculations by Hofmeister present a rather different picture. On the basis of her computations 10% of B2–B5 stars would have to have initial rotational velocities within 10% of the break-up value, if main-sequence evolution is to produce the emission phenomenon; this clearly requires no sharply-peaked initial distribution function. We are unable to comment on the origin of these discrepancies between the various computations but feel that they merit further investigation. We can, however, conclude that evolution under case A assumptions will maintain the Be star phenomenon throughout the main sequence and evolved phases. The question of whether this involves an excess of stars rotating at their break-up velocity on the ZAMS cannot be settled until discrepancies in the evolutionary models are resolved.

The conclusions drawn here do not apply to types earlier than B2. Table I shows that the increase of λ during the main-sequence phase is substantially larger as one goes to 9 or 15 solar masses. Anand and Sackmann (1970) computed the evolution of a 10 solar mass star, including the effects of rotation explicitly. They found that it is hard to prevent such a star from reaching the critical velocity (again under the assumption of rigid rotation all the way through): 250 km/sec at phase 1 leads to breakup as early as phase 2. For the validity of our conclusions it is therefore important to assign the right values of masses to the stars in question. It would be interesting to see Anand and Sackmann's method applied to the case of 5 $M_\odot$, in order to have a check on our rougher method.

One possible objection has still to be discussed: If we are right that Be stars are more or less in the same stage of evolution as non-emission B stars, why is it that they appear to lie *above* the main sequence for non-emission stars by approximately one magnitude? Meisel (1967) found that 10 Be stars in visual binaries lie an average $0^m.7$ above the mean MK-main sequence. This, however, is just the shift one expects from the effect of gravity darkening for stars at critical velocity (see, for example, Collins 1966). This need not therefore be an evolutionary effect, or even cannot be an evolutionary effect unless the amount is considerably larger than 1^m.

Finally we note that evolution under case B seems rather unlikely. Certainly the rotational parameter would decrease below its critical value as soon as the star left the ZAMS and the star would never become unstable again. In view of the fact that many Be stars lose their emission characteristics for long periods it seems to us that continuous replenishment of gaseous material in the disc is required to maintain the emission phenomenon. Since this is precluded under case B we feel that detailed conservation of angular momentum can probably be excluded. This is in agreement with results of stability analyses by Goldreich and Schubert (1967) and by Fricke (1967).

3. Angular Momentum Transport

We have assumed throughout this paper that there is sufficient transport of angular momentum to maintain rigid rotation during the main-sequence phase. For later type stars this has been shown to be likely by Faber and Danziger (1970). We did not dare to subdivide our small sample to do the same analysis for the earlier types, but in principle this could be done. In this appendix we merely want to summarize what mechanisms can be made responsible for maintaining rigid rotation.

TABLE II

Mechanisms for transport of angular momentum

Mode	Time scale	Years
1. Viscous stresses	R^2/ν	10^{12}
2. Dynamical instability	$(R/h)^{3/2}/\Omega$	10^2
3. Spin down	$(R^2/(\nu\Omega))^{1/2}$	10^4
4. Magnetic stresses	$R(4\pi\varrho)^{1/2}/H$	$10^4/H$

Table II lists four processes with their repective time scales (here $h=$ scale height, $\nu=$ microscopic viscosity, $\varrho=$ density, $H=$ magnetic field).

Clearly viscous stresses are too slow, whereas each of the remaining processes could act on a time scale short compared to the main-sequence lifetime. The dynamical instability is the one discussed by Goldreich and Schubert (1967) and by Fricke (1967); its time scale is rather uncertain. The spin-down process has been discussed for an analogous case by Howard *et al.* (1967). Processes 2 and 3 involve transport of material from the inner regions towards the surface and would therefore make for well mixed stars whose evolution is not in line with observations. However, process 2 could nevertheless be at work: Goldreich and Schubert have shown that dynamical instability is inhibited by a stable molecular weight gradient of the type established in the core region due to the shrinking of the convective core during evolution. If this type of dynamical instability is responsible for the outward transfer of angular momentum then we should expect the convective core not to participate in this transport. How would this fact alter the conclusions drawn in the preceding paragraph?

We recomputed the moment of inertia of the models, omitting those regions which were convective on the zero-age main sequence. The corresponding numbers for the rotation parameter λ^E of these exterior regions are given in Table I. The results are very similar to those calculated previously, because the core contributes too little to the moment of inertia. For this reason no observational test of whether the core is rotationally decoupled from the rest of the star seems possible at the present.

There could, however, be another reason why the convective core has to be excluded from angular momentum considerations: It has been suggested by Gough and Lynden-Bell (1968) that vorticity is expelled from convective regions. If this were the case, the convective core would not rotate at all underneath a rotating exterior. In cases like the sun with a convective exterior it would be the other way round: the exterior would expel its angular momentum to the inner regions, which might explain Dicke's oblateness measurements.

We cannot go into theoretical or experimental details of this process here but merely wish to list two more points of astronomical evidence for it. The first is the observed break in the main-sequence rotational velocity distribution at spectral type $\simeq$ F2. This coincides with the development of a strong surface convective zone. The second is Kraft's (1968) observation of rotational velocities of the four Hyades giants, as compared with what one expects from the velocities on the main-sequence: He observed $\leqq 8$ km/sec, whereas velocities of 40 km/sec would be expected on the rigid rotation assumption, and 20 km/sec with detailed conservation of angular momentum. The explanation could be that angular momentum is transferred to the interior radiative regions as soon as the star moves to the right in the HR diagram and develops a deep outer convection zone (we do not claim that this is the only possible explanation). Once these stars evolve further and eventually move to the left again, onto the horizontal branch, they cease to have a convective envelope. Then the angular momentum locked up in the interior could speed up the outer parts again on a time scale for dynamical instabilities. This might explain the high rotational velocities of horizontal branch stars in M67 even though these are presumably evolved G stars (Deutsch, 1967). Clearly, the mechanism of vorticity expulsion could be of considerable astrophysical importance. Further evidence, both observational and from laboratory experiments, is required before its true significance can be assessed.

Acknowledgements

We wish to thank Drs. Hofmeister, Iben and Kippenhahn for allowing us to use unpublished details of their models.

References

Anand, S. P. S. and Sackmann, I. J.: 1970, this volume, p. 63.
Collins, G. W. II: 1966, *Astrophys. J.* **146**, 914.
Crampin, J. and Hoyle, F.: 1960, *Monthly Notices Roy. Astron. Soc.* **120**, 33.
Deutsch, A. J.: 1967, *Astron. J.* **72**, 383.
Faber, S. and Danziger, J.: 1970, this volume, p. 39.
Faulkner, J., Roxburgh, I. W., and Strittmatter, P. A.: 1968, *Astrophys. J.* **151**, 203.
Fricke, K.: 1967, *Z. Astrophys.* **68**, 317.
Goldreich, P. and Schubert, G. 1967, *Astrophys. J.* **150**, 571.
Gough, D. O. and Lynden-Bell, D.: 1968, *J. Fluid Mech.* **32**, 437.

Hardorp, J. and Strittmatter, P. A.: 1968, *Astrophys. J.* **153**, 465.
Hofmeister, E.: 1967, *Z. Astrophys.* **65**, 164.
Hofmeister, E., Kippenhahn, R., and Weigert, A.: 1964, *Z. Astrophys.* **59**, 242.
Howard, L. N., Moore, D. W., and Spiegel, E. A.: 1967, *Nature* **214**, 1297.
Iben, I.: 1967, *Ann. Rev. Astron. Astrophys.* **5**, 571.
Kippenhahn, R., Kohl, K., and Weigert, A.: 1967, *Z. Astrophys.* **66**, 58.
Kraft, R. P.: 1968, Otto Struve Memorial Vol. (ed. by Herbig).
Meisel, D. D.: 1967, *Astron. J* **72**, 1126.
Morton, D. C. and Adams, T. F.: 1968, *Astrophys. J.* **151**, 614.
Popper, D. M.: 1967, *Ann. Rev. Astron. Astrophys.* **5**, 85.
Schmidt-Kaler, Th.: 1964, *Veröffentl.* Bonn **70**, 1.
Slettebak, A.: 1949, *Astrophys. J.* **110**, 498.
Slettebak, A.: 1954, *Astrophys. J.* **119**, 146.
Slettebak, A.: 1966, *Astrophys. J.* **145**, 121.
Slettebak, A. and Howard, R. F.: 1955, *Astrophys. J.* **121**, 102.

Discussion

Collins: (1) I feel, without further comment, that one should be careful of small number statistics, particularly when one of the cases agrees with the theory to be disproven.

(2) It is not at all clear that Be stars are rotating at the critical velocity. In the event they are not, the argument used to eliminate the several cases of small statistics disappears. This is a result of the fact that 'gravity-darkening' corrections to $v \sin i$ do not apply.

(3) Care must be exercised when one talks about 'gravity-darkening' corrections in the spectral type-luminosity plane. Indeed, the correction is in the other direction (i.e., to earlier types).

(4) If the Be stars lie less than 1.5 magnitude above the main sequence you cannot have the Crampin-Hoyle (or Schmidt-Kaler) result as they require the existence of rapid rotation and the rotation effects of about 1 magnitude added to the evolutionary effects of about 1 magnitude to imply a height of 2 magnitudes above the main sequence.

Hardorp: In answer to your second comment, if Be stars were not rotating at critical velocity, our conclusions would not be at all changed: we could not apply the gravity-darkening correction to $v \sin i$, which simply means that not $\frac{2}{3}$ but about $\frac{1}{2}$ of the B6–B9 stars are now too slow to ever rotate as fast as the Be stars. In the case of the B2–B5 stars, our point would even be strengthened since we are then justified in using the spherical models right up to the point where emission sets in.

Jaschek: I think you underestimated the proportion of Be stars because the figures are higher if you count as Be stars not only those which at the time of the survey show emission lines, but also all those which showed emission at any other time.

Hardorp: The quoted proportions are certainly underestimates, which strengthens our conclusions.

Roxburgh: We do not know what limits the angular velocity of a star. It may not be equatorial shedding, but an instability that sets in earlier, such as a pulsational instability. Could your arguments be reversed to calculate the maximum rotational velocity such that enough Be stars are produced by evolution?

Hardorp: No, because the maximum rotational velocity is taken from the observations and is therefore fixed.

Van den Heuvel: Could Be stars not be stars contracting toward the main sequence? Such stars are also expected to rotate with the break-up velocity.

Hardorp: Only a very small percentage of B stars can be expected to be contracting towards the main sequence – therefore, the Be stars cannot be identified with them. However, since all B stars pass through the contraction phase, the extended rings could be remnants of that phase.

Roxburgh: In the calculation of the increase in rotational velocity it is important to remember that the moment of inertia does not have to decrease by a large factor in order to increase the rotational velocity by a similar factor, as the equatorial radius increases very rapidly for a small change in moment of inertia if the star is near maximum rotational velocity. Have you included this effect in your calculations?

Hardorp: Yes, insofar as we showed that not enough stars ever come near the critical velocity through evolutionary effects, so that your argument does not apply. In fact, near critical velocity the equatorial radius blows up by a factor 1.45 which makes for an extra λ-increase by a factor $1.45^3 = 3$ near critical velocity. This just means that near this velocity our way of reasoning is not applicable in any case, because it relies on spherical models.

Dicke: This is a question addressed both to the speaker and the other participants. The problem of rotation on the horizontal branch is very important. What is the status of the observations concerning such a rotation?

Deutsch: The so-called 'blue stragglers', like those in M67, are rotating two or three times too slowly for A stars, but 40 or 50 times too fast for G stars. At least one of these objects is probably a close-binary remnant, in the sense of Van den Heuvel, and I think others are likely to be so. For this reason and others, I now disbelieve my earlier conjecture that these stars are metamorphs of red giants that have lost their outer layers but have retained their initially high angular momentum in the interior.

Relative to the misunderstanding between Dr. Ostriker and Dr. Hardorp, one should note that if a star rotates rigidly while its density profile changes as the result of evolution, then some process must indeed occur to transfer angular momentum within the star.

When stars are red giants, they have chromospheres and, probably, stellar winds. These can transport angular momentum outwards very effectively, as the work of O. C. Wilson and R. P. Kraft has shown for solar-type dwarfs. The angular momentum therefore need not be expelled into the interior to account for the rotation seen in any metamorphs of those stars that are found near the main sequence – if there are any such stars!

Hardorp: I agree. The expulsion of angular momentum from convective layers to the interior was only proposed as an alternative mechanism.

Ruben: Evolutionary time scales depend on chemical composition. How much are your theoretical values of 5% for B2–B5 and 8% for B6–B9 influenced by the composition of the models?

Hardorp: The higher the helium abundance, the harder it is to disprove Schmidt-Kaler's hypothesis. Not only does the relative duration of the secondary contraction phase rise with helium abundance, but there is also a larger increase of the rotation parameter during evolution.

Ostriker: Have there been any masses determined for Be stars?

Cowley: Sure!

EVOLUTION OF A ROTATING STAR OF NINE SOLAR MASSES

R. KIPPENHAHN, E. MEYER-HOFMEISTER, and H.-C. THOMAS

Universitätssternwarte Göttingen and Max-Planck-Institut für Physik und Astrophysik, München, Germany

To compute the evolution of a rotating star, the following approximations were used in order to obtain some of the main effects of rotation with simple models of spherical symmetry:

(a) The angular velocity distribution was assumed to be spherically symmetric: $\omega = \omega(r)$, and the stars assumed to be spherical.

(b) In the (spherical) stellar structure equations only the radial component $c_r = \omega^2 r \sin^2 \vartheta$ of the centrifugal force $\mathbf{c}$ per gram was taken into account, and in order to keep spherical symmetry was replaced by its mean value $\bar{c}_r = (\frac{2}{3}) \omega^2 r$ over a sphere. Oblateness effects were neglected.

One therefore gets the following equation for hydrostatic equilibrium:

$$\frac{dP}{dM_r} = -\frac{GM_r}{4\pi r^4} + \frac{\omega^2}{6\pi r}, \tag{1}$$

which replaces the normal condition of hydrostatic equilibrium in our stellar evolution program described elsewhere (see Kippenhahn *et al.*, 1967).

If one computes time sequences of stellar models, a prescription of how the function $\omega(r)$ or $\omega(M_r)$ changes with time is necessary. The computations have been started on the main sequence. Until now we have used only solid body rotation for the initial rotation on the zero age main sequence. The change of the angular velocity distribution with time was computed under the following assumption: either the angular momentum has to be conserved locally (i.e. in spherical shells) or over a certain region in the star which is rotating as a solid body. If one prescribes in which regions one has local angular momentum conservation and in which regions one has $\omega = \text{const.}$ then the evolution of the angular velocity distribution of the model is determined.

We followed the evolution of a star of nine solar masses with a chemical composition of $X = 0.739$, $Z = 0.021$. Two cases of rotation were computed:

Case (α): Convective regions are in solid body rotation, angular momentum is conserved locally in radiative regions;

Case (β): Regions which are chemically homogeneous are in solid body rotation; angular momentum is conserved in regions in which the molecular weight increases inwards.

The latter case is based on the picture that in a rapidly rotating star the Eddington-Vogt circulation and the Goldreich-Schubert-Fricke instability (Goldreich and Schubert, 1967; Fricke, 1968) produce a sufficiently strong mixing of angular momentum and since both types of motion are hindered by μ-gradients (Mestel, 1953; Goldreich and Schubert, 1967), the regions of varying molecular weight are the only regions where the angular momentum has to be conserved locally.

A. Slettebak (ed.), Stellar Rotation, 60–62. All Rights Reserved

For the main sequence model the maximum angular velocity possible for solid body rotation was used in both cases. The computations were terminated in a phase, where a carbon-oxygen core is surrounded by a helium burning shell which produces most of the energy radiated from the surface of the star. For comparison the evolution of the nonrotating model was also computed.

In the case of rotation the time scales for nuclear burning at the center of the star

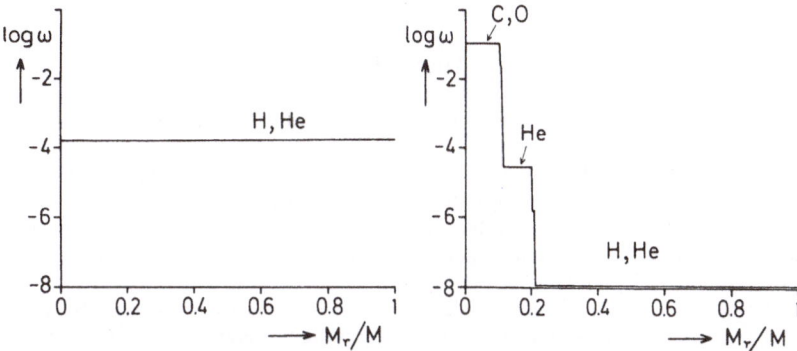

Fig. 1. The angular velocity distribution as a function of the depth in a star of 9 $M_\odot$. In the *left diagram* the star is in the zero-age main sequence stage rotating with break-up velocity. The angular velocity is assumed to be constant. In the *right diagram* the angular velocity distribution in an evolved stage is given. The star is now a red supergiant, a carbon-oxygen core has already been formed, which is surrounded by a helium envelope which again is surrounded by an envelope consisting of the original hydrogen-rich mixture. The evolution of the ω-distribution has been computed according to the assumptions of our case (β). ω varies only in regions of variable chemical composition. In the evolved stage the star has a rapidly spinning core with a rotational period of about 60 sec. With this period the star has reached a stage where the centrifugal force balances gravity in the equatorial plane at the boundary of the carbon-oxygen core.

were longer: about 4% for hydrogen burning and about 12% for helium burning, with slight differences between cases (α) and (β) (see Figure 1).

Two problems follow from these computations. In case (β) the rotation law near the main sequence will be governed by the fact that the convective core and the envelope, both in solid body rotation, are separated by a zone of varying molecular weight, which prevents the transport of angular momentum from the core into the envelope. But the structural changes in the envelope are such that the ratio of centrifugal to gravitational force at the equator increases with time, causing a mass loss due to rotation in and close to the main sequence band, since the star was assumed to be fully rotating at the beginning. This might give an explanation for Be-stars like Pleione without the assumption of solid body rotation for the entire star. The question is how much mass will be lost before the star becomes a red giant.

The second problem is raised by the formation of rapidly rotating cores, as indicated in Figure 1. After central helium burning the material in the carbon-oxygen core reaches an angular velocity such that in the equatorial plane the centrifugal force balances gravity. This is true for both cases, although the details of the rotation law for these two series differ from each other. This raises the question of what happens if the core of an evolved star spins faster and faster.

References

Fricke, K.: 1968, *Z. Astrophys.* **68**, 317.
Goldreich, P. and Schubert, G., 1967, *Astrophys. J.* **150**, 571.
Kippenhahn, R., Weigert, A., and Hofmeister, E.: 1967, in *Methods in Computational Physics*, vol. 7, (ed. by B. Alder, S. Fernbach, and M. Rothenberg), Academic Press, New York and London, p. 129.
Mestel, L.: 1953, *Monthly Notices Roy. Astron. Soc.* **113**, 716.

Discussion

Deutsch: Is it possible to scale these solutions to approximate stars starting from lesser rotational velocities?

Thomas: It may be possible, but in case (β) I would prefer to re-do the computations near the main sequence with a smaller starting value for the angular velocity.

Roxburgh: The solutions you have can be scaled at least approximately since the effect of rotation on evolution is small and the change in angular velocity can be calculated from the spherical evolution, with the assumptions you have made.

Ostriker: Some of the very evolved models ($\alpha_c \simeq 0.1$) would appear to be unstable to the Kelvin modes which are thought to lead to fission. This leads to the intriguing possibility of the creation of white dwarf-like binary stars *within* red giants!

EVOLUTION OF RAPIDLY ROTATING B-TYPE STARS

I. JULIANA SACKMANN* and S. P. S. ANAND

David Dunlap Observatory, University of Toronto, Canada

(Paper read by M. J. Clement)

1. Introduction

It has become clear in recent years through the work of Collins (1963, 1965, 1966), Collins and Harrington (1966), Hardorp and Strittmatter (1968a, b) and Roxburgh and Strittmatter (1965) how seriously the spectrum of a star may be affected by rotation. However, before atmospheric models for rotating stars can be computed, the results of corresponding interior models have to be available. For polytropic configurations a great many rotating interior models have been constructed (Chandrasekhar, 1933; James, 1964; Stoeckly, 1965; Monaghan and Roxburgh, 1965; and Anand, 1968), but for rotating main-sequence stars few computations exist (Sweet and Roy, 1953; Roxburgh *et al.*, 1965; Ostriker and Mark, 1968; Faulkner *et al.*, 1968). In fact for B- and Be-type stars for which the most rapid rotation along the main sequence is observed, not a single interior model has yet been computed except in this conference reported by Kippenhahn and Thomas (1970). It is this gap that the present paper attempts to close.

Sixty-four interior main-sequence models of slowly and rapidly rotating B-type stars are presented. The mass of the models ranges from 5 to 15 $M_\odot$ and the rotation is taken as uniform. The contribution of radiation pressure is fully included in the basic equations. The basic set of equations are solved by applying a Henyey type of iteration procedure as developed for non-rotating stars by Larson and Demarque (1964). The details of all these calculations are under publication (Sackmann and Anand, 1970).

2. Results

Tables Ia and Ib define the characteristics of the 64 main-sequence models. Table II gives some of the most important results for various models at zero and at critical rotation. Most of the parameters in these tables have the usual meanings (Sackmann and Anand, 1970).

A. THE LUMINOSITY

The luminosity as used here means the total bolometric interior luminosity. The decrease of this luminosity as the rotation increases when measured by the usual η-criterion, is displayed in Figure 1 for models of 5, 10, and 15 $M_\odot$. The first feature to notice is the very close coincidence of the curves for the different mass models.

* Present address: Universitäts Sternwarte, Göttingen, Germany.

A. Slettebak (ed.), Stellar Rotation, 63–72. All Rights Reserved

TABLE Ia

Model characteristics

Model	$M/M_\odot$	Chemical composition		Radiation pressure
		X	Z	
I	15	0.67	0.03	included
II	12	0.67	0.03	included
III	10	0.67	0.03	included
IV	9	0.67	0.03	included
V	7	0.67	0.03	included
VI	5	0.67	0.03	included
VII	5	0.67	0.03	omitted
VIII	5	0.71	0.02	omitted

TABLE Ib

Model characteristics

Model	α_s	η	v	R_p/R_0	R_e/R_0	r_i/R_0
1	0	0	1.000	1.0000	1.0000	1.00
2	0.05	0.0781	1.002	0.9756	1.0137	1.00
3	0.10	0.1642	1.007	0.9524	1.0306	1.00
4	0.15	0.2620	1.019	0.9302	1.0521	1.00
5	0.20	0.3797	1.040	0.9091	1.0817	0.94
6	0.25	0.5386	1.079	0.8889	1.1283	0.87
7	0.30	0.9371	1.193	0.8696	1.2770	0.82
8	0.300692	1.0000	1.200	0.8693	1.3040	0.82

Although such a coincidence is observed for no other physical parameter, it must be related to the similarity of the structure of these stars.

The second feature to notice is that the maximum decrease in the luminosity due to rotation is extremely small, being only 6.7%±0.1% for the models from 5 to 15 $M_\odot$. This maximum decrease is much smaller than that of 24.7% obtained for critical rotation by Roxburgh *et al.* (1965) or that of 16.8% obtained for only slow rotation by Sweet and Roy (1953). Because they were the only ones available until recently, these two results have often been used in atmospheric calculations.

The third feature to notice is the detailed behavior of the luminosity change as the rotation builds up. This decrease in luminosity is certainly not linear with η. About 80% of the total change has already occurred when η is only 0.5 of the critical value. A quadratic polynomial

$$L(\eta) = L(0) \left[0.9977 - 0.1661\eta + 0.1028\eta^2 \right]$$

is found to describe the 10 $M_\odot$ curve better than 6%.

B. THE POLAR RADIUS

The decrease of the polar radius with increasing rotation is displayed for the 64 main-sequence models in Figure 2. The first feature to notice is that a separation with mass occurs. The maximum percentage decrease at 15 $M_\odot$ is only 1.2% while that of a

TABLE II
Main-sequence models

Model	M_b interior	$\log T_{e0}$	$\dfrac{R_0}{R_\odot}$	$\log T_c$	$\log P_c$	$10^4\,\Omega$ (rad/sec)	V (km/sec)	$\dfrac{\Delta L\,(\alpha_s)}{L(0)}$ %	$\dfrac{\Delta R_p\,(\alpha_s)}{R_p(0)}$ %
I, 1	−6.160	4.480	5.42	7.511	0.699	0	0	0	0
I, 8	−6.086	4.460	6.16	7.510	0.705	1.07	597	6.5	1.3
II, 1	−5.452	4.438	4.75	7.495	0.794	0	0	0	0
II, 8	−5.377	4.418	5.38	7.493	0.800	1.17	571	6.7	1.5
III, 1	−4.842	4.401	4.26	7.480	0.876	0	0	0	0
III, 8	−4.767	4.381	4.82	7.478	0.883	1.26	551	6.7	1.7
IV, 1	−4.477	4.378	3.99	7.471	0.927	0	0	0	0
IV, 8	−4.402	4.359	4.51	7.469	0.934	1.32	540	6.7	1.8
V, 1	−3.570	4.321	3.42	7.449	1.052	0	0	0	0
V, 8	−3.495	4.302	3.86	7.447	1.060	1.47	515	6.7	1.8
VI, 1	−2.278	4.238	2.77	7.417	1.230	0	0	0	0
VI, 8	−2.203	4.219	3.13	7.415	1.239	1.70	484	6.6	1.9
VII, 1	−2.329	4.243	2.77	7.421	1.209	0	0	0	0
VII, 8	−2.250	4.224	3.11	7.418	1.218	1.72	485	7.0	2.3
VIII, 1	−2.143	4.233	2.66	7.425	1.267	0	0	0	0
VIII, 8	−2.062	4.214	2.99	7.423	1.276	1.82	495	7.1	2.3

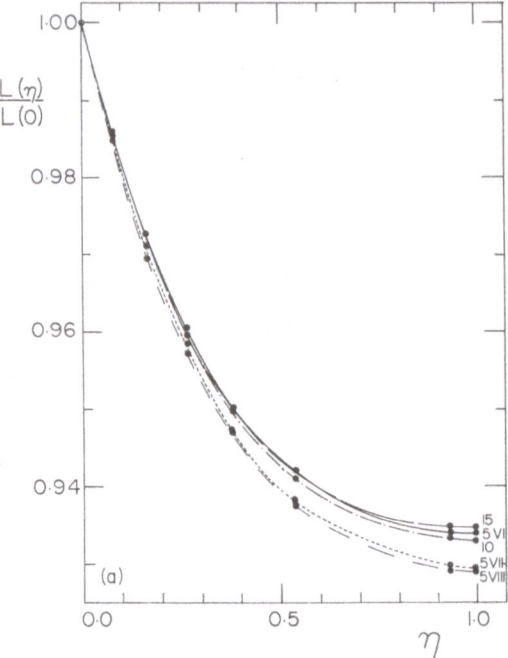

Fig. 1.   The detailed variation, as the rotation builds up, of the luminosity of a rotating star relative to that of a non-rotating star of the same mass. The Roman numeral refers to the different models, while the Arabic numeral gives the mass of the models in solar units (reprinted courtesy of the *Astrophysical Journal*).

corresponding 5 $M_\odot$ model is 1.9%. Notice how small these changes are, especially when compared to those frequently used, namely those of Roxburgh *et al.* (1965), and Sweet and Roy (1953); the former obtained 10.87% at critical rotation and the latter 5.40% for slow rotation. Morover, the change of the polar radius when plotted against η is highly non-linear. The quadratic polynomial

$$R_p(\eta) = R_p(0)\left[0.9987 - 0.0486\eta + 0.0336\eta^2\right]$$

is found to describe the 10 $M_\odot$ curve to better than 9%.

The effects of rotation on other physical quantities are also investigated (Sackmann and Anand, 1970).

C. THE EVOLUTION

To allow for evolution, three modifications have to be applied to the basic structure equations given for main-sequence stars. The change of chemical composition as hydrogen is burned into helium is computed as in Larson and Demarque (1964). An additional release of energy due to gravitational contraction is included when the central hydrogen content has fallen to a very low level. For rotating stars the change of the rotational velocity during evolution has to be considered as well. To compute

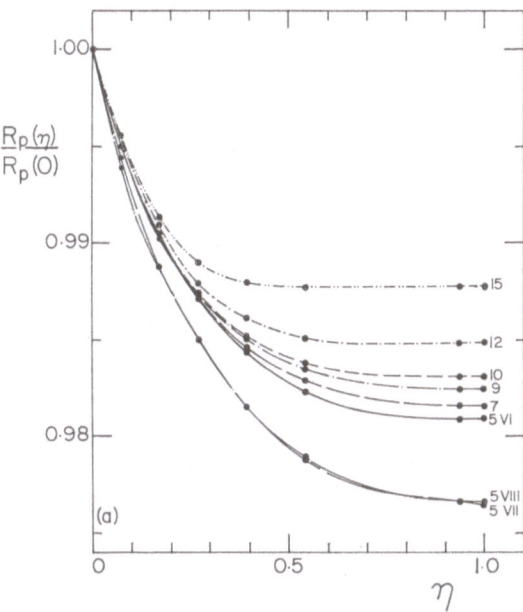

Fig. 2.   The variation of the polar radius of a rotating star, relative to that of a non-rotating star having the same mass, as the rotation is increased. The Arabic and Roman numerals have the same meaning as in Figure 1 (reprinted courtesy of the *Astrophysical Journal*).

this change, one has to know whether angular momentum is conserved or lost during evolution. There are some recent observations by Kraft (1967) that suggest that late-type stars lose an appreciable amount of angular momentum on a time scale of 10^8 years. However, since the present work is the first quantitative attempt to compute the evolutionary tracks of rotating stars, the case where angular momentum is conserved ought to be considered first.

There are two limiting cases in which angular momentum can be conserved. In the one case, angular momentum is conserved by cylindrical shells. This will give rise to differential rotation even if the star starts off rotating rigidly on the zero age main-sequence. The other limiting case is that in which the star as a whole, like a solid body, conserves its angular momentum. Thus, a star starting off with uniform rotation will continue to rotate uniformly during evolution. The latter case is of course the simplest and is considered in this work. The change in angular velocity with evolution is related to the change in the moment of inertia in an obvious way.

Because of the large amount of computer time needed, the evolution of only one mass was considered. A mass of 10 $M_\odot$ was selected as being fairly representative of Be stars and also as being right in the middle of the 5 to 15 $M_\odot$ range considered for main-sequence models. A number of evolutionary tracks for different initial rotational velocities were computed.

During evolution, the moment of inertia (I) increases, causing angular velocity (Ω) to decrease if the star's total angular momentum is conserved. In Figure 3, the evolutionary change of the rotation parameters α_s and α_c and of the equatorial velocity V and the angular velocity are shown.

It is very interesting to compare the equatorial velocities at critical rotation for evolved and main-sequence models. For a 10 $M_\odot$ main-sequence model, the equatorial

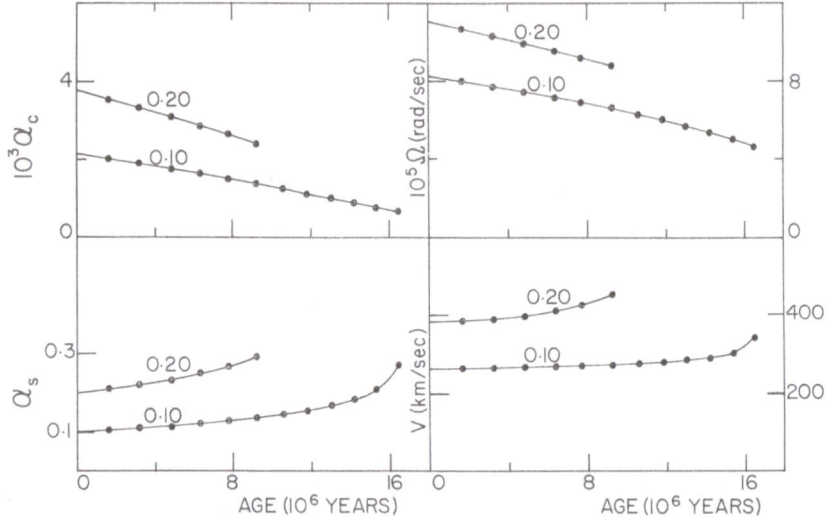

Fig. 3.   For two rotating 10 $M_\odot$ models starting their evolution off the main-sequence with $\alpha_s = 0.10$ and $\alpha_s = 0.20$, respectively, the evolutionary change of the rotation parameters α_s and α_c and of the equatorial velocity V and the angular velocity Ω are shown (reprinted courtesy of the *Astrophysical Journal*).

velocity at critical rotation is 551 km/sec, while for a 10 $M_\odot$ evolved model with moderate initital rotation ($\sim \frac{1}{2}$ critical) the equatorial velocity at critical rotation is only 344 km/sec. Hence, Be stars that are evolved could be expected to rotate with a critical equatorial velocity of the order of 200 km/sec lower than that of corresponding main-sequence Be stars.

The most interesting result of the present computations are the evolutionary tracks of rotating stars plotted in the H-R diagram. Two evolutionary tracks of 10 $M_\odot$ models rotating with different initial velocities are shown in Figure 4 together with the evolutionary track of the non-rotating model. Also illustrated are the main-sequences of non-rotating and critically rotating stars.

The evolutionary tracks of the two rotating stars are approximately parallel to each other and to the non-rotating case.

The last point plotted for the non-rotating model occurs just before the gravitational contraction phase. The last point plotted for the rotating models is the last output value from the computer before α_s becomes greater than its critical value. The most interesting result of all the work here is that the evolved rotating models reach critical rotation long before gravitational contraction sets in unless the initial rotation is very slow. This behavior is in contrast to Schmidt-Kaler's (1964) suggestion, that the

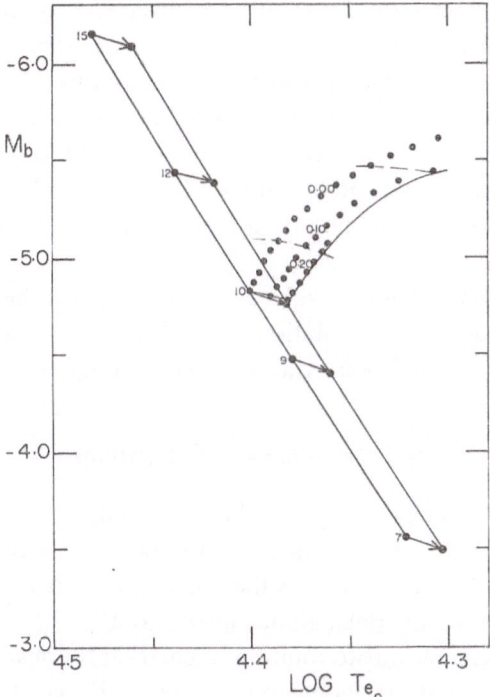

Fig. 4.   The evolutionary tracks of 10 $M_\odot$ models starting with different rotational velocities from the main-sequence. The number on a track refers to the initial α_s value. The last point plotted for the tracks of rotating stars refers to the last output value from the computer before critical rotation sets in. The dashed curve refers to a locus of constant time. The solid curve to the right indicates the locus of 10 $M_\odot$ stars at critical rotation. For comparison, the main-sequences of non-rotating and critically rotating stars are also shown. The numbers thereon refer to the masses in solar units (reprinted courtesy of the *Astrophysical Journal*).

extreme Be stars, sitting tightly grouped about a magnitude above the main-sequence in the H-R diagram, are those rotating stars which are in their evolutionary stage just before, during, or immediately after the gravitational contraction phase. If Schmidt-Kaler's suggestion is indeed the correct one, then one must change the basic assumptions which led to the present evolved rotating models; that is, rotating stars either do not obey the momentum conservation law assumed here and/or they do not rotate uniformly. They would then be able to evolve up to the gravitational contraction phase before critical rotation sets in.

The time needed to reach critical rotation after the zero age mean sequence phase is 9.4×10^6 and 16.6×10^6 years for the models of initial $\alpha_s = 0.20$ and $\alpha_s = 0.10$, respectively.

The locus of evolved stars at critical rotation is also shown. This locus is very similar to the evolutionary tracks themselves.

Also interesting are the loci of constant time for the rotating models. These loci slope slightly upward and to the left in the H-R diagram. This is in contrast to the constant time locus obtained for non-rotating, evolved stars of different masses where the slope is upward to the right. Hence, an imaginary cluster consisting of rotating stars all of the same mass but having different initial rotational velocities would tend to show evolutionary effects by exhibiting a band of rapidly rotating stars parallel to the main-sequence and above it in accordance with age. Indeed Schild (1966), found for the h and χ Persei cluster a band of ordinary Be stars, parallel and above the main-sequence, while the extreme Be stars occupy a much more localized section of this same band.

It should be pointed out that the evolved models computed here with zero rotation agree very well with independent evolutionary models of non-rotating stars. This gives some confidence in the evolutionary tracks of the rotating models.

3. Conclusions and Suggestions

Let us summarize the principal results of this investigation:

(1) The interior effect of uniform rotation on upper main-sequence stars is smaller than usually thought. For a given mass the maximum decrease in the interior luminosity is 6.7% and that of the polar radius by 1% to 2%.

(2) The luminosity change due to rotation is remarkably constant as the mass varies from 5 to 15 $M_\odot$; the corresponding polar radius change decreases as the mass increases.

(3) The assumption of a Roche model in the outer layers of a rotating star is a very good one, the distortion from a Roche model being no more than 0.05%.

(4) At critical rotation, equatorial velocities vary from 484 km/sec to 597 km/sec and angular velocities from 1.70×10^{-4} radians/sec to 1.07×10^{-4} radians/sec as the mass increases from 5 to 15 $M_\odot$.

(5) The evolution of rotating stars, computed by assuming conservation of total angular momentum and uniform rotation, causes the angular speed of rotation as well as the central rotation parameter α_c to decrease. But the surface rotation parameter α_s and the equatorial velocity increase with evolution. Critical equatorial velocities of evolved rotating stars can be of the order of 200 km/sec smaller than those of corresponding main-sequence stars.

(6) The most important result found is that stars starting off from the main-sequence with a slow or intermediate rotation reach critical rotation very early in their evolution and still long before their gravitational contraction phase. Only for very slow initial rotation may a rotating star reach its gravitational contraction phase before critical rotation sets in. Schmidt-Kaler's (1964) suggestion that the extreme Be stars are evolved rotating stars in or close to their gravitational contraction phase can therefore not be supported with the present models. The present evolutionary track of a model with the rotation set equal to zero agrees well with that computed in independent work.

(7) The critical rotation locus in the H-R diagram is roughly parallel to the evolutionary tracks.

(8) In the H-R diagram the locus of constant age for evolved rotating models of the same mass but with different initial rotational velocities has a slope of opposite sign to that produced by non-rotating stars differing in mass.

Acknowledgements

This research was supported in part by the National Research Council of Canada and the Department of University Affairs of the Province of Ontario.

Note added in proof: Recently Sanderson *et al.* (*Astrophys. J.* **159**, L69, 1970) found the internal inconsistency in the calculation of the luminosity change by Roxburgh *et al.* (1965).

The corrected value of the luminosity change calculated by Sanderson *et al.* is in good agreement with the results of the present authors.

References

Anand, S. P. S.: 1968, *Astrophys. J.* **153**, 135.
Chandrasekhar, S.: 1933, *Monthly Notices Roy. Astron. Soc.* **93**, 390.
Collins, G. W., II: 1963, *Astrophys. J.* **138**, 1134.
Collins, G. W., II: 1965, *Astrophys. J.* **142**, 265.
Collins, G. W., II: 1966, *Astrophys. J.* **146**, 914.
Collins, G. W., II and Harrington, J. P.: 1966, *Astrophys. J.* **146**, 152.
Faulkner, J., Roxburgh, I. W., and Strittmatter, P. A.: 1968, *Astrophys. J.* **151**, 203.
Hardorp, J. and Strittmatter, P. A.: 1968a, *Astrophys. J.* **151**, 1057.
Hardorp, J. and Strittmatter, P. A.: 1968b, *Astrophys. J.* **153**, 465.
James, R. A.: 1964, *Astrophys. J.* **140**, 552.
Kippenhahn, R. and Thomas, H. C.: 1970, this volume, p. 20.
Kraft, R. P.: 1967, *Astrophys. J.* **150**, 551.
Larson, R. B. and Demarque, P. R.: 1964, *Astrophys. J.* **140**, 524.
Monaghan, J. J. and Roxburgh. I. W.: 1965, *Monthly Notices Roy. Astron. Soc.* **131**, 13.
Ostriker, J. P. and Mark, J. W.-K.: 1968, *Astrophys. J.* **151**, 1075 (Paper I).
Roxburgh, I. W., Griffith, J. S., and Sweet, P. A.: 1965, *Z. Astrophys.* **61**, 208.
Roxburgh, I. W. and Strittmatter, P. A.: 1965, *Z. Astrophys* **63**, 15.
Sackmann, I. J. and Anand, S. P. S.: 1969, *Astrophys. J.* **155**, 257.
Sackmann, I. J. and Anand, S. P. S.: 1970 (in press).
Schild, R. E.: 1966, *Astrophys. J.* **146**, 142.
Schmidt-Kaler, Th.: 1964, *Bonn. Veröffentl.* **70**, 1.
Stoeckly, R.: 1965, *Astrophys. J.* **142**, 208.
Sweet, P. A. and Roy, A. E.: 1953, *Monthly Notices Roy. Astron. Soc.* **113**, 701.

Discussion

Jordahl: What were the assumptions in the form of the equations of structure used to produce these models (i.e., the modifications to allow for rotation effects)?

Anand: In the basic structure equation, we have assumed that the system is rotating uniformly but neglected meridian circulation. The effect of radiation pressure is fully included in the basic equations. The opacity and the nuclear energy-generation formulae are those which have been used often for non-rotating stars. For more details, see Sackmann and Anand (1970, *Astrophysical Journal*, in press).

Roxburgh: What is the lowest rotational velocity on the main sequence that will give equatorial breakup during the nuclear evolution phase?

Anand: 200 km/sec.

Roxburgh: This means that there is no problem with the Be stars as there will be many stars at equatorial breakup for a substantial fraction of their evolutionary lifetime so that the statistics on Be stars will be O.K.

Jaschek: But really the majority of the Be stars look slightly overluminous – mostly IV, sometimes even III.

Hardorp: (1) It has been said in the conclusion that the effects of rotation are smaller than hitherto thought: at critical velocity, luminosity, and polar radius drop only by 6% and 2%, respectively, as compared to the non-rotating case. As far as the radius is concerned, this is a *larger* effect than found by Roxburgh *et al.*, because it means an increase of the equatorial radius by 47%, as compared to the 31% of these authors.

(2) For stars of 10 solar masses it has now been proven that they do reach critical rotational velocity already during their core hydrogen-burning phase, provided they start out faster than 250 km/sec on the zero-age main sequence and rotation is always rigid. From the statistics of rotational velocities of the B2–B5 stars, P. Strittmatter and I concluded, by means of much rougher arguments, that most of the breakup stars are *not* created during this phase. The question whether these different answers are due to the fact that the B2–B5 stars have masses smaller than 10 solar masses and therefore show less change in moment of inertia, or whether our arguments are indeed too rough, could be settled by appelying Anand and Sackmann's computations to a star of 5 solar masses.

Deutsch: Miss Faber and Danziger have found strong observational evidence that most stars do evolve with uniform rotation. Can this evidence be dismissed as coincidental?

Slettebak: In this paper, as in the earlier paper by Hardorp and Strittmatter this afternoon, we have heard that Be stars are not expected to be located very far above the main sequence. Observations of Be stars in clusters and visual binary systems show that these objects are located in a band approximately one magnitude above the main sequence. Such displacements can be explained as an effect of rotation, according to the work of Collins and others. On the other hand, if one uses line ratios to estimate the luminosities of Be stars, the majority of the Be's seem to be of luminosity class IV and, in some cases, III. This is rather puzzling, in view of the recent results of Collins that rotation should not significantly affect the line ratios used to estimate luminosity, and suggests that the location of the Be stars on the H-R diagram is due in part to evolutionary effects in addition to aspect effects.

Collins: My conclusion relating to effects of rotation upon luminosity class are firm concerning stars of moderate to large rotational velocity. Concerning stars rotating with $\omega = \omega_c$ the conclusions are more tentative but are still of a similar nature. If the Be stars are not rotating exactly at critical velocity what you say is true. If they are then the situation with regard to luminosity class may be much more complicated.

ON THE LOSS OF ANGULAR MOMENTUM FROM STARS IN THE PRE-MAIN SEQUENCE PHASE

ISAO OKAMOTO

The International Latitude Observatory of Mizusawa, Iwate, Japan

(Paper read by K. Nariai)

Abstract. The braking of stellar rotation in the wholly convective phase in the pre-main sequence is numerically discussed. The structure of stars in that phase is expressed by a rotating polytrope with an index of 1.5 and the Schatzman-type mechanism is used as the means of loss of angular momentum. The magnetic energy is assumed to change with evolution as $H_0^2/8\pi(R/R_0)^s$, where H_0 and R_0 are initial magnetic field and radius, and s is a free parameter. The changes of angular momentum, rotational velocity, etc. with contraction are calculated from the initial state, which is taken to be the state when the stars flared up to the Helmholtz-Kelvin contraction. It is shown that the exponent s must be in the range from -1 to -3 so that the stars with adequate strength of the initial magnetic field may lose almost all of their angular momenta in a suitable rate if they are initially in the state of rotational instability.

Stellar rotation from the time of star formation to the main sequence stage is discussed. Also, the formation of the solar system and other planetary systems is discussed, with respect to the braking.

1. Introduction

It is well known that the axial rotation of the main sequence stars stops quite abruptly at about F5. Schatzman (1962) suggested an efficient mechanism due to mass loss through stellar magnetic activity related to the surface convection zone. Mestel (1968a, b) constructed an elaborate theory about magnetic braking by a stellar wind. Nariai (1968) discussed braking of stellar rotation in terms of acoustic energy supplied to coronae. Huang (1965b, 1967) investigated rotational behavior of the main sequence stars and its plausible consequences concerning formation of planetary systems.

It was shown by Hayashi (1961) that the stars in the pre-main sequence contraction stage are wholly-convective and contract downward almost vertically on the H-R diagram until the radiative core develops appreciably. Poveda (1964) presented a theory of flare stars on the basis of Schatzman's theory and calculations of stellar evolution by Hayashi and his collaborators (1962, 1963). Huang (1965a) considered the sequence of events in the early phase of the solar system and suggested that Schatzman's mechanism operated in the wholly convective phase and the sun thereby lost its angular momentum.

We connect Schatzman's loss mechanism to the theory of evolution of Hayashi *et al.* and calculate numerically the variation of angular momentum of the sun and stars with their evolution.

2. The Rotation of Stars in the Wholly Convective Phase

The angular momentum and equatorial velocity of a wholly convective star with the polytropic index of 1.5 are expressed by the following relations;

$$\omega C = (\omega C)_A \sqrt{GM^3 R} \tag{1}$$

and

$$v_{eq} = \omega(A) \sqrt{\frac{GM}{R}}, \tag{2}$$

where $A = \omega^2/8\pi G\varrho_c$ (see Okamoto (1969) (Paper I) for the derivation of Equations (1) and (2)). The dimensionless quantities $\omega(A)$ and $(\omega C)_A$ are functions of A only and shown in Figure 1. The rotation equation which contains both terms of gravitational

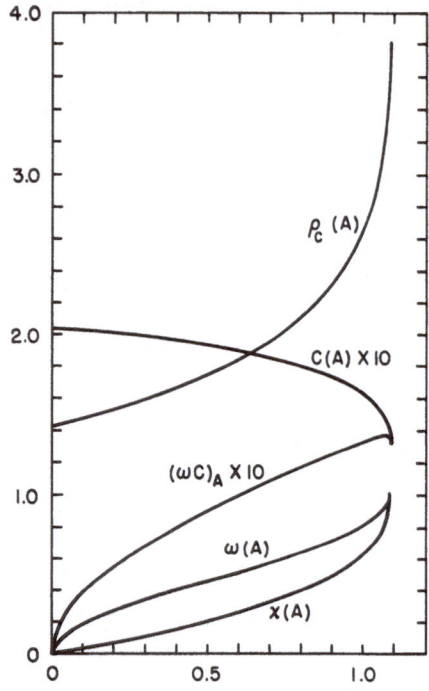

Fig. 1. $C(A), \rho_c(A), \omega(A), (\omega C)_A$ and $\chi(A)$ are plotted as functions of A for the polytrope with $n = 1.5$. The values multiplied by a factor 10 are plotted for $C(A)$ and $(\omega C)_A$.

contraction and magnetic braking is obtained, by equating the time derivative of ωC to the magnetic torque, as follows;

$$\frac{\mathrm{d}}{\mathrm{d}A}(\omega C)_A \cdot \frac{\mathrm{d}A}{\mathrm{d}t} = (\omega C)_A \cdot \frac{1}{2}\left(-\frac{1}{R}\frac{\mathrm{d}R}{\mathrm{d}t}\right) + \frac{1}{\sqrt{GM^3R}}\frac{1}{4\pi}\int H_\phi r \sin\theta (\mathbf{H}\,\mathrm{d}\mathbf{S}). \tag{3}$$

The Helmholtz-Kelvin contraction time is expressed in terms of the stellar luminosity;

$$\left(-\frac{1}{R}\frac{\mathrm{d}R}{\mathrm{d}t}\right)^{-1} = \frac{3\gamma - 4}{3(\gamma - 1)}\cdot\frac{3}{5 - n}\cdot\frac{GM^2}{RL} = \frac{3}{7}\frac{GM^2}{RL}, \tag{4}$$

where the values $n = 1.5$ and $\gamma = \frac{5}{3}$ are inserted. For the mechanism of magnetic braking we use a modified form of the Schatzman-type. Then, we can equate approximately the second term of Equation (3) in the following way;

$$\frac{1}{4\pi} \int H_\phi r \sin\theta (\mathbf{H} \, d\mathbf{S}) = \frac{dM}{dt} \omega r_c^2 \quad (\text{g}\cdot\text{cm}^2\cdot\text{sec}^{-1}/\text{years}), \tag{5}$$

where mass loss dM/dt and critical distance r_c are given by the following Equations (6) and (7) respectively.

$$\frac{dM}{dt} = -4.93 \times 10^{32} \times \frac{H_*^2}{4\pi V^2} \left(\frac{L}{L_0}\right)^{1/2} \left(\frac{M_0}{M_\odot}\right)^{19/24}$$

$$\times \left(\frac{R_0}{R_\odot}\right)^{27/24} \left(\frac{M}{M_0}\right)^{-5/24} \left(\frac{R}{R_0}\right)^{53/24} \omega(A)^{3/4} \quad (\text{g/years}) \tag{6}$$

and

$$r_c = R \left[\frac{1}{\sqrt{2}} \frac{1}{\omega(A)} \left(\frac{V_A}{V}\right)^2\right]^{1/3} \tag{7}$$

The velocity of ejected matter and its change along the evolutionary path cannot be definitely determined from observations and theory. Therefore, for simplicity, we take the escape velocity of the star for the ejection velocity of matter;

$$V = V_{\text{esc}} = \left(\frac{2GM}{R}\right)^{1/2} = 617.7 \left(\frac{M}{M_\odot}\right)^{1/2} \left(\frac{R}{R_\odot}\right)^{-1/2} \quad \text{km/sec}. \tag{8}$$

The initial radii of stars when they flare up from the pre-opaque dynamical contraction and reach the state of quasi-static contraction are given by the relation

$$\frac{1}{2} \frac{3}{5-n} \frac{GM^2}{R} = \chi \cdot \frac{M}{m_H},$$

where m_H is the mass of the hydrogen atom and $\chi = 15.84X + 19.75Y$, X and Y being the concentrations by mass of hydrogen and helium. If we take the stars of Population I, putting $X = 0.61$ and $Y = 0.37$, we obtain the initial radii in terms of stellar mass as follows;

$$R/R_\odot = 49.88 (M/M_\odot). \tag{9}$$

Then, from Equation (2) the rotational velocity at the initial radius (9) depends only upon A and not upon the stellar mass. The radius given by Equation (9), however, corresponds to the non-rotating state. In order to carry out the integration of Equations (3) and (4), we must know the initial value of A also. But we are not able to determine both R_0 and A_0 at the same time. Thus, we assume the value of 1.069×10^{-2} for A_0, at which stars are near to the limit of rotational instability.

The path of the evolution from the initial state (9) is approximated by the following relation

$$L/L_\odot = Q(R/R_\odot)^{1.5}, \tag{10}$$

where $Q = 0.154, 0.269, 0.624$ and 1.161 for stars of $0.4, 0.6, 1.0$ and $2.0 \, M_\odot$, respectively. The radii where radiative cores just begin to grow at the centers are $0.776, 1.318,$

2.570 and 11.75 in units of the solar radius for stars of these masses (see Table I in Okamoto (1967)). Though Iben (1965) has calculated the evolutionary paths of the pre-main sequence stars in detail, the paths themselves do not have an important effect upon the present calculations. Also, the effects of rotation and the magnetic field are not taken into account in his calculations. Thus, the approximate relation (10) is sufficient for our purpose.

Since the magnetic energy appears in the equations for dM/dt and r_c, we need knowledge of the change of stellar magnetic field with evolution. But we do not know accurately the origin of the stellar magnetic field and its change with evolution at present. Here we consider the dynamo mechanism in the stellar convection zone (Parker, 1955) and assume for the change with evolution the following:

$$\frac{H_*^2}{8\pi} = \frac{H_0^2}{8\pi}\left(\frac{R}{R_0}\right)^s,\tag{11}$$

where H_0 is the initial field at $R = R_0$. If the magnetic field is frozen in the stellar material during the contraction, we should take $s = -4$. Provided that the primordial

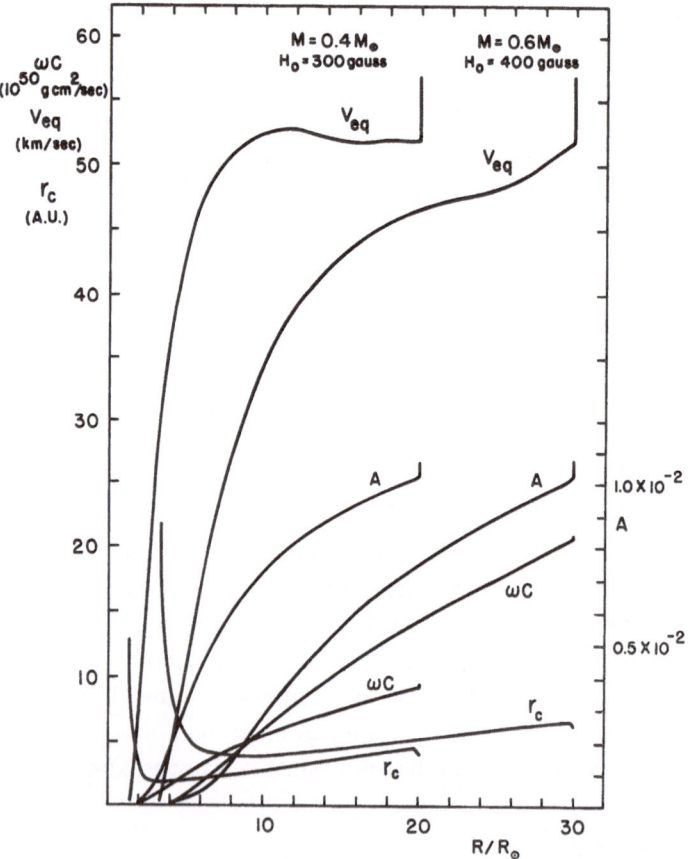

Fig. 2. The change of angular momentum ωC, equatorial velocity V_{eq}, critical distance r_c and the parameter A with the contraction of the stars of 0.4 and 0.6 $M_\odot$ for $H_0 = 300$ and 400 G, respectively, when $s = -1$.

field decays with contraction, s is positive. We take $-4 \leqslant s \leqslant 1$ in the following numerical calculations.

Numerical integrations of simultaneous differential Equations (3) and (4) are carried out, starting from the initial radius (9) through the evolutionary path (10), where relations (5), (6), (7) and (8) are used. If we know R and A through integrations, we can know the angular momentum ωC and the equatorial velocity V_{eq} of the star from Equations (1) and (2). The quantities A, ωC, V_{eq} and r_c are plotted in Figures 2 and 3 as functions of the radius in terms of the solar radius. For $s = -1$ we take $H_0 = 300$ G for 0.4 $M_\odot$, $H_0 = 400$ G for 0.6 $M_\odot$, $H_0 = 400$ G for 1.0 $M_\odot$ and $H_0 = 600$ G for 2.0 $M_\odot$ (see Okamoto (1969) for results for other values of s and H_0). Figures 2 and 3 show the sharp decreases of A, ωC, V_{eq} at the initial state. It is because $(\omega C)_A$ has the maximum value at the point $A_{max} = 1.08 \times 10^{-2}$ slightly smaller than $A_{cr} = 1.0906$, where the rotational instability sets in according to James (1964), and $d(\omega C)_A/dA = 0$ at $A = A_{max}$. If the right-hand side of Equation (3) is non-zero, we have $dA/dt = \pm \infty$. We have avoided the maximum of $(\omega C)_A$ in the numerical calculation and selected

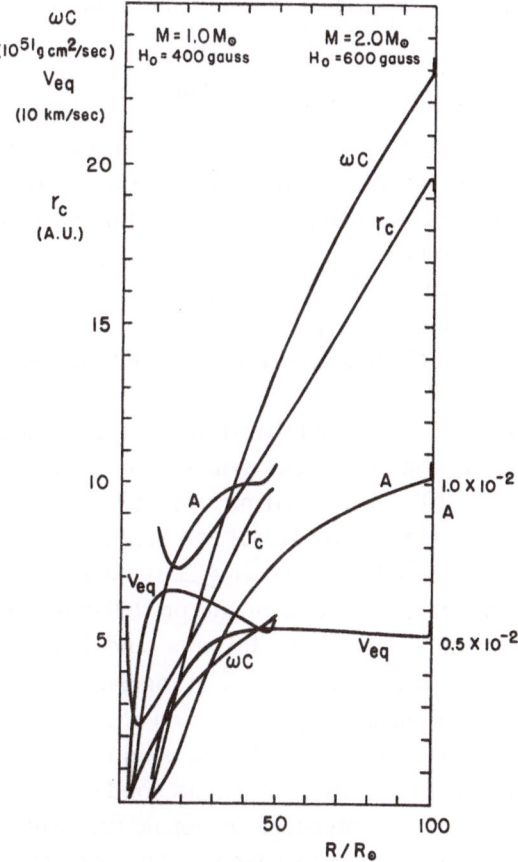

Fig. 3.   The change of angular momentum ωC, equatorial velocity V_{eq}, critical distance r_c and the parameter A with the contraction of the stars of 1.0 and 2.0 $M_\odot$ for $H_0 = 400$ and 600 G, respectively, when $s = -1$.

1.069×10^{-2} instead of A_{cr} for the initial value of A. However, $d(\omega C)_A/dA$ is still so small that A, ωC and V_{eq} show a rapid decrease.

In the above calculations, we have taken the escape velocity of a non-rotating star for the ejection velocity of matter due to flare activity. The escape velocity of a rapidly rotating star is considerably smaller than given by Equation (8), owing to the centrifugal force. Then, as we see from Equations (6) and (7), this mechanism for throwing out stellar angular momentum would be much more effective.

The ratios of the total mass loss necessary to throw away almost all the stellar angular momenta to the stellar mass are smaller than, or at most equal to 2×10^{-4}. This degree of mass loss lies well within the range deduced from observations of T Tauri stars (Kuhi, 1964, 1966).

We can also guess the change of magnetic energy of stars in the wholly convective phase from the braking of stellar rotation. The protostars will have a reasonable strength of the initial field H_0 resulting from the interstellar magnetic field. Then, from the calculations we know that the exponent s must be negative for protostars to be braked during the wholly convective phase. We suggest a value from -1 to -3 for s so that the protostars may be braked with a suitable speed.

3. Discussion

We now conjecture about the rotations of stars from the time of star formation to the main sequence stage. According to recent theories of star formation (Mestel, 1965), an interstellar cloud starts contracting owing to gravitational instability and protostars are formed by fragmentation of a collapsing cloud. The subsequent evolution is divided into two main phases, one is the phase of dynamical contraction which proceeds more or less rapidly after star formation, and the other is the following phase of slow contraction (Helmholtz-Kelvin contraction) toward the main sequence (Hayashi, 1966).

In the previous section we assumed that stars are near the limit of rotational instability when they flare up from the state of dynamical contraction to the Hayashi phase and we adopted 1.069×10^{-2} for A_0. At present we have little knowledge about the rotational state of protostars. Struve (1945) and Huang and Struve (1954), however, showed from a statistical study that rotating stars have not acquired their angular momentum from the galactic rotation of the prestellar gaseous medium and their angular momenta must have been derived from a random process, such as a random spin in fragmentation of a collapsing cloud. The angular momentum and the magnetic field may be formidable obstacles for condensation of new stars. Here we assume that protostars may succeed in evolving into the pre-opaque dynamical contraction stage. The magnetic energy of a gravitationally-bound condensation can never exceed the gravitational energy. Thus, the interstellar magnetic field will not be able to become strong enough to transfer all of the stellar angular momentum. As the contraction proceeds, the rotational instability may take place in this stage and break up equatorially to shed the stellar matter. Therefore, when the protosun and protostars flare up to the stage of quasihydrostatic equilibrium, they will be at least nearly rotationally unstable.

The angular momentum of the solar system (3.156×10^{50} g·cm²/sec) is possessed almost entirely by the orbital motion of the giant planets, with the sun having only 0.5% of the whole as its spin angular momentum. If the angular momentum of the present solar system had concentrated in the sun when it flared up, we obtain $A_0 = 0.008 \times 10^{-2}$ and the ratio of the centrifugal force to gravity at the equator $\chi(A_0) = 0.003$ from Figure 1. If these values for the protosun were true, the protosun would have had to suffer a great degree of braking during the dynamical contraction stage. As discussed above, however, the magnetic field could not effectively brake the protosun to the state of such a slow rotation during this stage. Thus, from Equations (1) and (9) and using the value of 1.069×10^{-2} for A_0, the protosun had the values of 58×10^{50} g·cm²/sec as the angular momentum at the moment of the flare-up, which is about twenty times larger than the present value of the solar system. The chemical composition of the protoplanetary cloud would be the same as that of the sun, and the lower limit on the initial mass of the cloud is estimated by diluting planetary material with volatile substances until the solar composition is reached. The extra mass over the present total mass of the planets would be lost out of the solar system because of the increasing centrifugal force that resulted as the protosun began to transfer its angular momentum.

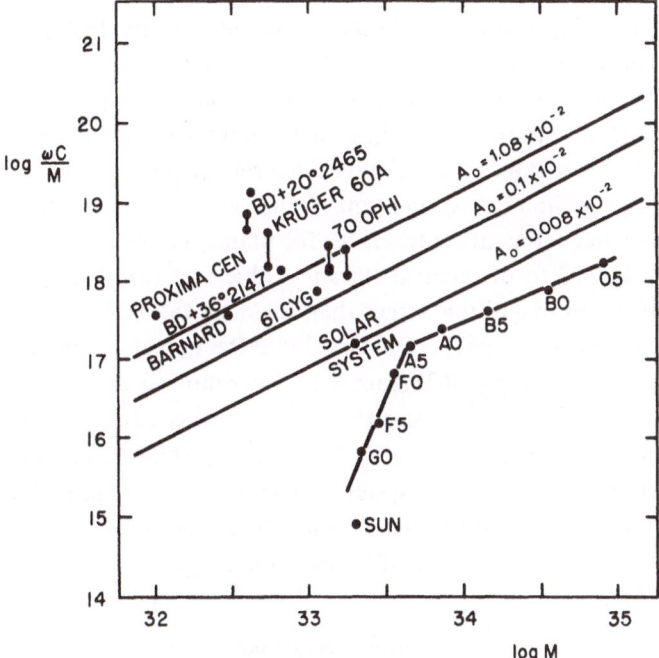

Fig. 4.   The logarithm of angular momentum per unit mass versus the logarithm of stellar mass. The values for the main sequence stars are taken form McNally (1965) and the values for the 'planetary systems' from van den Heuvel (1966).

We plot in Figure 4 the logarithm of the stellar angular momentum per unit mass versus the logarithm of the stellar mass. The straight line $A_0 = 1.08 \times 10^{-2}$ is drawn from Equations (1) and (9) and shows the angular momentum per unit mass at the limit of the rotational instability. The line $A_0 = 0.008 \times 10^{-2}$ corresponds to the case where the present total angular momentum of the solar system is concentrated in the protosun at the flare-up. We also plot the angular momentum per unit mass for the main sequence stars from the calculation of McNally (1965). The line for $A_0 = 1.08 \times 10^{-2}$ and the curve for the main sequence stars of early type are nearly parallel. The difference between both lines shows that even early type stars suffered a considerable degree of braking. Actually, as Huang (1965b) suggested, the braking is not limited only to stars of spectral type later than F5, but extends to O, B, A and early F stars as well. These early-type stars are considered to have suffered braking during the Hayashi phase and no braking afterwards because of the retirement of the surface convection zone in the succeeding post-Hayashi phase. Further contraction caused them to rotate rapidly because of conservation of angular momentum. The slow rotation of late-type stars, including the sun, may be regarded as a result of their having continued to receive braking from the wholly convective phase to the main sequence.

Finally we discuss the possible occurrence of planetary systems around stars. Huang (1965, 1967) suggested that the formation of planetary systems is closely related to the braking of stellar rotation, and defined the following three characteristics for any planetary system; (1) a small mass ratio of the total mass of the entire planetary system to the mass of the central star; (2) a large angular momentum; (3) the majority of its members moving in nearly coplanar and circular orbits. In Figure 4 we plot the angular momentum per unit mass for 'planetary systems' from van den Heuvel's paper (1966). While more than half of them lie beyond the line for $A_0 = 1.08 \times 10^{-2}$, the remainder are near to that line. These facts imply that contrary to the case for the solar system these 'planetary systems' have a nearly equal or larger amount of angular momentum than the angular momentum of the central stars at their flare-up. They seem to have angular momentum too large for planetary systems. This is because the mass ratio of 'planets' to the central stars in Table I of van den Heuvel's paper are more than an order of magnitude larger than the mass ratio of the solar system. Also, Kumar (1964) and Huang (1967) emphasized that the unseen companion of Barnard's star inferred by van de Kamp (1963) from the proper motion should not be regarded as forming a planetary system, because the eccentricity is equal to 0.6, a fairly large value. Consequently van den Heuvel's 'planetary systems' are different from the definition of Huang for planetary systems. These may be regarded as intermediate between binary systems and planetary systems. Nevertheless we expect that planetary systems must be common in space. We may not as yet have discovered them with today's instruments.

Acknowledgements

The author wishes to thank Dr. K. Nariai for reading this manuscript at the Colloquium on Stellar Rotation. His thanks are also due to Dr. T. Osaki for suggesting to him that the paper be sent to the Colloquium and to Professor Slettebak for various arrangements.

References

Hayashi, C.: 1961, *Publ. Astron. Soc. Japan.* **13**, 450.
Hayashi, C.: 1966, *Annual Rev. Astron. Astrophys.* **4**, 171.
Hayashi, C. and Nakano, T.: 1963, *Prog. Theoret. Phys.* **30**, 460.
Hayashi, C., Hoshi, R., and Sugimoto, D.: 1962, *Prog. Theoret. Phys. Suppl.* **22**, 165.
Huang, S.-S.: 1965a, *Publ. Astron. Soc. Pacific.* **77**, 42.
Huang, S.-S.: 1965b, *Astrophys. J.* **141**. 985.
Huang, S.-S.: 1967, *Astrophys. J.* **150**, 229.
Huang, S.-S. and Struve, O.: 1954, *Ann. Astrophys.* **17**, 85.
Iben, I.: 1965, *Astrophys. J.* **141**, 993.
James, R. A.: 1964, *Astrophys. J.* **140**, 552.
Kuhi, L. V.: 1964, *Astrophys. J.* **140**,1409.
Kuhi, L. V.: 1966, *Astrophys. J.* **143**, 991.
Kumar, S. S.: 1964, *Z. Astrophys.* **58**, 248.
McNally, D.: 1965, *Observatory* **85**, 166.
Mestel, L.: 1965, *Quart. J. Roy. Astron. Soc.* **6**, 161, 265.
Mestel, L.: 1968a, *Monthly Notices Roy. Astron. Soc.* **138**, 359.
Mestel, L.: 1968b, *Monthly Notices Roy. Astron. Soc.* **140**, 177.
Nariai, K.: 1968, *Astrophys. Space Sci.* **3**, 150.
Okamoto, I.: 1967, *Publ. Astron. Soc. Japan* **19**, 384.
Okamoto, I.: 1968, *Publ. Astron. Soc. Japan* **20**, 25.
Okamoto, I.: 1969, *Publ. Astron. Soc. Japan* **21**, 350.
Parker, E. N.: 1955, *Astrophys. J.* **122**, 293.
Poveda, A.: 1964, *Nature* **202**, 1319.
Schatzman, E.: 1962, *Ann. Astrophys.* **25**, 18.
Struve, O.: 1945, *Pop. Astr.* **53**, 201, 259.
Van den Heuvel, E. P. J.: 1966, *Observatory* **86**, 113.
Van de Kamp, P.: 1963, *Astron. J.* **68**, 515.

PART II

THE EFFECTS OF ROTATION ON STELLAR ATMOSPHERES

THE EFFECTS OF ROTATION ON THE ATMOSPHERES
OF EARLY-TYPE MAIN-SEQUENCE STARS

(Review Paper)

GEORGE W. COLLINS, II

Perkins Observatory, The Ohio State and Ohio Wesleyan Universities, U.S.A.

1. Introduction

In discussing the status of the theory of rotating stellar atmospheres, it is necessary to draw upon the contributions of many well established aspects of astrophysics and to interconnect them in a cohesive pattern structured so as to provide insight into a rather specific problem – namely, the structure and characteristics of a surface of the star undergoing axial rotation. Many different connections are possible having varying degrees of emphasis and, of necessity, those given here represent only one such presentation. The discussion could be much simplified if it were not necessary to test the efficacy of the theoretical development by referring to observations. Unfortunately, such a comparison is necessary and the results are at the moment somewhat inconclusive. This unhappy situation arises from the retrospectively obvious fact that axial rotation does not play a dominant role in determining the directly observable properties of stars. Indeed, if rotation were a dominant factor, earlier attempts at describing stellar structure and evolution would have met with little success. However, it is becoming increasingly clear that the structure and final evolution of highly evolved stars are greatly influenced by the total angular momentum which they retain from their earlier history. In order to understand the angular momentum distribution present in the final state, it is necessary to understand the effects of stellar evolution on the total angular momentum and its distribution. But even before this step can be taken, one must first successfully describe the rotational structure of the main-sequence phase as it is this state which provides the initial conditions necessary for any further study. It is further appropriate that we attempt to describe this period of a star's life as the largest body of observational material with which we must test our results exists for these stars. In addition, we should expect a study of the atmospheric structure to be the most fruitful as this is the region of the star which provides the final modification of the radiation we observe.

The observations will be based on the integrated contributions from regions on the surface of the star which differ greatly from one another with respect to the parameters that define the local radiation field. Thus, unlike the theory of stellar atmospheres, our task is not just to determine the variation of state-variables with depth at a specific point on the star, but rather at any point on the star. In principle, we are therefore required to construct many model atmospheres representing the local structure of the star, recognizing that to some extent these models will interact with each other. In addition, it is clear that the parameters defining the atmospheric structure will be determined by the underlying interior structure of the star. We shall not attempt to describe the state of the theory of rotating interiors, but only present the results of several investigators and indicate the extent to which observed parameters may be sensitive to their results.

We shall then assume that the theory of stellar interiors provides us with not only the variation of effective temperature and gravity over the surface, but also with the shape of that surface. With this information, we may then estimate the extent to which various points on the surface will affect the model structure at other points. If the extent of this interaction is not large, we may employ the existing theory of stellar atmospheres to determine the state of the local radiation field resulting in both con-tinuous and line radiation. Then, taking into account the Doppler effect, the contri-bution of the local field to the observed field may be integrated over the surface of the star yielding observed parameters. Many different detailed approaches have been taken to this problem and we shall herein undertake to describe some of the more successful ones and indicate their results.

In the first section, we shall describe the results of interior studies insofar as they enable us to estimate the variations of the effective temperature, surface gravity and shape and then using these results, indicate the extent of the horizontal interaction that may be expected between models. Following this, we shall undertake a brief description of the methods used to determine the continuum and bolometric properties of the radiation field indicating some of the results of these studies. Since the numerical techniques required to investigate the effects of rotation on the formation of spectral lines are essentially the same as those required for the investigation of the continuum, we shall conclude this section with a review of those studies relating to spectral lines. In the next section, we shall undertake to investigate the success which theory has had in describing observational results. Finally, we shall make some concluding remarks about what may be inferred at the present time about the nature of the early-type stars.

2. Specification of Atmospheric Parameters

As we have already indicated, it is necessary to inquire of the results of rotating interior investigations to ascertain the run of effective temperature and gravity over the sur-face of the star. In addition, we must determine the extent to which atmospheric structure at one point on the surface will influence the structure at another. These problems are not unrelated as the resolution of the latter depends largely on the former.

Ever since the internal structure of stars became of interest to astrophysicists, researchers have demonstrated concern over the effects of rotation upon this structure. A detailed account of the development and present status of model rotating interiors here would be inappropriate. However, it must be noted that in all such models made to date, the distribution of angular momentum within the star is either specified in an ad hoc manner or, within certain constraints, is arbitrary. The central reason for this situation is that the nature of the forces which bring about redistribution of angular momentum tending to an equilibrium value, is not well understood. Indeed, it may well be that if radiative and kinematic viscosity are the only agents available for this re-distribution, the star may not reach an equilibrium state in a time which is less than its main-sequence life-time. The implication of this for the computation of model stars is simply that the apparent arbitrariness now present in existing models may be seen reflected in nature.

In any event, we may specify the local gravity and the surface shape of the star from the combined gravitational and rotational potential. Thus

$$\mathbf{g} = + \nabla \Phi = + \nabla \left[\Phi_0 + \tfrac{1}{2} \omega^2 r^2 \sin \theta \right] \tag{1}$$

where Φ_0 is the gravitational component of the potential while r and θ are defined by Figure 1.

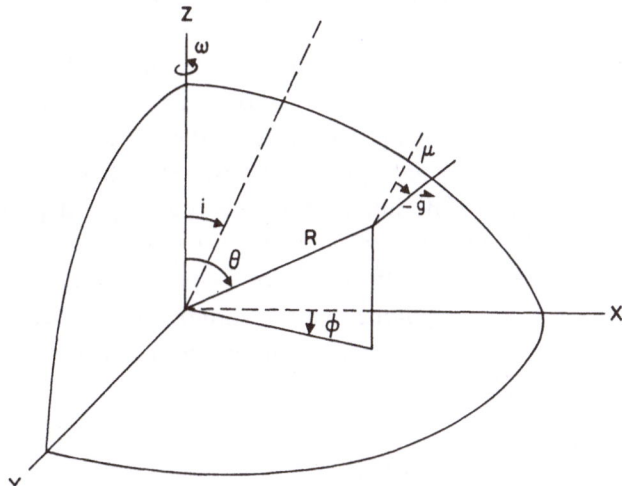

Fig. 1.   One octant of a star rotating at critical velocity, indicating the defining coordinate system, local surface gravity, etc. (reprinted courtesy of the *Astrophysical Journal*).

Also

$$R(\theta) = R \left[\Phi = \text{const} \right]. \tag{2}$$

The remaining parameter required to specify an atmosphere is the effective temperature. In order to do this, we must make use of the often misunderstood theorem of von Zeipel (1924), the most lucid presentation of which is due to Eddington (1926). In light of the modern theory of stellar structure, this theorem can be rigorously viewed as a proof by contradiction that stars in both hydrostatic and radiative equilibrium cannot rotate as rigid bodies. However, the more interesting and more used corollary that, a rigidly rotating star in both hydrostatic and radiative equilibrium will have a variation of radiative flux which is proportional to the local gravity, may well hold to a high order of approximation in stars exhibiting differential rotation. Indeed, if the following two conditions are met, then the corollary will hold rigorously in any distorted star:

(a) the state variables P, T, and ϱ are functions of the potential alone,

(b) the radiation pressure is a function of the local state variables alone.

The first of these conditions may be violated by the existence of any circulation currents generated by differential rotation as can be seen from the equation of hydro-dynamic flow for a medium in steady state:

$$\varrho \nabla \Phi + \nabla P = \varrho \left[\mathbf{u} \cdot \nabla \right) \mathbf{u} \right]. \tag{3}$$

Here local u is the stream velocity in the rotating coordinate frame and the right-hand side of (3) represents essentially the forces exerted on a volume element by the circu-lation flow. Since the flow is driven by the rotation, we should expect the magnitude of the right-hand side to be small compared to either term on the left-hand side, even in the lower atmosphere. However, the magnitude of the effect of the right-hand side of (3) on the von Zeipel condition that $F \sim g$ still remains to be investigated in a rigorous manner.

The dependence of the radiation pressure on the local state variables is nicely satisfied in the lower regions, but of necessity must break down to an increasing extent as one enters the atmosphere. This is evident from the more or less global nature of the equation of transfer appropriate for the atmosphere. That is, in the atmosphere the radiation field, and hence its higher moments, no longer depends on the local physical parameters of the gas but is strongly influenced by the neighboring conditions. The extent to which this condition affects the von Zeipel condition has also never been thoroughly investigated. We may hope that the variation of physical parameters con-tributing to the majority of the emergent radiation is sufficiently small to allow the use of the von Zeipel result that

$$F = -\frac{4c}{\kappa \varrho} \frac{dP_r}{d\Phi} g = -f(\Phi) g \tag{4}$$

which implies $F \sim g$ on a surface of constant Φ.

Slettebak (1949) has observed that the constant of proportionality appropriate for the integrated flux may then be obtained by integrating the gravity over the surface of the star and equating the result to the total luminosity of the star. So that

$$L(\omega) = \sigma C(\omega) \int_A g \, dA \tag{5}$$

yielding the effective temperature variations over the surface through

$$T_e^4(\omega, \theta) = C(\omega) g. \tag{6}$$

This procedure has been used since by Collins (1963, 1965, 1966, 1968a, b), Roxburgh and Strittmatter (1965), and Rubin (1966) while Hardorp and Strittmatter (1968a) preferred to replace the surface integral in (5) with a volume integral of $\nabla^2 \Phi$.

It is clear from Equation (5) that any study of rotating stellar atmospheres must rely on interior calculations to provide knowledge relating to the variation of the total luminosity of a star of given mass $L(\omega)$ with ω. Many authors have investigated this variation (Chandrasekhar, 1933; Sweet and Roy, 1953; Roxburgh *et al.*, 1965; Limber and Roberts, 1965; Epps, 1968; Sackmann, 1968; and Roxburgh *et al.*, 1968) by means of perturbation theory and have found the intuitively expected result that at least to first order in ω^2 the luminosity decreases linearly by an amount depending slightly on the detailed nature of the model. At present the effect and indeed even the sign of the second order term is somewhat in doubt. Roxburgh and Strittmatter (1965) find them to be small and of the same sign as the first order terms, while Sackmann (1968) finds them to be comparable to the first order term, but of opposite sign. Her results for $L(\omega_c)$ seem in agreement with Roxburgh *et al.* (1968).

Having discussed the relationship of the effective temperature variation to the total luminosity and the surface gravity, we still must turn to the interior models for the detailed evaluation of Equation (1). The geometrical shape of the equipotential surface is governed only by the internal mass and angular momentum distributions. Most investigators agree that to better than 0.1% the gravitational potential can be represented at the surface by that of a point mass. The apparent arbitrariness of the angular momentum distribution would indicate that greater accuracy than this is unnecessary. A notable exception to this is presented by Mark (1968) wherein he shows that appreciable differential rotation may lead to extreme distortion of the inner core. The scale of the equipotential surface (i.e., that level surface which can be said to define the surface of the star) is given by the boundary conditions of the interior models. The variation of this parameter with rotational velocity is, like the total luminosity, determined by perturbation theory. Again, there is general agreement concerning the behavior of the first order terms, but the nature of the second order terms is subject to some controversy. There does appear to be near general agreement that both $L(\omega)$ and the scale factor (e.g., the polar radius $R_p(\omega)$) are affected in the same sense by a variation of ω. That is an increase in ω for models of constant mass brings about a decrease in both $L(\omega)$ and $R_p(\omega)$. This appears to be true for a large range of models and angular momentum distributions. As pointed out by Collins (1965), this type of behavior for $L(\omega)$ and $R_p(\omega)$ will result in a reduced effect on the distribution of effective temperature over the surface of the star. In essence then, the variation of effective temperature over the surface is rather insensitive to the change in the model parameters $L(\omega)$ and $R_p(\omega)$. Unfortunately, this is not true for the surface gravity or the integrated flux.

Finally, in order to construct a local atmosphere appropriate for any point on a rotating star, we must estimate the extent to which one atmosphere model will interact with its neighbor. This interaction could arise from two causes:

(1) Disruption of radiative equilibrium by circulation currents which convey energy in and out of a given region.

(2) Radiative flux being transported horizontally through the atmosphere driven by a latitudinal temperature gradient.

Von Zeipel's Theorem assures us that both these conditions will prevail, but does not allow us to estimate the effect they will have on atmospheric structure. However, we may crudely estimate the extent to which radiative equilibrium can be disturbed by the meridional circulation by estimating the ratio of energy flow that might be expected from circulation currents to the energy carried by radiation. This ratio is very roughly given by

$$r \ll \tfrac{1}{2}\varrho u^3 / \pi \sigma T_e^4 \tag{7}$$

where u is the stream velocity and T_e is some typical effective temperature. It is unlikely that the stream velocity could reach much more than 10% of the equatorial rotational velocity as these currents are driven by the rotation itself. Thus, choosing the generous value for the density in the atmosphere of a rotating early-type star to be 10^{-9} g/cm^3 and a maximum equatorial velocity of 500 km/sec, together with a mid-latitude temperature of 10^4 K, we arrive at $r < 3\%$. Since we have been generous on the side of circulation in our choice of numbers, it is most probable that we can safely ignore the effects of circulation currents on the atmospheric structure. A more sophisticated argument reaching a similar conclusion is given by Strittmatter (1969).

A 'Devil's Advocate' for the case for circulation currents, at this point, would point out that in the equatorial region of Be stars the situation may well be different due to the lower effective temperature and possibly higher u. Although this may be true, the relevance of this argument to the computation of the theoretical spectrum of such a star is diminished by: (a) the substantially lower density, (b) the reduced luminosity of the equatorial region which contributes to the integrated spectrum. A far more serious problem is the sensitivity of the surface temperature distribution in early-type stars to small departures from radiative equilibrium. It may well be that departures from radiative equilibrium of 0.1% that might result in early-type stars may be sufficient to upset the calculation of relatively strong lines. Unfortunately, a thorough study of this effect has never been made.

The final possibility for horizontal interaction of atmosphere models lies in the existence of a horizontal surface temperature gradient. The magnitude of this gradient compared to the radial gradients will provide an estimate of the degree of coupling to be expected between adjacent points on the stellar surface. The radial gradient may be estimated by noting that in most atmosphere models of early-type stars, the temperature drops from 10^5 K to 2×10^4 K in a physical distance of about 10^4 km. Thus the radial gradient should be on the order of 10^{-4} K/cm. However, the horizontal gradient cannot exceed that resulting, ala von Zeipel, from a temperature of 2×10^4 K at the pole to zero at the equator. This yields a temperature gradient of less than 10^{-7} K/cm or three orders of magnitude less than the radial gradient. Thus, we may simply ignore this form of horizontal coupling.

It is also to be pointed out that magnetic fields do exist in most stars and in some may well be strong enough to affect the stellar structure. For a detailed discussion of the influence of such fields, as well as a more comprehensive picture of the present state of rotating stellar interior studies, the reader is referred to review articles by Strittmatter (1969) and Roxburgh (1970).

3. Calculation of Observable Parameters

Having established that we can locally approximate the surface of a rotating star with atmospheres made in accord with present atmospheric theory, we may now turn to the calculation of those parameters accessible to observation. To do this, we must first describe the local radiation field of the atmosphere at any point on the surface of the star. The only major difference between this procedure and the one normally followed in stellar atmospheres is that we shall require knowledge of the emergent specific intensity rather than the local flux. This is necessitated by the fact that the star is no longer spherical and the conditions defining the local radiation field vary over the surface. Thus

$$F_0 = 2\pi \int_A I\mu \, dA \neq 4\pi^2 R^2 \int_{-1}^{+1} I\mu \, d\mu$$

as is the case for spherical stars. As a result, the left-hand integral must be evaluated numerically.

The basic approaches for evaluating the integral of the specific intensity over the surface of a star are independent of whether or not one is dealing with radiation arising in the continuum or in a region dominated by spectral lines. Indeed, calculation of the net state of polarization of the integrated radiation may be done using basically the same techniques if care is taken always to refer the state of polarization to an external coordinate frame. The details of the techniques used vary from author to author, but usually involve the use of some general two-dimensional quadrature technique. The exact nature of the quadrature scheme is not terribly important unless computer time is a scarce commodity. In this case, the Gauss-type quadrature schemes are to be preferred due to their much higher efficiency. Regardless of the precise nature of the chosen quadrature scheme, one must be careful to use a sufficient number of points to inspire confidence in the accuracy of the result.

One numerical technique is worthy of note as it greatly simplifies the formulation of the integration limits and reduces the number of atmospheres which must be calculated. Collins (1965) noted that if the function to be integrated exhibits both axial and equatorial plane symmetry, the integral over the apparent surface is formally equal to the integral having limits of 0 and π in the polar angle and $\pm\pi/2$ in the azimuthal angle. This is true regardless of the angle of inclination. Of course, since the angle of inclination explicitly enters into the calculation of μ, the integral, $\int_{-1}^{+1} I$ $(\mu)\mu \, d\mu$, will be angle dependent. However, a corollary of this symmetry property is that any scalar function which has the proper symmetry and is uniquely defined on the surface of the star will have an area-average which is independent of the angle of inclination. This is the case for such functions as the effective temperature and the scalar value of the local gravity. Since this result is contra-indicated by both theoretical calculation of the spectral energy distributions and observations of rotating stars, one may conclude that arithmetical area-average values of T_e and g are inappropriate for describing the observed properties of these stars.

One may divide the results of studies of the theoretical properties of the radiation field of rotating stars into two broad areas:

(a) Results relating to the continuum.

(b) Results relating to line profiles and strengths.

We shall discuss the results from these two areas separately and in the last section, attempt to relate them to observational results.

A. CONTINUUM RESULTS

The effects of rotation upon the continuum radiation of stars have been studied by a number of authors since 1963. However, the earliest effort appears to be due to Sweet and Roy (1953) where, on the basis of interior models, they estimated the effect on the bolometric magnitude to be on the order of one magnitude. Their calculation relates to the total energy output of the star and not to that quantity that would be determined by an observer. The difference lies in the fact that the radiation field of such a star is not isotropic and any 'absolute' quantity measured by an observer will depend on the orientation of the star.

Attempts to measure the effects of varying the aspect of the star upon the observed bolometric magnitude were first made by Zhu (1963) and Collins (1963). In the absence of the interior model studies, both authors made assumptions about the variation of total luminosity and polar radius with angular velocity. Subsequent interior studies have shown that these variations are not important insofar as the estimation of the aspect effect is concerned. Zhu (1963), assuming that $F = \sigma T_e^4$, calculated the variation in the observed bolometric magnitude due to aspect effects for stars rotating at the critical velocity. Collins (1963) investigated this case while also studying the effect on the spectral energy distribution when the atmosphere was assumed to be locally gray. In addition, he considered the variation of these effects with rotational velocity. These studies, as well as all others since (with the exception of Rubin, 1966), have assumed that the gravitational part of the effective potential is given by that of the Roche Model. Rubin (1966) carried out calculations similar to Zhu (1963) with a homogeneous mass distribution rather than a Roche model. The resulting Maclaurin Spheroids, being more highly distorted than a Roche Model with corresponding angular velocity, yielded a larger maximum aspect effect of 1.36 magnitude as opposed to about 0.80 magnitude for the Roche Model. The surprising result is the close agreement between these two vastly different models. Thus, one may expect about 1 magnitude variation in the observed bolometric magnitude to be present in the most rapidly rotating stars as a result of aspect effects alone.

Roxburgh and Strittmatter (1965) repeated some of the calculations of Collins (1963) including the variation of total luminosity and polar radius with rotational velocity which are predicted by the interior calculations of Roxburg et al. (1965). Allowing for the variation of total luminosity, the two studies are in excellent agreement and indicate that a star will be displaced to be right and above the main sequence on a color-magnitude diagram (see Figure 2). Strittmatter (1966) states that this displacement will be proportional to the square of the equatorial velocity. However,

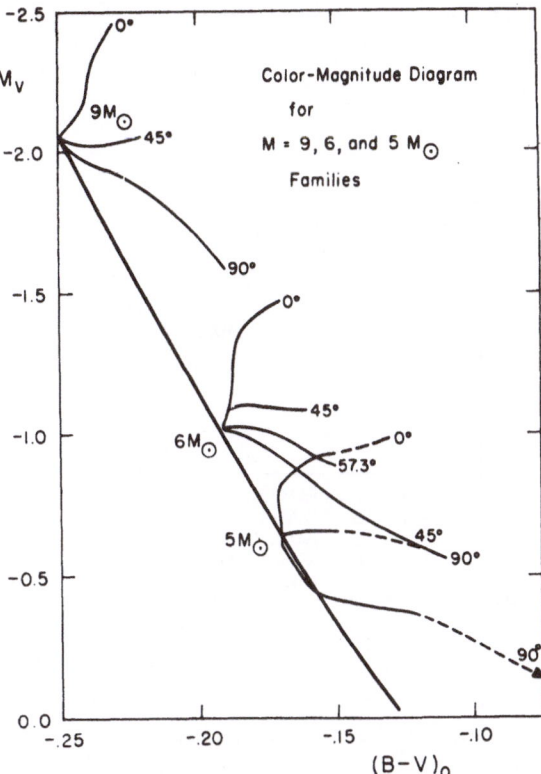

Fig. 2.   A color-magnitude diagram indicating the effect of rotation upon early-type main-sequence stars. A model moves along any given line a distance roughly proportional to ω^2 with the end points representing models for which $\omega = \omega_c$. The dotted lines depict models in which there is some uncertainty (reprinted courtesy of the *Astrophysical Journal*).

Mander (1968) finds observationally that the displacement is proportional to V^α where α varies with spectral type. Golay (1968) finds that the constant of proportionality is also a function of spectral type. Replacing the gray atmosphere with a non-gray atmosphere and attempting to reconcile the critical angular velocity models of Roxburgh *et al.* (1965) with the first-order perturbation theory of Sweet and Roy (1953), Collins (1965) concludes that higher-order terms may defy any simple description and in addition, that the amplitude of the displacement is color-dependent. Collins and Harrington (1966), on the basis of an extended non-gray atmosphere study, find that rotation has little or no effect on photo-electric color-color diagrams. This is clearly indicated in Figure 3. This leads to the result that reddening effects may be distinguished from color changes introduced by rotation on a color-color diagram.

All work on the continuum which has attempted to predict the spectral energy distribution has lead to the conclusion that the largest effects will occur in the vicinity of the energy maximum. Thus, in the B stars, one would expect rather large effects in the far ultraviolet. However, as Hardorp and Strittmatter (1968a) point out, not only is the ultraviolet flux sensitive to aspect, it is also very sensitive to $(B-V)_0$. Thus,

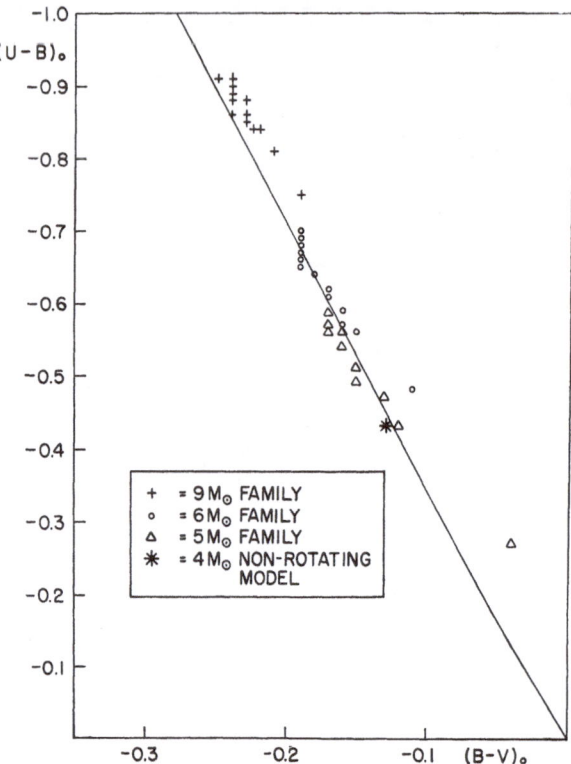

Fig. 3. A color-color diagram for various rotating models indicating the displacement along the Johnson-Morgan zero-age main sequence (reprinted courtesy of the *Astrophysical Journal*).

they predict no more than 0.5 mag. variation in m_{1350} due to rotation for stars of the same $(B-V)$. In this study, they also carry out theoretical calculations for A and early F stars which heretofore had been largely neglected. They found, as did Kraft and Wrubel (1965), that a change in the Strömgren c_1 index of about 0.2 magnitudes may be expected to result from rotational effects.

We may therefore summarize the effects of rotation upon the spectral energy distribution of stars as predicted by the numerous theoretical studies carried out since about 1963 as follows:

(i) large effects (about 1 mag.) in the observed bolometric magnitudes (and hence in bolometric corrections) may be expected.

(ii) rotation may generate between 0.5 and 1 mag. changes in M_v depending on V, aspect and $(B-V)_0$.

(iii) for small rotation, we may expect these changes to be roughly proportional to V^2.

(iv) rotation has little or no effect on the color-color diagram.

(v) the effect of rotation on color is such as to displace the star to the right and above the main sequence.

(vi) approximately 0.2 magnitude increase in the Strömgren c_1 index for a star may be introduced by extreme rotation.

Before we turn to the effects of rotation upon spectral lines, there is one further aspect of the continuum radiation field that should be discussed. Chandrasekhar (1946) and later Code (1950) indicated that polarization would arise in atmospheres of early-type stars due to the presence of free electrons and the anisotropy of the radiation field near the surface of the star. Normally, this polarization would average to zero when the contribution from all parts of a spherical star were summed up. However, as this zero result arose from the spherical symmetry and not from any physical properties of the atmosphere, it was to be expected that a rotationally distorted star should show a net intrinsic polarization. This possibility was investigated by Harrington and Collins (1968) using the gray atmosphere development of Chandrasekhar (1946) and it was found that up to 1.8% net polarization might be expected in the continuum of stars undergoing extreme rotation and seen nearly equator-on. However, further investigation by Collins (1970) employing non-gray atmospheres ascertained that the wavelength dependence of the polarization was such that it could only be detected in the line-free sections of the far ultraviolet and that the amount to be expected in the visible is essentially zero. There appears to be some difficulty in reconciling this result with recent observations and we shall return to this problem in the last section. Both Harrington (1969) and Collins (1970) point out that polarization arising from Rayleigh Scattering in the atmospheres of late-type giants may yield an amount considerably in excess of that to be expected from the gray studies of Chandrasekhar (1946) and Code (1950). Indeed, if these stars are distorted by some mechanism (i.e., rotation, magnetic fields, or non-radial pulsation), intrinsic polarization of an amount comparable with that presently observed may well be present.

B. ROTATIONAL EFFECTS ON LINE PROFILES AND EQUIVALENT WIDTHS

In contrast to the rather short time during which the rotational effects upon the continuum have been studied, the history of the investigation into the effects of rotation upon spectral lines dates back 50 years. The theoretical study of rotational profiles initiated by Shapley and Nicholson (1919) was expanded and formalized by Carroll (1928, 1933) and later applied to actual stars by Carroll and Ingram (1933) and others. An essentially geometrical formulation of the problem was given by Shajn and Struve (1929) and it is this form which has been most frequently used.

However, the problem as formulated by both Carroll (1928, 1933) and Shajn and Struve (1929) neglects effects of limb-darkening, 'gravity-darkening' (resulting from von Zeipel's condition), and shape distortion. Unsöld (1955) describes an analytic method for including the effects of limb-darkening in the method of Shajn and Struve (1929). Slettebak (1949) using essentially this method did include effects of limb-darkening, shape distortion and gravity darkening separately, but did not estimate the effects of varying aspect. For a more detailed account of these first efforts to analyze rotationally-broadened lines, the reader is referred to the very excellent review article by Huang and Struve (1960).

All of the initial efforts quite properly focused primary attention on the effects of the rotational Doppler broadening alone. Thus, the nature of the line profile that is formed locally on the surface of such a star received little or no attention. This somewhat historical breakdown of the problem also serves as a useful astrophysical delineator for its investigation.

To investigate the effects of rotational Doppler broadening alone, one has usually assumed a form for the local line profile and then, correcting for the rotationally induced Doppler shift, has convoluted these profiles over the surface of the star to obtain the observed profile. Initially, the locally assumed profile had some simple analytic form such as a delta function or a Doppler profile. Except for the limb-darkening correction of Unsöld (1955), the profile was assumed to be the same at all points on the disk. Slettebak (1949) and later Stoeckley (1968a, b, c) assumed that the local profile could be accurately determined by using carefully measured empirical profiles for known sharp-line stars. This has the advantage that in using the theoretically resulting profiles to interpret observed profiles for rotating stars, the instrumental broadening is automatically included in both results. It has the disadvantage that one cannot be sure that a profile obtained from a sharp-line star will be at all related to the profile generated locally on the surface of a rotating star. For determining first-order effects such as the value of $V_e \sin i$, the approach is probably justified for all but the most rapidly rotating stars. However, for investigating second-order effects such as differential rotation, there appears to be some doubt as to the effect of such an assumption.

More recently a variety of authors (e.g., Collins and Harrington 1966, and Hardorp and Strittmatter, 1968b) have attempted to remove the assumption of the form of the local line profile shape. This is done by calculating a non-gray atmosphere appropriate for the local surface condition and then, in accordance with the theory of radiative transfer, calculating a specific intensity profile at each point on the disk. Thus, provided with the local profile shape, the results are then integrated over the apparent disk so as to yield a flux profile. This provides a self-consistent means for estimating the importance of various effects resulting from the rotation of the star on a resulting profile.

Rather than delineate the findings of each of the many investigators in this field, it is perhaps more appropriate to indicate the nature of the effects and discuss what is presently known about the importance of each upon the resulting profile. Starting from a purely theoretical base, one naturally inherits all of the uncertainties inherent in calculating a line profile for an ordinary star. In addition, the importance of these uncertainties will vary over the surface of the star and therefore the ultimate effect on the final result is more difficult to ascertain than in the non-rotating case. As the nature of the assumptions which result in these uncertainties have been widely discussed in the literature, we shall only briefly list them with some qualitative comments relating to their estimated importance.

Considerable emphasis has been placed on the effects that departures from L. T. E. may have on the formation of spectral lines (cf. Underhill, 1966, 1969). In all likelihood, we need not be concerned about non-L.T.E. influences on the continuum field as Mihalas (1967) has shown that these departures are negligible for non-rotating models. However, as the core of strong lines are formed much higher up in the atmosphere, departures from L.T.E. may be important in their formation. Among main-sequence stars, the most likely candidates for having a physical condition which might lead to departures from L. T. E. are the early B-type stars – particularly those which are rapidly rotating. These departures have two basic origins which are somewhat related. Firstly, a breakdown in the Saha-Boltzman equation resulting from a distortion of the free electron velocity distribution from that expected of a gas in thermal equilibrium. This results in a variation in the population of the various atomic levels from the thermal equilibrium values. This effect, although investigated for non-rotating stars, has never been dealt with for rotating atmospheres. However, it must be pointed out that those regions where the pressure and density are low enough to bring about significant departures from L.T.E. are the equatorial regions where due to the von Zeipel condition, the radiative flux is relatively low. Thus, the contribution to the surface integral of the intensity should be minimized. It is also this region where one would expect effects of meridional circulation, a breakdown of the von Zeipel condition, etc., to have the largest effect.

Secondly, a deviation of the line source function from the Planck function resulting from non-coherent resonance scattering is important when investigating the effects of rotation upon the low-order members of the Balmer series of hydrogen as well as any resonance line. This effect was included by Collins and Harrington (1966) and it was found to lead to approximately a 20% difference in the intensity of the core region of Hβ. Since it is the core radiation that dominates most of a rotating hydrogen line profile, the inclusion of this effect appears to be important and may explain some of the qualitative differences between the calculations of Collins and Harrington (1966) and Hardorp and Strittmatter (1968b). Strom and Kalkofen (1966, 1967) have considered the effects of departure from L.T.E. upon the spectral energy distribution and suggest an observational test for determining the extent to which they are present. They found (1967) that the Paschen- to Balmer-jump ratio was insensitive to rotation, but systematically affected by departures from L.T.E. However, a very carefully designed observational program would be required to detect the effect, since it is possible that other factors may influence this ratio.

In addition to departures from L.T.E., one should also be concerned with the possible existence of small departures from radiative equilibrium which can lead to large variations in the surface temperature distribution in B stars, effects of line blanketing, and the choice of broadening theory to use for hydrogen lines. Although the relevance of none of these problems to rotating line profiles has been investigated, one may take some solace in the fact that at least the last two do not seem to have a major effect on the profiles of single stars. However, since the former may only arise because of the presence of axial rotation, its importance will be much harder to assess. Additional problems may arise from the exclusion of an important opacity source in the theoretical models. Strom and Strom (1969) have been lead to the conclusion that silicon is important in determining not only the shapes of hydrogen line profiles, but also the continuum energy distribution. If this is so, then great care must be exercised in understanding the importance of heretofore ignored elements in early-type stars.

Most of the current investigation has been directed toward estimating the global effects of rotation upon the line shape. The early work of Shajn and Struve (1929) showed that the profile resulting from a uniformly rotating and undarkened star of an infinitely sharp line would be elliptical or 'dish-shaped'. This basic form will be modified by limb-darkening, 'gravity-darkening', shape distortion, differential rotation, the variation of ionization equilibrium over the surface and the aspect which the star presents to the observer. The inclusion of limb-darkening as formulated by Unsöld (1955) is only appropriate for spherical stars not affected by 'gravity-darkening'. The more recent studies of Collins and Harrington (1966) and Hardorp and Strittmatter (1968b) automatically include this effect as well as 'gravity-darkening' through the nature of the locally non-gray atmospheres. Two papers of Collins (1968a, b) were specifically aimed at estimating the relative importance of limb-darkening and aspect effects, 'gravity-darkening' and the variation of ionization equilibrium over the surface. Although the simplified expression for the center-limb variation of pure absorption and scattering lines resulting from the Milne-Eddington atmosphere approximation together with an ad-hoc dependence of the line absorption coefficient on temperatures were used, the results were sufficient to indicate the relative importance of these three effects on the equivalent width. Even in cases where the temperature dependence of the line absorption coefficient was large (i.e., $\eta_\nu \sim T_{\mathrm{e}}^{10}$) the dominant factor in determining the variation of line strength with rotational velocity and aspect appeared to be the nature of the center-limb variation. However, the sign of the temperature dependence was of importance in that a line with a mass absorption coefficient decreasing with increasing temperature was less affected by other rotational effects than one with the opposite dependence on temperature (see Figure 4). Thus, one would expect that for B stars, the equivalent width of lines such as the Balmer lines to show fewer effects of rotation than say He I lines. This result seems to be borne out by the more detailed studies of Collins and Harrington (1966) and Hardorp and Strittmatter (1968b). However, it must be pointed out that all line effects seem to have a somewhat stronger dependence on angular velocity than ω^2 which might have been expected from the first-order interior studies. Indeed, noticeable effects on the equivalent width of spectral lines appear to be present only for stars rotating faster than 80% of their critical velocity. This led Collins (1968b) to conclude that only stars undergoing extreme rotation might be assigned a spectral type that was different from a similar non-rotator. It was also found that the qualitative effect of rotation upon the equivalent width was almost totally insensitive to the choice of the temperature gradient required by the Milne-Eddington model. This would lead one to hope that some of the problems mentioned earlier leading to errors in the surface temperature distribution (e.g., departure from L.T.E.) would not affect the equivalent width of the spectral lines of similar stars differing only in rotational velocity and aspect. This will probably not be the case for the line profiles.

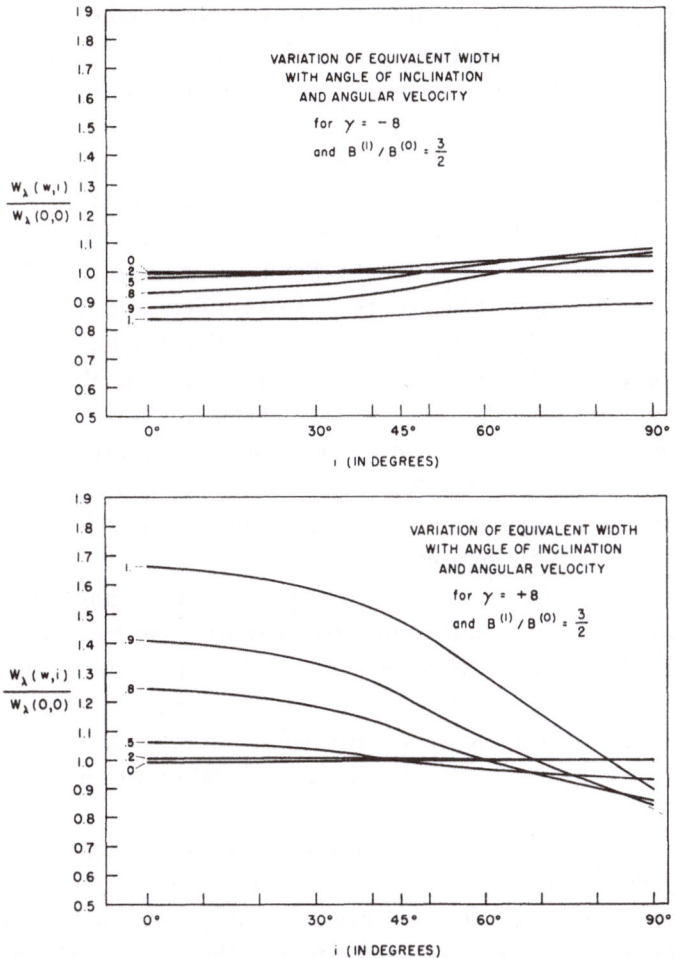

Fig. 4.   Variation of equivalent width with fractional rotational velocity and angle of inclination for lines where $\eta_\nu \sim T^{\gamma}_e$ and $\gamma = \pm 8$. The factor $B^{(1)}/B^{(0)}$ specifies the source function gradient for the Milne-Eddington type atmosphere (reprinted courtesy of the *Astrophysical Journal*).

The effects of limb-darkening and 'gravity-darkening' on the line profile itself have been investigated by several authors. Slettebak (1949) found that neglecting both limb-darkening and 'gravity-darkening' led to a change in the shape of the line profile such as to lead to an underestimate of $V_e \sin i$. However, he also noted that only in the case of extreme rotation was the effect of 'gravity-darkening' important. Hardorp and Strittmatter (1968b) find a much larger discrepancy (about 40%) between $V_e \sin i$'s determined from gravity-darkened and undarkened profiles for stars rotating within 5% of their critical velocity. Nothing can be said about the far larger number of more slowly rotating objects as their study does not include models of moderate rotation. Similar results are indicated by Friedjung (1968) with a maximum error attributable to gravity-darkening at the critical velocity of 25%. However, his calculations extend to lower velocities and it appears that the effect quickly disappears. At least for $V_e < 300$

km/sec., he shows that the errors introduced by his assumptions regarding the nature of the local line formation exceed the difference introduced by gravity-darkening.

Since it is guaranteed by the von Zeipel Theorem that stars rotate differentially and it is observed to be the case with the sun (Abetti, 1955), it is logical to inquire into the effects of differential rotation on the shape of spectral lines. This problem has been investigated by a number of people from several different points of view.

Slettebak (1949) concluded that the effect of differential rotation upon line profiles would be difficult to detect observationally. Similar conclusions have been reached by other authors (see Huang and Struve, 1960; also Qin-Yur and Jia-bing, 1964). More recently, Stoeckley (1968a), using the assumption that the local line profile shape could be empirically determined, presented a theoretical method for determining $V_e \sin i$, and a differential rotation parameter he calls S. However, even with observational line profiles which surpass those of Slettebak by nearly two orders of magnitude in accuracy, he is forced to conclude that "No general conclusion can be reached at this time, because many of the observations, although accurate by normal standards, should be much improved".

The importance to astrophysics of observationally determining the extent to which rotating stars differentially rotate cannot be minimized, but great care must be exercised when that determination rests on a theory where the nature of the second-order effects are not well understood. One cannot help but wonder if the uncertainties in the theoretical structure of rotating stellar atmospheres previously mentioned justify the pursuit of such an elusive quantity as differential rotation which apparently requires observational accuracy better than 0.1% in the line profile.

There remain other uncertainties than those already mentioned which cannot be ignored if a highly accurate theory of rotational line broadening is to be formulated. For example, it is not clear that turbulence, induced by rotation, might not be present in stars which rotate. This added complication to the velocity field might vitiate all other attempts to account for second-order effects. Thus great care must be exercised in drawing conclusions based on these second-order effects and indeed only those of a qualitative nature can be given any chance of surviving future inspection. In spite of the uncertainties, we may attempt the following qualitative conclusions resulting from line investigation:

(i)   no unambiguous effects other than geometrical broadening are apparent in spectral lines except for the most rapidly rotating stars (i.e., $\omega > 0.8\omega_c$).

(ii)   the determination of $V_e \sin i$ for the most rapid rotators is probably underestimated by between 10% and 40% by neglecting numerous 'second-order' effects.

(iii)   for the most rapidly rotating stars, the center-limb variation is probably the dominant parameter in determining the line strength and the effect is most noticeable for lines with a mass absorption coefficient which increases with temperature.

(iv)   in spite of its importance to astrophysics, no definitive conclusions can be drawn regarding the extent or even the existence of differential rotation. Due to the impressive increases in observational accuracy now possible in determining observational line profiles, it would appear that the greatest insight into this problem could be gained from an improved theoretical study of the conditions determining the surface velocity field in these stars.

4. Comparison of Theoretical Effects with Observations

In this section, we shall review the attempts that have been made to verify the predictions of the theory of rotating stellar atmospheres in the wealth of observational material which exists for these stars. Due to the volume of material and the existence of several other review papers in this colloquium dealing with this subject, we shall omit from discussion all of the material relating to multiple-star systems and most of the material on clusters.

There are several difficulties to be encountered in relating theory to observation. Firstly, except for the rotational Doppler broadening of spectral lines, all the directly observable effects we would hope to detect are small and quite near the limit of detection. Secondly, virtually all of the observational material which is at our disposal has been gathered for other purposes. Thus, it is not surprising that rotational effects have been largely ignored in developing systems to observationally specify stars. For instance, we shall find photometric systems whose standards are chosen without regard to rotation. Since a set of standards is by definition without systematic error, we have the possibility then that rotational effects will be largely lost from the system or at most appear as a systematic error in non-standard stars. Only the morphological system of spectral classification deliberately attempts to avoid a systematic bias introduced by rotational broadening by choosing a dispersion sufficiently low that rotational broadening is difficult to detect in any but the most rapidly rotating stars.

It would appear logical to divide our discussion of observational effects in much the same way as we did the theoretical predictions, but such is not the case. The nature of a star is usually specified by referring exclusively to either properties of the continuum such as M_v and color or to properties of absorption line spectra such as luminosity class and spectral type. Once the specifications have been made, then one looks for variance within that specification attributable to axial rotation. However, this is at best misleading as it is most likely that rotation will affect the nature of the specification to a degree equal to or larger than the variance to be expected within that classification. Thus, it is most appropriate that we first consider the evidence for effects of axial rotation on the parameters normally used to observationally specify an early-type star. The method of specifying the nature of a star for which there is a vast amount of observational material and which is probably the oldest is spectral type. In spite of the high degree of reliability and self-consistency that an individual investigator may acquire, the basic nature of the classification procedure is highly subjective and in many cases not fully specified. Indeed, it appears that the original specification of the system by Morgan et al. (1943) is not exactly followed by many investigators. Thus, without knowing in advance which line ratios and line strengths will be used by a given investigator, it is impossible for the theoretician to predict precisely what the effect of rotation may be upon the spectral class assigned any given star. In addition, it is not obvious that the concept of line strength is related to the predictable quantities of equivalent width and central depth, although if the dispersion is low enough, one would expect it to be more strongly correlated with the former than the latter.

The only hope of using this information for rotation studies arises from the result that equivalent widths of stellar lines will only be affected at a detectable level for spectral classification in the most rapid rotators (i.e., $\omega < 0.8\,\omega_c$) and thus we may assume that the spectral class assigned a star will be essentially unaffected by aspect or rotation effects even for stars of moderately large rotational velocity. However, Collins (1968b) did ascertain that if the original Morgan *et al.* (1943) criteria were adhered to, that the most rapidly rotating stars in the middle B region might well be classified a tenth of a class earlier than their non-rotating counterparts. Thus, he considers that a systematic difference between spectral type and photometric color should exist for the Be stars. A careful search for this effect has never been carried out as the photometric color must be corrected for the presence of any emission lines lying within the filter band pass.

A similar investigation led Collins (1968b) to conclude that again, except for the most rapid rotators, the luminosity class of a B star should be relatively unaffected by rotation. Unfortunately, his investigation did not extend to the A and F stars as the classification criteria became too complicated, but it would not be surprising if such a study resulted in similar conclusions for these stars as well.

A second and increasingly common method of specifying the nature of a star is through the use of various photometric indices. We shall not attempt to discuss the effects of rotation upon all of the numerous photometric systems in use today, but rather concentrate upon two types of systems in the hope that they will be illustrative of the types of effects that might be expected in other photometric systems of similar nature.

First, let us consider the rotational effects on the Johnson UBV system. As we have already seen, a theoretical investigation into these effects demonstrates that both the absolute magnitude M_v and photometric colors $(U-B)_0$ and $(B-V)_0$ will be affected both by the degree of rotation and the aspect presented by the star. This will be true even for moderate rotational velocities (e.g., $\omega = 0.5\,\omega_c$). The investigation of these effects is severely complicated by three different problems. Firstly, and most fundamental is that effects of rotation on the continuum radiation field are very similar to effects produced by evolution. Thus to clearly separate the two effects, an observational program should be designed to investigate stars in the same state of evolutionary development with varying rotational characteristics, or vice versa. Secondly, the UBV standard stars have not, in general, been selected with any regard to their rotational properties. Finally, the absolute characteristics of the UBV system seem to be difficult to reproduce (see Code, 1960) and it seems likely that the published definition of the filter transmission functions is in error. In light of these difficulties, the only clear statements that can be made are:

(i) The V magnitude should be affected by rotation by an amount varying with spectral type of between 0.8 and 1.5 mag. This appears to be verified from cluster studies, notably Roxburgh *et al.* (1966), Strittmatter and Sargent (1966) and Strittmatter (1966).

(ii) The effect of the rotation upon stars plotted on a color-color diagram is to move them down the main sequence with respect to their non-rotating counterparts. This theoretical result has at the present time not been verified primarily due to the intrinsic difficulties of the *UBV* system and small size of the effect.

We turn now to the observed effects on an intermediate band system such as that defined by Strömgren (1956). The effects on the *uvby* colors should theoretically be essentially the same as those predicted for the Johnson system with the possible exception of the *u* filter. Accurate theoretical estimates of rotational effects on this part of the spectrum require atmospheres which accurately represent the Balmer decrement. An analysis of this nature has recently been carried out by Golay (1968).

Hardorp and Strittmatter (1968a) compute the change of the c_1-index with rotation and aspect. They find that for A and early F-type stars, one should expect changes due to aspect and extreme rotation of as much as 0.16 in c_1 above the average value to be expected for these stars. Kraft and Wrubel (1965) calculated a similar effect and indicated that they felt an anomalously large spread in c_1 on a c_1-$(b-y)$ diagram existed for stars in the Hyades. This they attributed to rotation. The more careful study of Golay is in qualitative agreement with these earlier studies, but indicates that they may have overestimated the size of the effect. However, Strömgren (1967) points out that at that time, no evidence existed for the presence of any rotational effects on c_1, m_1 or the β index as defined by Crawford (1958, 1960, 1964) for early F and A stars. It remains to be seen how much of this difference is attributable to theory and how much to observation. For stars earlier than spectral type A, Hardorp and Strittmatter (1968a) predict a spread of 0.2 in c_1 to be introduced by rotation.

It is worth noting one remaining aspect of the continuum radiation field. The work of Harrington and Collins (1968) implied the possible existence of intrinsic polarization in rapidly rotating B stars. The extension of this study to non-gray atmospheres by Collins (1970) showed that this was not the case in the visible and this result has been confirmed observationally by Serkowski (1968, 1969). However, several observers (Coyne and Gehrels, 1967 and Serkowski, 1968, 1969), have inferred the existence of intrinsic polarization in Be stars and the results have been directly confirmed by observation of differing polarization between two members of several visual binary systems by Bottemiller (1969). This result, combined with that of Collins (1970) implies that the polarization present in these stars must arise above the atmosphere, presumably in a circumstellar shell.

Finally, to conclude our discussion of observed rotational effects, let us turn to effects which appear to be present in the line spectrum of rotating stars. As previously mentioned, the most obvious observational aspect of stellar rotation is the Doppler broadening of lines. This broadening has been used by numerous workers to determine the $V \sin i$ appropriate for many stars. A catalogue of $V \sin i$'s for 2558 stars as determined by various investigators has been compiled by Boyarchuk and Kopylov (1964) and recently extended to 3951 stars by Uesugi and Fukuda (1970). Since the method of determination of $V \sin i$ varies from author to author, some care must be exercised in the use of these catalogues and one must always be watchful for systematic differences arising from both observational sources and methods of reduction. It is probably safe to say that the best determinations of $V \sin i$ have an internal consistency of no better than 5% while 10% is far more common. Indeed, for the most rapidly rotating stars, it has been suggested that 20% might not be an unreasonable observational error to expect. We have already seen that there exists considerable doubt concerning the interpretation of the $V \sin i$ measures for the most rapidly rotating stars and that systematic errors of 10% to 40% exist. It is therefore surprising that Slettebak (1966) finds such a smooth relationship between the maximum rotational velocity of stars and spectral type. This could only result if the measured parameter was determined with a relatively high degree of internal consistency and represented something fundamentally connected with the structure of these stars. This latter fact must be true regardless of the precise interpretation of the term $V \sin i$. However, it is true that the measured 'critical velocities' do fall below those to be expected from existing main-sequence models. Whether this results from the presence of second-order effects as suggested by Hardorp and Strittmatter (1968b), or from the absence of stars rotating at the verge of instability is not clear. This specific point is of extreme importance for future observational studies of stellar rotation.

A possible resolution of this problem could perhaps be found in the detection of the presence of second-order effects on the equivalent width of absorption lines. Guthrie (1963) suggested that such effects were present in the photoelectric indices related to the equivalent widths of Hβ and Hγ. Collins and Harrington (1966) showed that their results were perfectly consistent with Guthrie's (1963) estimates of the behavior of Hβ and Hγ for rotating stars provided the observed $V \sin i$ scale was somewhat increased. However, Strömgren (1967) indicated that there were no rotation effects evident in the A and F stars studied with this filter system. Crawford and Mander (1966) also conclude that no rotation effects were present in the measurements of the β index for the sample of B stars they investigated. Petrie (1964), in revising his Hγ equivalent width-luminosity classification (Petrie, 1958), indicates that no rotational effects on Hγ are present in line studies. Bappu et al. (1962) also found good agreement between Hγ luminosities and independent determinations for Orion and NGC 2362 stars.

How may this apparently conflicting observational material be understood? Theory (e.g., Collins and Harrington, 1966; Hardorp and Strittmatter, 1968b), unambiguously indicates that the equivalent width of Hβ and Hγ should decrease roughly as $(V \sin i)^2$. Collins (1968b) indicates that the helium lines should show a strong positive correlation with $V \sin i$ as was found by Deeming and Walker (1967a, b). It is not likely that this ambiguity between theory and some observations and the apparent conflict within the observations themselves has a simple solution. Indeed, in order to understand this result, one must examine each case with great care in order to ascertain the reason why rotation effects were or were not detected. However, several general considerations must be kept in mind when examining any of these observational studies.

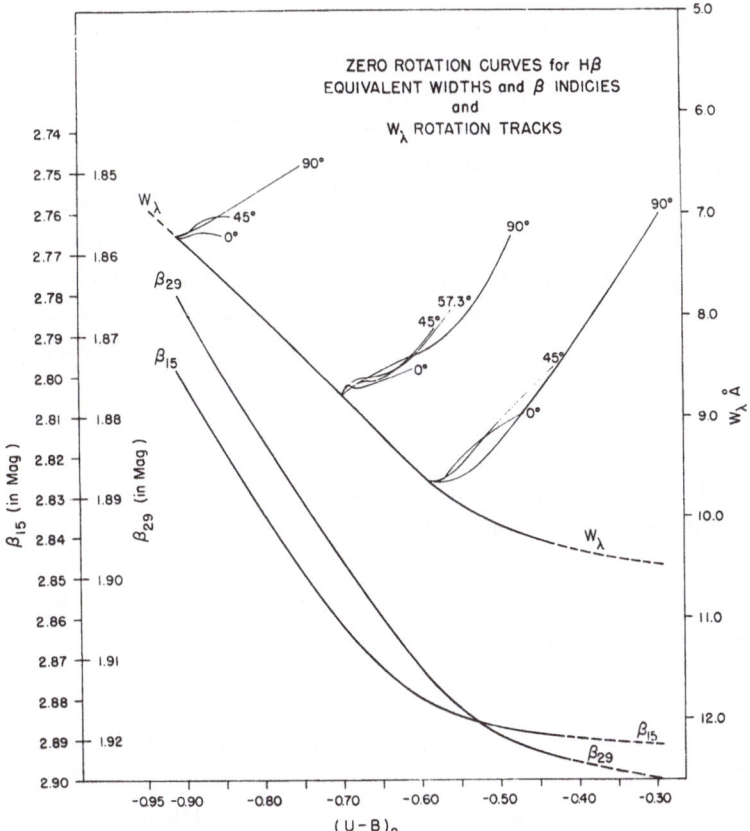

Fig. 5. Rotational effects predicted for the photometric β index for three model stars seen at different angles of inclination. The displacement along the various curves is roughly proportional to $(V \sin i)^2$ (reprinted courtesy of the *Astrophysical Journal*).

Firstly, the effect of rotation upon the hydrogen line strengths is not terribly large unless one includes the most rapid rotators (see Figure 5). In that event, the situation may be greatly confused by the presence of emission. Secondly, the effect as predicted by Collins and Harrington (1966), may not be as obvious when the comparison is made between line strengths and spectral type rather than intrinsic color due to the subjective nature of spectral classification. Although the photoelectric line-strength measures (e.g., γ and β indices) admit the possibility of great accuracy, the precision obtained in practice often falls short of that expected. All observing systems now in use suffer from the difficulty that the defining standards of the system have been chosen without regard to rotation. Depending on the distribution function of $V \sin i$ for the standards, this may introduce a systematic error as well as a variance in studies carried out with the system. Although one would expect both the systematic error and the variance to be small, it could be expected to be of the same size as the rotational effects in question and thus mask them in any statistical search.

The disagreement existing between observers of Hγ equivalent widths is somewhat more difficult to understand. However, it must be remembered that equivalent widths are very difficult to measure, with the results depending strongly on the dispersion used. Buscombe (1969), using high-dispersion spectra, finds a clear effect in the M_v vs. $W_{\gamma\delta}$ relationship due to the value of $V \sin i$ and presents a detailed discussion relating to the difference between his result and that of Petrie (1964).

Finally, some mention must be made regarding observations of the helium lines where the effect of rotation should be larger (Collins, 1968b). Unfortunately, very little information is available on this matter. However, what is available appears to support the theory. Deeming and Walker (1967a), find a positive correlation of HeI (4471) equivalent width and $V \sin i$. Strittmatter and Sargent (1966) find that certain sharp-line stars in Orion exhibit weak helium lines and on grounds other than the low value of $V \sin i$, feel that they are intrinsically slow rotators. In addition, Buscombe (1969) finds that a correlation between weak helium lines and slow rotation is compatible with his results.

In light of this discussion, we may draw the following conclusions:

(i)   The subjective criteria of spectral classification are probably less affected by rotation than differences between observers.

(ii) *UBV* photometry is of limited usefulness in studying rotational effects; however, the effect of rotation upon the observed absolute magnitude predicted by theory appears to exist.

(iii) Narrow band photometry should provide an excellent tool for studying rotation effects, but to date, the results have been partly negative. However, this does not conclusively demonstrate that the effects do not exist.

(iv) An observed discrepancy exists between the theoretical value for the critical rotational velocity and the largest observed values of $V \sin i$. However, the existence of second-order effects on line profiles is more than sufficient to explain the difference.

(v)   Present photometric data on the hydrogen and helium lines tend to confirm the theoretical predictions and hence the existence of second-order effects on line profiles due to rotation.

5. Concluding Remarks

In this paper we have attempted to describe the present status of the theory of rotating stellar atmospheres. The central question to which all our discussion points is: 'To what extent are the effects of extreme axial rotation present in stars?'. It is imperative that this question be answered if one wishes to proceed to the next logical question of: 'How do these stars evolve?'. This is necessary from two points of view. Firstly, the angular momentum will be important in determining the evolved structure of the stellar interior. Secondly, the extent to which observational anomalies existing among early-type stars can be explained as evolutionary effects depends on the existence and importance of rotational effects.

Considering the uncertainties present in the theory and the difficulties encountered in observing second-order effects, it appears that excellent agreement now exists between theory and observation at least on a qualitative basis. Thus, we may conclude that rotational effects must be considered wherever observations are to be interpreted in light of the theory of stellar evolution. The question that remains is a quantitative one. The answer to the quantitative problem will require substantial advances in both theory and observation. Theoretically, we need to know the most probable distribution of velocities that should be expected on the surface of a rapidly rotating star. Observationally, we require carefully designed programs specifically aimed at accurately measuring differences between similar stars resulting from rotation. If rotational effects in equivalent widths and line profiles are present, then stars exist which are rotating at or near their critical velocity and most of the anomalies in early type stars may largely be due to rotation or rotationally induced effects. If these effects are not present, then we may infer the existence of processes preventing stars from ever arriving at a state where they would rotate at or near their critical velocity. In addition, most all of the observed anomalies and differences existing in early-type stars would have to result from differences in their evolutionary state.

Unfortunately, at the moment, there is not one piece of observational evidence which unambiguously confirms the existence of stars rotating at the critical velocity. Within their inherent errors, all present observations are consistent with either of two pictures. Firstly, very rapidly rotating stars exist but their detection is made difficult by the presence of second-order effects on line profiles. Alternatively, such stars do not exist and the second-order effects on line profiles are not present. Either view is tenable but the implications of each for astrophysics are quite different.

References

Abetti, G.: 1955, *The Sun* (transl. by J. B. Sedgwick), Faber and Faber, London.
Bappu, M. K. V., Chandra, S., Sanwal, N. B., and Sinuhal, S. D.: 1962, *Monthly Notices Roy. Astron. Soc.* **123**, 35.
Bottemiller, R. L.: 1969, private communication.
Boyarchuk, A. A. and Kopylov, I. M.: 1964, *Isz. Krym. Astron. Obs.* **31**, 44.
Buscombe, W.: 1969, *Monthly Notices Roy. Astron. Soc.* **144**, 1.
Carroll, J. A.: 1928, *Monthly Notices Roy. Astron. Soc.* **88**, 548.
Carroll, J. A.: 1933, *Monthly Notices Roy. Astron. Soc.* **93**, 478, 680.
Carroll, J. A. and Ingram, L. J.: 1933, *Monthly Notices Roy. Astron. Soc.* **93**, 508.
Chandrasekhar, S.: 1933, *Monthly Notices Roy. Astron. Soc.* **93**, 390.
Chandrasekhar, S.: 1946, *Astrophys. J.* **103**, 351.
Code, A. D.: 1950, *Astrophys. J.* **112**, 22.
Code, A. D.: 1960, *Stars and Stellar Systems* **6**, 59.
Collins, G. W., II: 1963, *Astrophys. J.* **138**, 1134.
Collins, G. W., II: 1965, *Astrophys. J.* **143**, 265.
Collins, G. W., II: 1966, *Astrophys. J.* **146**, 914.
Collins, G. W., II: 1968a, *Astrophys. J.* **151**, 217.
Collins, G. W., II: 1968b, *Astrophys. J.* **152**, 847.
Collins, G. W., II: 1970, *Astrophys. J.* **159**, 583.
Collins, G. W., II and Harrington, J. P.: 1966, *Astrophys. J.* **146**, 152.
Coyne, G. V. and Gehrels, T.: 1967, *Astron. J.* **72**, 887.
Crawford, D. L.: 1958, *Astrophys. J.* **128**, 185.
Crawford, D. L.: 1960, *Astrophys. J.* **132**, 66.
Crawford, D. L.: 1964, in *I.A.U. Symposium* 24.
Crawford, D. L. and Mander, J.: 1966, *Astron. J.* **71**, 114.
Deeming, T. and Walker, G. A. H.: 1967a, *Nature* **213**, 479.

Deeming, T. and Walker, G. A. H.: 1967b, *Z. Astrophys.* **66**, 457.
Eddington, A. S.: 1926, *The Internal Constitution of the Stars,* Dover Publ., New York.
Epps, H. W.: 1968, *Proc. Nat. Acad. Sci.* **60**, 51.
Friedjung, M.: 1968, *Astrophys. J* **151**, 779.
Golay, M.: 1968, *Pub. Geneva Obs.* **75**, 105.
Guthrie, B. N. G.: 1963, *Pub. Roy. Obs. Edinburgh* **3**. 84.
Hardorp, J. and Strittmatter, P. A.: 1968a, *Astrophys. J.* **151**, 1057.
Hardorp, J. and Strittmatter, P. A.: 1968b, *Astrophys. J.* **153**, 465.
Harrington, J. P.: 1969, *Astrophys. Letters* **3**, 16.
Harrington, J. P. and Collins, G. W., II: 1968, *Astrophys. J.* **151**, 1051.
Huang, S. S. and Struve, O.: 1960, *Stars and Stellar Systems* **6**, 321.
Hyland, A. P.: 1967, Thesis, Australian National Observatory, Canberra.
Kraft, R. and Wrubel, M.: 1965, *Astrophys. J.* **142**, 703.
Limber, D. N. and Roberts, A. H.: 1965, *Astrophys. J.* **141**, 1439.
Mark, J. W.-K.: 1968, *Astrophys. J.* **154**, 627.
Mander, A.: 1968, *Pub. Geneva Obs.* **75**, 125.
Mihalas, D.: 1967, *Astrophys. J.* **149**, 169.
Morgan, W. W., Keenan, P. C., and Kellman, E.: 1943, *Atlas of Stellar Spectra*, 1st ed., University of Chicago Press, Chicago, Ill.
Petrie, R. M.: 1958, *Astron. J.* **63**, 181.
Petrie, R. M.: 1964, *Pub. Dom. Astrophys. Obs.* **12**, 317.
Qin-Yur, Q. and Jia-bing, Z.: 1964, *Acta Astron. Sinica* **12**, 29.
Roxburgh, I. W.: 1970, this volume, p. 9.
Roxburgh, I. W. and Strittmatter, P. A.: 1965, *Z. Astrophys.* **63**, 15.
Roxburgh, I. W. and Strittmatter, P. A.: 1966, *Monthly Notices Roy. Astron. Soc.* **133**, 345.
Roxburgh, I. W., Griffith, J. S., and Sweet, P.: 1965, *Z. Astrophys.* **61**, 203.
Roxburgh, I. W., Sargent, W. L. W., and Strittmatter, P. A.: 1966, *Observatory* **86**, 118.
Roxburgh, I. W., Faulkner, J., and Strittmatter, P. A.: 1968, *Astrophys. J.* **151**, 203.
Rubin, R.: 1966, *Monthly Notices Roy. Astron. Soc.* **133**, 339.
Sackmann, I. J.: 1968, Thesis, University of Toronto, Toronto, Canada.
Sargent, W. L. W. and Strittmatter, P. A.: 1966, *Astrophys. J.* **145**, 938.
Serkowski, K.: 1968, *Astrophys. J.* **154**, 115.
Serkowski, K.: 1969, *Astrophys. J.* (in press).
Shajn, G. and Struve, O.: 1929, *Monthly Notices Roy. Astron. Soc.* **89**, 222.
Shapley, H. and Nicholson, S. B.: 1919, *Comm. Nat. Res. Sci., Mt. Wilson Obs.* **2**, 65.
Slettebak, A.: 1949, *Astrophys. J.* **110**, 498,
Slettebak, A.: 1966, *Astrophys. J.* **145**, 126.
Stoeckley, T. R.: 1968a, *Monthly Notices Roy. Astron. Soc.* **140**, 121.
Stoeckley, T. R.: 1968b, *Monthly Notices Roy. Astron. Soc.* **140**, 141.
Stoeckley, T. R.: 1968c, *Monthly Notices Roy. Astron. Soc.* **140**, 149.
Strittmatter, P. A.: 1966, *Astrophys. J.* **144**, 430.
Strittmatter, P. A.: 1969, *Ann. Rev. Astron. Astrophys.* **7**, 665.
Strittmatter, P. A. and Sargent, W. L. W.: 1966, *Astrophys. J.* **145**, 130.
Strom, S. and Kalkofen, W.: 1966, *Astrophys. J.* **144**, 76.
Strom, S. and Kalkofen, W.: 1967, *Astrophys. J.* **149**, 191.
Strom, S. and Strom, K.: 1969, *Astrophys. J.* **155**, 17.
Strömgren, B.: 1956, *Vistas in Astronomy*, Vol. 2 (ed. by A. Beer), Pergamon Press, New York and London, p. 1336.
Strömgren, B.: 1967, 'The Magnetic and Related Stars', in *Proc. AAS-NASA-Symp. on Peculiar and Metallic-Line A Stars* (ed. by R. C. Cameron), Mono Book, Co. Baltimore, p. 461.
Sweet, P. and Roy, A. E.: 1953, *Monthly Notices Roy. Astron. Soc.* **113**, 700.
Uesugi, A. and Fukuda, I.: 1970, in press.
Underhill, A.: 1966, *The Early Type Stars*, Reidel, Dordrecht, The Netherlands.
Underhill, A.: 1969, *Observatory* **89**, 22.
Unsöld, A.: 1955, *Physik der Sternatmosphären*, 2nd ed., Springer-Verlag, Berlin.
Von Zeipel, H.: 1924, *Monthly Notices Roy. Astron. Soc.* **84**, 665.
Zhu, C.-S.: 1963, *Acta Astron. Sinica* **11**, 41.

Discussion

Jaschek: I am a little bit surprised that you omitted entirely a phenomenon which after all is well known to exist – namely, the emission lines in rapidly rotating B stars. How do they come into the picture?

Collins: Although it is clear that the Be phenomenon is correlated and probably closely related to rotation, the emission lines are most likely generated outside the atmosphere. Thus I considered the phenomenon itself to be outside the scope of this paper.

Mark: I would like to comment that where a star is differentially rotating the effects of the aspect angle would be very much reduced because the effective gravity does not vary by more than four as compared to the infinite ratio of values for the Roche Model. If we allow a variation of angular velocity comparable to the observed differential rotation of the sun, the stellar models differ markedly from a uniformly rotating model. In particular, the luminosity changes by up to two magnitudes for emission stars as compared to non-rotating ones. However, luminosity and effective temperatures averaged over aspect angle stay near to the non-rotating main sequence. Aspect effects are smaller as mentioned. Thus, small changes in photometric calibrations are expected.

Roxburgh: That cannot be so. If you just give the relative variation in angular velocity to be the same as observed on the sun, then there will still be a limit to the amount of angular momentum that can be stored in a star. Indeed, there will be very little change in the observed properties over the effects of uniform rotation. A 20% variation in ω, particularly to make ω larger at the equator, can do very little.

Collins: I tend to agree with Dr. Roxburgh. However, even should your picture be correct and the second-order effects be reduced, this would only tend to support the picture I have presented where the second-order effects are already small. Your comment regarding ratio of polar gravity to equatorial gravity is at best misleading. Although it is true that this ratio formally goes to infinity as ω approaches ω_c, if one accepts the von Zeipel condition, the equatorial region contributes nothing to the integrated properties of the star. Thus one should consider not this ratio but rather some mean value of $\bar{g}$. I think you will find most reasonable mean gravities to be similar for both the differential and rigid models. This is partially borne out by the variation of 2 magnitudes in bolometric magnitude for your model. The Roche model gives a value of between 1 and 1.5 magnitudes depending somewhat on the spectral type.

Deutsch: I want to address the question of aspect determination from high-dispersion spectrograms, as distinguished from ordinary spectral classification. Consider two sharp-line A stars with weak, zero-volt FeI lines that have the same equivalent widths. Suppose that one star has $V = 300$ km/sec and $i \simeq 0°$, while the other has $V = 10$ km/sec and $i \simeq 90°$. Now consider the behavior of weak lines from high levels of FeI, or from FeII or SiII. Can one predict that these lines also will have equivalent widths that are the same within a few percent?

Collins: The question is not simply answered for it requires a detailed knowledge of the mass absorption coefficient dependence on temperature for the lines in question. A velocity of 300 km/sec is only about 50% or 60% of the critical velocity; thus, it might well be the case that rotational effects would only be a few percent. Unless the velocity is considerably higher I feel the effect would be difficult to detect.

ON THE EFFECT OF ROTATION ON STELLAR
ABSORPTION LINES

ILKKA V. TUOMINEN and OSMI VILHU

Astrophysics Laboratory, University of Helsinki, Helsinki, Finland

Abstract. The effect of rapid uniform rotation on equivalent widths of spectral lines is studied in the case of a star seen pole-on. Contrary to the situation for the strong lines Mg II 4481 and Si II 4128, the effect may be significant for the weak ones, especially those of the Eu II-type. Because the mean atmospheric parameters of a rapidly rotating star seen pole-on are very similar to those of a non-rotating star of the same mass, this may cause significant errors in abundance determinations, especially for Eu-type elements. The aspect effect on the central intensity of two Eu II lines of different strengths is also studied.

1. Introduction

The effect of rapid uniform rotation on radiation from stars has been studied extensively, e.g., by Hardorp and Strittmatter (1968a, b). One of their results is that the pole-on cases (at least when $\omega \leqq 0.99$) show colors and spectral types differing very little from the values for non-rotating stars of the same mass. Also the mean atmospheric parameters θ_e and $\log g$ are very similar if deduced from the Balmer discontinuity and the width $D(0.2)$ of $H\gamma$. When these atmospheric parameters are used to determine the abundances for individual stars, errors may be caused if the star is rotating and seen pole-on. Since in the linear section of the curve of growth the abundance A is proportional to the equivalent width W, the error $\Delta \log A$ in the abundance, when weak lines are used, is equal to $\log W$ (rotating and pole-on) $-$ $\log W$ (non-rotating).

The results of Hardorp and Strittmatter (1968b) show that for Mg II 4481 the effect of rotation is small. In the present report we present our results for this same line and for the similar one Si II 4128 in the mass-range 1.55–3.55 solar masses, the star being viewed pole-on. Furthermore we have calculated the effect of rotation on weak lines of Fe II, Cr II and Eu II. In addition to abundance determinations generally, the results may be interesting in connection with the overabundances of peculiar stars. Aspect effects are not considered in the present communication in greater detail. Our preliminary result for the central intensity of the Eu II profile is presented. The visibility of the line is important from the point of view of the pole-on and fast-rotator hypothesis of peculiar stars.

2. Methods of Calculation

The physics involved is practically the same as that used by Hardorp and Strittmatter (1968a, b). The interior model of the rotating star was taken from Roxburgh *et al.* (1965), and the surface gravity calculated from the Roche model. The effective temperature is given by von Zeipel's law. The atmospheric parameters θ_e and $\log g$, as a

A. Slettebak (ed.), Stellar Rotation, 110–114. All Rights Reserved

function of latitude, thus depend only on the mass and a dimensionless parameter ω, representing the angular velocity in units of the break-up angular velocity. The radii and luminosities of the non-rotating stars were those given by Faulkner and taken from Hardorp and Strittmatter (1968b). The temperature and pressure at each latitude, as functions of the optical depth, were interpolated from the tables of Mihalas (1965). The corresponding opacity was interpolated from Bode's (1965) opacity tables. In the specific intensity

$$I_v(0, \mu) = \int_0^\infty S_v(\tau_v) \exp\left(- \tau_v/\mu\right) d\tau_v/\mu$$

the Planck function was used as the source function. In the pole-on case the integration over the surface was finally performed using 11 latitude points. For the aspect effect 15 latitude and 17 longitude points with varying stepwidth were used.

To study the effect of rotation on weak lines, three singly ionized elements (Fe$_{II}$, Cr$_{II}$, Eu$_{II}$) were chosen. The lines were calculated at 4000 Å for two different values of the excitation potential. In addition to zero, the values 3.25, 3.9 and 1.5 were used for Fe$_{II}$, Cr$_{II}$ and Eu$_{II}$, respectively. The latter values represent mean values in the spectrum of α^2 CVn between $\lambda\lambda$ 4950–3600, as taken from Burbidge and Burbidge (1955). The abundances and the oscillator strengths were chosen so that the lines were sufficiently weak, i.e. on the linear part of the curve of growth.

For comparison, the equivalent widths of the doublet Mg$_{II}$ 4481 and of Si$_{II}$ 4128.05, using the abundances $\log(A_{Mg}/A_H) = -4.0$ and $\log(A_{Si}/A_H) = -4.50$, were also calculated. The same values were used also by Hardorp and Strittmatter (1968b) and by Mihalas and Henshaw (1966). For Mg$_{II}$ 4481 $\log(gf\lambda)$ was put equal to 4.63 and for Si$_{II}$ 4128.05 equal to 4.18, as was done also by Mihalas and Henshaw (1966). The line profiles were assumed to be given by the Voigt profile and the damping constants calculated taking into account both radiation damping and collisional damping. The damping constants were calculated from the formulae given by Mihalas and Henshaw (1966).

3. Results

The results are shown in Figures 1–3. The effective temperature and spectral class are given on the abscissa axis. The masses for which the calculations have been performed are also given. The ordinate is $\log W$ (rotating and pole-on) $- \log W$ (non-rotating). The possibility of using the effective temperature or the spectral class, the latter taken from Morton and Adams (1968), as the abscissae is based on the fact that these are practically the same for a non-rotating star and for a rotating star of the same mass when seen pole-on (at least when $\omega \leq 0.99$, Hardorp and Strittmatter, 1968b).

Figure 1 shows the results for strong Mg$_{II}$ 4481 and Si$_{II}$ 4128 lines when $\omega = 0.99$. The effect of rotation is seen to be very small ($< 10\%$). For comparison, the dashed line shows the results of Hardorp and Strittmatter (1968b) for Mg$_{II}$ 4481. Figure 2 represents the results for weak lines of Fe$_{II}$, Cr$_{II}$ and Eu$_{II}$ when $\omega = 0.99$. The numbers in parentheses indicate the values of excitation potentials. The maximum

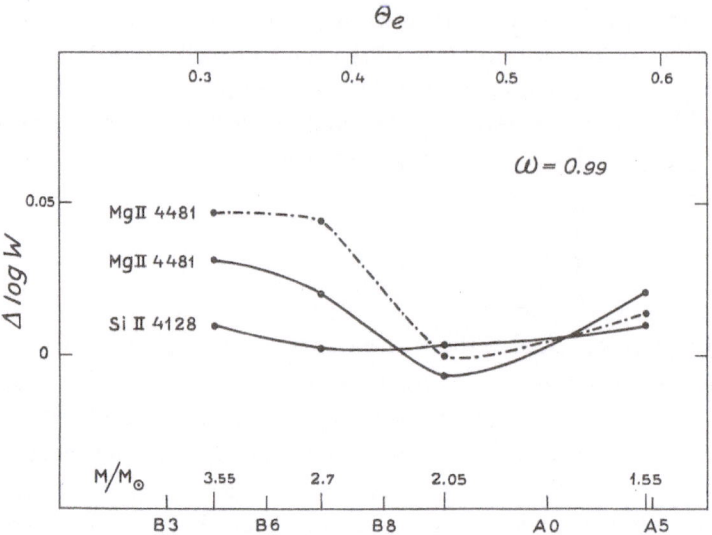

Fig. 1.   Effect of rotation on the equivalent width (W) of Mg II 4481 and Si II 4128.05 as a function of spectral type (or effective temperature) when $\omega = 0.99$ and the stars are seen pole-on ($\Delta \log W = \log W$ (rot.) $- \log W$ (non-rot.)). The masses for which the computations have been performed are also given. The dash-dot line gives the results of Hardorp and Strittmatter (1968b) for Mg II 4481.

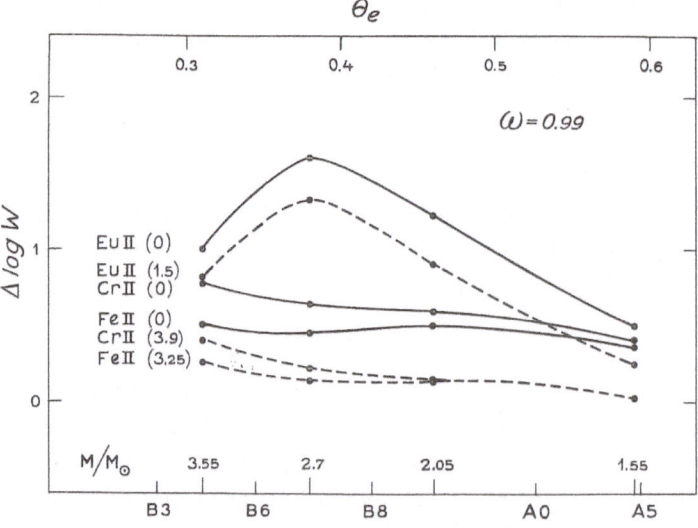

Fig. 2.   The same as in Figure 1 but for weak lines of Fe II, Cr II and Eu II at 4000 Å for two different excitation potentials given in parentheses.

amplification occurs for Eu II at $\theta_e = 0.4$ and amounts to 40 for the zero excitation potential. There are also considerable differences between different elements, and clearly the rapid rotation significantly affects the abundance determinations from weak metallic lines, especially for elements like Eu and Sr, both having similar ionization

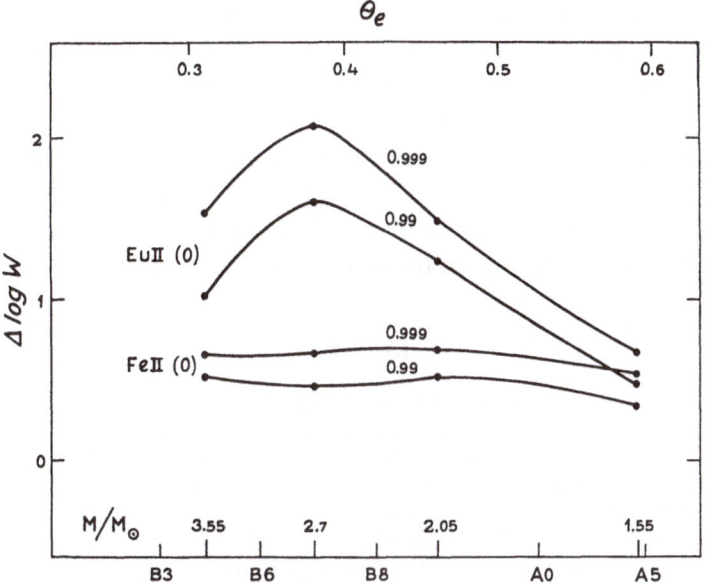

Fig. 3.   The same as in Figures 1 and 2 for Fe II and Eu II lines with zero excitation potentials and
two values of the rotation ($\omega = 0.99$ and 0.999).

TABLE I

Equivalent width W in mÅ and central intensity I_c
in percent of Eu II for three inclinations and two
line strengths. $M = 2.7\ M_\odot$, $\omega = 0.99$ and excitation
potential $= 0$.

	$i = 0°$	$i = 30°$	$i = 60°$
W	13	15	27
I_c	12	0.17	0.26
W	92	99	128
I_c	65	1.3	1.4

potentials. In Figure 3 the results for the rotation $\omega = 0.999$ are shown together with
those for $\omega = 0.99$, for the zero excitation lines of Eu II and Fe II. The maximum amplification for Eu II is now as great as two orders of magnitude. The reason for the
behavior of the Eu II line follows from the powerful growth of the amount of Eu II
ions with decreasing effective temperature. This growth attains its maximum at
$\theta_e \approx 0.4$. This in turn follows mainly from the small (11.2 eV) second ionization
potential.

Also it may be useful to remember that the present study is based on interior models
in uniform rotation. Eu-type lines are very sensitive to effective temperature and gravity. Deviations from the Roche model can easily strengthen the lines greatly. In such
a case, however, the equivalent width for a rotating star cannot be directly compared

to the corresponding value of a non-rotating star of the same mass, but instead to the same spectral type. The reason is that the rotation may significantly change the mean atmospheric parameters and spectral types derived from observations. This may be true also for the case $\omega = 0.999$ in Figure 3.

Table I gives the central intensity of the profile of Eu II for two different strengths for the mass for which the amplification was greatest. Although calculated for three inclinations only, it can be clearly seen that the reasonably strong line ($W \approx 100$ mÅ) is also greatly flattened even when the inclination is relatively small.

References

Bode, G.: 1965, *Abh. Inst. Theor. Phys. Sternwarte Kiel* **10**.
Burbidge, G. R. and Burbidge, E. M.: 1955, *Astrophys. J. Suppl.* **1**, 431.
Hardorp, J. and Strittmatter, P. A.: 1968a, *Astrophys. J.* **151**, 1057.
Hardorp, J. and Strittmatter, P. A.: 1968b, *Astrophys. J.* **153**, 465.
Mihalas, D.: 1965, *Astrophys. J. Suppl.* **9**, 321.
Mihalas, D. and Henshaw, J. L.: 1966, *Astrophys. J.* **144**, 25.
Morton, D. C. and Adams, T. F.: 1968, *Astrophys. J.* **151**, 611.
Roxburgh, I. W., Griffith, J. S., and Sweet, P. A.: 1965, *Z. Astrophys.* **61**, 203.

Discussion

Collins: What are the non-rotating values for Table I?

Tuominen: Table I was calculated only to demonstrate the aspect effect. The stronger line is no longer on the linear part of the curve of growth. To estimate the non-rotating value for that line we need in addition to Figure 2 also the curve of growth, which is different for a non-rotating and a rotating star.

Collins: Have you made any calculation of the models with a lower rotational velocity?

Tuominen: Not yet. These first calculations were done in order to find out how large an effect it is possible to obtain when the star is rotating uniformly.

Hardorp: These huge effects of rotation on the Eu II lines do not help us in explaining the peculiar A stars: as far as I remember the hypothesis that these stars are rapid rotators seen pole-on died four years ago at the conference on Magnetic and Related Stars. I am rather suspicious about the magnitude of the effects you found. They are in direct conflict to what Dr. Collins told us this morning.

Tuominen: Two of the objections against the pole-on hypothesis are that rotation can not produce spectral peculiarities and that these peculiarities should also be observed in stars with broader lines. What these results suggest is only that these two objections may not necessarily be correct.

Preston: How was the ionization equilibrium of Eu established? In particular, how did you determine the partition functions for successive stages of ionization as a function of temperature?

Tuominen: For Eu we have used the partition functions of La given by Aller (*Stars and Stellar Systems* **6**, 232) and by Allen (*Astrophysical Quantities*, 1958, p. 33). For curves like those in Figures 2 and 3 the influence of the partition functions is small. To test this we put all the partition functions of Eu equal to unity, and the resulting curves differed by less than 5 % from those in Figures 2 and 3. In addition, for Sr II, whose partition functions are better known and whose ionization potentials are similar to those of Eu, the resulting curves were also within 5 % of the curves of Eu.

OBSERVATIONS OF STELLAR ROTATION AND LINE STRENGTHS FOR OB STARS

WILLIAM BUSCOMBE

Northwestern University, Evanston, Ill., U.S.A.

For 121 southern stars of spectral types between O5 and A3, including 35 supergiants, equivalent widths and values of $v_e \sin i$ have been measured for absorption lines on direct-intensity tracings of coudé spectrograms from Mount Stromlo, with original dispersion 150 microns per angstrom. For stars of a particular temperature and luminosity class, the strength of the hydrogen lines is weaker and the triplet helium lines stronger in fast rotators than in sharp-lined objects which are rotating more slowly. Full details are in print (*Monthly Notices Roy. Astron. Soc.* **144** (1969), 1): subsequent papers in press present detailed profiles of the Balmer lines for 23 slowly rotating stars and details of emission features shown by several fast rotators.

The investigation is being continued with spectrograms of northern B stars secured at the Dominion Astrophysical Observatory, supported in part by grant GP-13544 from the National Science Foundation.

Discussion

Slettebak: I notice that for a number of stars in common, your rotational velocities are generally higher than mine. Do your velocities contain corrections for limb or gravity darkening?

Buscombe: No: the profiles are treated as approximately Gaussian, with a half-width parameter $= v_e \sin i \, (\log_e 2)^{1/2}$. However, for stars in which the empirical $v_e \sin i$ term < 60 km/sec, an adjustment was made for an instrumental profile of half-width 20 km/sec in the same parameter.

Slettebak: I believe that it is difficult to find any stars earlier than O9 with really sharp lines, which would imply that macroturbulence must play a role in these stars. Did any of your O-type stars show sharp lines?

Buscombe: The star HD 57682, with $V = 6.4$ and $v_e \sin i = 40$, has not quite as sharp lines as 10 Lac.

Slettebak: The fact that the O9.5 V star, Zeta Ophiuchi, has exceedingly broad lines and yet has never been observed to show emission lines is interesting, as you pointed out. The critical rotational velocity for break-up rises rather sharply as one moves from the B-stars into the O-stars, however – the star may not be rotating quite rapidly enough to produce an emitting ring.

Buscombe: Since the value of $v_e \sin i$ determined from half-widths on my microphotometer tracings substantially exceeds that quoted by you and other investigators, it will be of interest to attempt a convolution of the line profiles for a slow rotator in the hope of fitting the observed distribution of intensity.

Collins: I would like to point out that this work is one of the strongest justifications for the existence of rotational effects on equivalent widths of spectral lines. The qualitative behavior of all lines mentioned is in accord with theory. However, the magnitude of the effects is somewhat less than one would expect if there were many stars rotating at or very near the critical velocity.

Buscombe: Although I was aware of the results of Deeming and Walker, I had not studied your predictions in detail before reducing my own measurements. It is remarkable that the selection of material analyzed by Hyland did not make the effects more noticeable.

ROTATING STELLAR ATMOSPHERES AND THE
ENERGY DISTRIBUTION OF ALTAIR

PETER R. JORDAHL

The University of Texas at Austin, Austin, Tex., U.S.A.

Abstract. Models of rotating stellar atmospheres have been constructed for masses appropriate to middle and late A stars, and some of the photometric properties of these models are discussed. A specific model for the rapidly rotating star Altair has been constructed matching the observed radius, luminosity, and projected rotational velocity. It is found that the energy distribution predicted for the model agrees well with that observed for Altair, after the latter has been corrected for line blanketing.

This investigation was begun in the hope of comparing predictions from rotating model atmospheres and observations of a relatively rapidly rotating bright star, and thus, to some degree, testing the validity of the theory. The star Altair (α Aquilae) was chosen for the following reasons: (1) the spectral type (A7) indicates that most of its flux is emitted in the accessible regions of the spectrum. (2) Collins (1965) has suggested that the observable effects of rotation on the continua of A stars might be larger than for the B stars which he studied, due in part to (1). (3) Altair has a $V_e \sin i$ of 250 km/sec (Slettebak, 1966) determined by the standard method; Stoeckly (1968) has carefully studied line profiles in Altair photoelectrically and has derived a similar value based on a more complete model for rotation effects. (4) Altair has an excellent parallax ($\pi = 0.194 \pm 0.004$ (Jenkins, 1963)) and an interferometric angular diameter determination (Hanbury Brown *et al.*, 1967); thus we can attempt to match the total luminosity and radius of the star with our model.

The major disadvantage of the use of Altair as a test case is that it is a field star, and therefore (1) we cannot independently determine its approximate age by the cluster turnoff method, and (2) we cannot say anything about its composition from similar but sharp-lined stars which are members of the same group.

An attempt was made to associate Altair with one of the stellar groups isolated and identified by Eggen (1962). In the $U - V$ plane Altair cannot be identified with any of these groups, although it is much closer to the Hyades group than to any other, and is significantly distant from the sun-Sirius group. Therefore, we might *assume* (for lack of any better information) that the composition of Altair is similar to that of the Hyades, and that therefore blanketing corrections to the energy distribution of Altair (which of course cannot be measured directly for a star with a large $V_e \sin i$) may be obtained from the work of Oke and Conti (1966) on the Hyades.

Following Collins (1965), we assume uniform rotation, a Roche model for the potential, and von Zeipel's theorem for the relation between local gravity and local radiative flux. The polar radius and total energy generation are assumed to vary with angular velocity according to the prescription given by Faulkner *et al.* (1968) for a 2.0 $M_\odot$ interior model (computed with uniform rotation). (This variation is much less than that suggested earlier by Roxburgh *et al.* (1965); the newer results make a fit between model and observations possible over a much smaller range in the latter). The stellar interior models used to obtain the non-rotating luminosity and radius for a given mass, age and chemical composition were from the evolutionary sequences of Kelsall (unpublished; see Kelsall and Strömgren, 1966) which were computed for a range of composition and mass, and which incorporated modern opacities, including the bound-bound contribution.

The composition selected for the interior models was $X = 0.67$, $Z = 0.03$, these values agreeing well with those determined by Kelsall and Strömgren to give the best fit between the zero-age Kelsall models and the observations of the zero-age main sequence. The evolutionary sequences are given for a sufficient range of composition and age that a mass-luminosity-radius (MLR) relation can be obtained for the above composition for ages of 0, 1, and 2×10^8 years.

Thus, for a given age, mass, angular velocity (here expressed in terms of a ratio $v = \omega/\omega_c$, where ω_c is the 'critical' or 'break-up' angular velocity), and inclination i, we may construct a rotating model which gives the shape, size, and variation of gravity (and hence temperature) over the surface.

To obtain the radiation distribution from the model, the following procedure was employed: A grid of 21 model atmospheres was generated for $6500\,\mathrm{K} \leqslant T_e \leqslant 9500\,\mathrm{K}$, $3.5 \leqslant \log g \leqslant 4.5$, using a program kindly supplied by Robert Kurucz of Harvard-SAO. The atmospheres were constructed with $X = 0.7$ and normal solar metal abundances and incorporated all the usual sources of continuous opacity; also included were the absorption from neutral Mg and Si (of great importance in the rocket ultraviolet) and an approximation to the blended wings of the Balmer lines, as suggested by Strom and Avrett (1964; their model 5). This latter makes the theoretical energy distributions much more realistic in the region between Hδ and the Balmer limit.

For each atmosphere, emergent intensities were calculated for 62 frequencies and for $1.0 \geqslant \mu \geqslant 0.05$, and these data were incorporated into the rotation program. For a specific rotating model (age, M, v, i given) the gravity g (and hence the effective temperature T_e) and μ are determined for a set of points (θ, φ) on the surface. For each point (θ, φ) and each frequency v, the intensity $I_v(0, \mu[\theta, \varphi]; T_e[\theta], \log g[\theta])$ in the direction of the observer is obtained from the model atmosphere intensities by quadratic interpolation in T_e, $\log g$, and μ. The monochromatic flux F_v per unit solid angle in the direction of the observer is then obtained by a numerical quadrature over the surface of the model.

If these values of F_v are integrated over all frequencies and multiplied by 4π steradians, we obtain the total luminosity L which would be *inferred* for the model if it were observed *at the given inclination*. Note that for rapidly rotating models this value can vary by a factor of 2 or more with inclination, while of course the true total luminosity of the model remains constant.

Energy distributions were calculated as above for models of $M = 1.7$, 1.8, and 1.9 $M_\odot$, $i = 0°$ and 90°, and a range of $v (= \omega/\omega_c)$ from 0. to 0.99. These were then subjected to another program which computes the parameters of the Strömgren 4-color (*uvby*) photometric system (based on a preprint from Matsushima, 1969). Some of the results

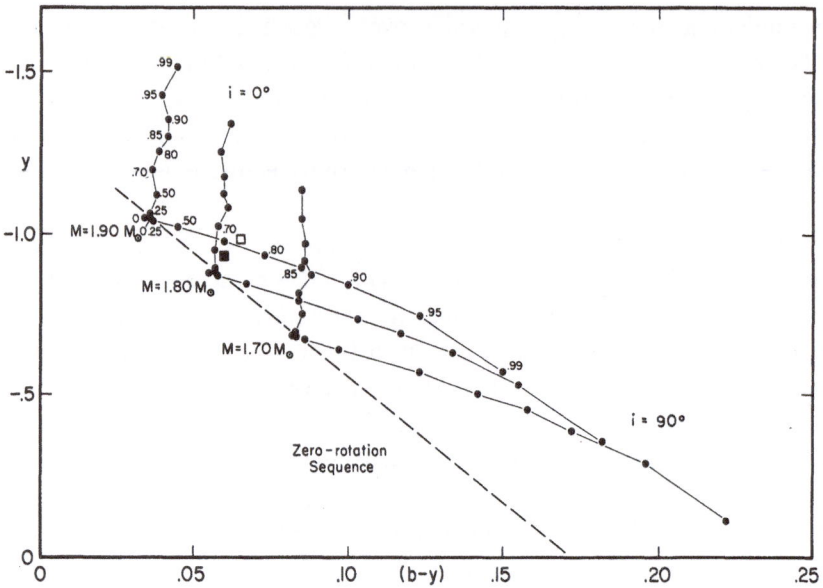

Fig. 1.   Pole-on ($i = 0°$) and equator-on ($i = 90°$) sequences of zero-age rotating models in the [y, $(b-y)$] color-magnitude diagram. Points for $M = 1.90\ M_\odot$ are labeled by the ratio of angular velocity to critical angular velocity. The squares show the position of non-rotating models for ages 1×10^8 years (filled) and 2×10^8 years (open) for $M = 1.80\ M_\odot$. (Note: magnitude y has an arbitrary zero point).

are shown in Figure 1, which demonstrates the variation in magnitude (in the y filter; *arbitrary* zero point) and color ($b - y$) with the parameters mentioned. The results for $1.9\ M_\odot$ are labeled by the value of v. Also shown in Figure 1 are the positions of non-rotating models for $1.8\ M_\odot$ and ages of 1 and 2×10^8 years (filled and open squares, respectively). The important thing to notice is that for the model with $v = 0.85$ (corresponding to $V_e = 260$ km/sec, about the largest observed in this mass range) the change in color (equator-on sequence) or magnitude (pole-on sequence) from the non-rotating model is only about half the maximum change predicted (for $v = 0.99$). Hence, although these theoretical models predict a rather large effect for 'near-break-up' stars, we shall probably not find such large effects. Taking the opposite view, *if* such large photometric effects are found and definitely attributed to rotation, then there is a greater impetus for incorporating a more complex model of rotation.

Another parameter which can be derived from the tilted, distorted rotating model is the average apparent radius $\bar{R}$. This is an average of the 'radii' of the object along one quadrant of the projected disk. These radii were not weighted by limb darkening and gravity darkening in taking the average. (However, the values of angular diameter taken from Hanbury Brown *et al.* incorporated *their* approximate correction for the effects of *limb* darkening.)

We now have three parameters with which we may compare models and stars: $V_e \sin i$, $\bar{R}$ and L. $\bar{R}$ and L, however, may be determined with reasonable accuracy only for the nearest stars, since both depend directly on distance; in addition, the angular diameter has been measured for only a few stars. Altair satisfies both requirements. Therefore a program has been written which performs a differential corrector procedure on the variables M, v, and i to fit the observed $V_e \sin i$, $\bar{R}$, and L, starting with a reasonable first guess for the variables. The derivatives required for the differential corrector procedure were calculated separately for each iteration, thus requiring four complete rotating models per iteration. For the age-zero Kelsall MLR relation the values (for a good fit) of these variables are:

$$V_e \sin i = 250 \text{ km/sec} \qquad M = 1.80 \, M_\odot$$
$$\bar{R} \qquad = 1.65 \, R_\odot \qquad v = 0.85$$
$$L \qquad = 11.0 \, L_\odot \qquad i = 73°.5.$$

No fit was possible for ages of 1.0 and 2.0×10^8 years, because the non-rotating model radius was already too large for these ages.

Observations of the energy distribution of Altair were made with the Ebert scanner on the 36″ telescope at McDonald Observatory. Those covering 3200 Å to 7000 Å

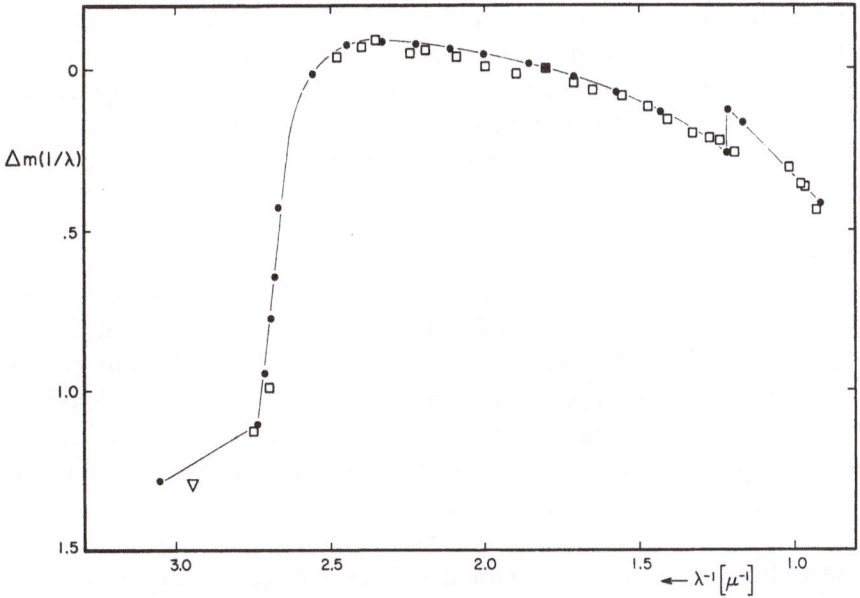

Fig. 2.   The energy distribution for a rotating model fitted to Altair ($V_e \sin i$, $\bar{R}$, and L) (filled circles and line) and deblanketed observations of Altair (open squares). The open triangle is derived from the observed point at 3400 Å and the blanketing for 3636 Å. The squares are 0.^{m}025 high; formal probable errors in the comparison between Altair and Vega are 0.^{m}005 to 0.^{m}015, but errors in the absolute calibration of Vega and in the assumed blanketing will increase these somewhat.

were made with the scanner operating in a rapid-scan, multiple pass mode incorporated recently by M. MacFarlane. The infrared observations were made earlier. The scans were reduced relative to Vega; the absolute calibration of Hayes (1967) corrected for the change in the gold point temperature (Labs and Neckel, 1968) was used to place the data on an absolute scale relative to 5556 Å. Blanketing corrections for

$\lambda \geqslant 3636$ Å were taken from the table in Oke and Conti (1966) interpolated to $(B-V)$ $= 0.22$. The resulting continuum energy distribution is shown in Figure 2 by the open squares. The open triangle at $\lambda^{-1} = 2.94$ (3400 Å) was obtained by assuming the blanketing at 3400 Å to be same as at 3636 Å, probably an underestimate.

The energy distribution from the fitted rotating model calculated *without regard* to the observed energy distribution is shown in Figure 2 by the filled circles and the smooth curve. The agreement between theoretical and observed energy distributions appears good. This might indicate that our assumptions are adequate for calculated models to represent the continuum observations. However, in view of the large number of assumptions concerning the specific test case, further models will be constructed to determine the sensitivity of the energy distribution fit to these assumptions and choices.

Another possibility for improvement (suggested by Dr. Collins at this Colloquium, this volume, p. 85), would be to calculate the 'average radius' with limb darkening and gravity darkening included. However, what really should be done is to determine the monochromatic intensity profile projected onto 'diameters' ranging from polar to equatorial for the wavelength used by Hanbury Brown and determine the radius from the Fourier transform of this profile in the same manner as the observational procedure. This would in addition show the degree of sensitivity of the radius so determined to the shape of the star; one might also be able to check the possibility of detecting (as mentioned at this Colloquium by Dr. Mark) a very elongated object, such as the rotating models constructed by Mark (1968), by use of interferometric observations.

Acknowledgements

I should like to thank Dr. Frank Edmonds and Dr. Terence Deeming for reading a preliminary version of this paper, and Kelly MacFarlane and Catherine Hilburn for their excellent work on the figures. The computing facilities and time provided by the University of Texas Computer Center are gratefully acknowledged. Special thanks are due the Director and staff of McDonald Observatory for making it possible to obtain these observations, and to M. MacFarlane and D. C. Wells III for assistance at the telescope and in the photometric reductions.

Note added in proof: Sanderson, Smith and Hazlehurst (1970, *Astrophys. J.* **159**, Part II, L69) offer an explanation of the discrepant results of Roxburgh *et al.*, on the variation of R_p and L with rotation.

References

Collins, G. W., II: 1965, *Astrophys. J.* **142**, 265.
Eggen, O. J.: 1962, *Roy. Obs. Bull.*, No. 51.
Faulkner, J., Roxburgh, I. W., and Strittmatter, P. A.: 1968, *Astrophys. J.* **151**, 203.
Hanbury Brown, R., Davis, J., Allen, L. R., and Rome, J. M.: 1967, *Monthly Notices Roy. Astron. Soc.* **137**, 393.
Hayes, D. S.: 1967, unpublished dissertation, University of California, Los Angeles.
Jenkins, L. F.: 1963, *General Catalogue of Trigonometric Stellar Parallaxes*, Yale University Observatory, New Haven.

Kelsall, T. and Strömgren, B.: 1966, *Vistas in Astronomy* **8**, 159.
Labs, D. and Neckel, H.: 1968, *Z. Astrophys.* **69**, 1.
Mark, J. W.-K.: 1968: *Astrophys. J.* **154**, 627.
Oke, J. B. and Conti, P. S.: 1966, *Astrophys. J.* **143**, 134.
Roxburgh, I. W., Griffith, J. S., and Sweet, P. A.: 1965, *Z. Astrophys.* **61**, 203.
Slettebak, A.: 1966, *Astrophys. J.* **145**, 126.
Stoeckly, T. R.: 1968, *Monthly Notices Roy. Astron. Soc.* **140**, 121.
Strom, S. E. and Avrett, E. H.: 1964, *Astrophys. J.* **140**, 1381.

Discussion

Collins: I would suggest that instead of using geometrically determined $\bar{r}$ to compare with an interferometer measurement, that an intensity-weighted mean be used so as to include effects of limb-darkening and gravity-darkening.

Jordahl: The results given by the interferometer group are given in two forms, the first being for the case of no limb darkening, the second for the case where limb-darkening is derived from model atmosphere data for the appropriate spectral type and wavelength, and the values so obtained are used to correct the derived radius. I am using the second value. The effect of gravity darkening for the case most nearly like Altair appears to be relatively small even for a model near break-up, according to Hanbury Brown.

Faber: Have the published radii measured by interferometry been derived by incorporating some correction for limb darkening?

Collins: Yes, the values of the radius derived from interferometric measurements do depend on the intensity distribution over the apparent disk. This is true for both the intensity interferometer and the phase interferometer. Thus radii must be corrected for both limb and perhaps 'gravity darkening'.

Mark: Concerning Dr. Collins' comment on gravity darkening effects on stellar interferometry, the Hanbury Brown group (Johnston, I. D., and Warbing, N. C.: 1969, Preprint CSUAC No.177, 'On the Possibility of Observing Interferometrically the Surface Distortion of Rapidly Rotating Stars'), has reported that for solid body rotation, oblateness effects are almost cancelled by gravity darkening. But since differential rotation results in small gravity darkening, it is perhaps still possible to detect oblateness if stars rotate in this way.

Jordahl: One would presumably have to put a model for the gravity darkening into the interferometer data reduction. Hanbury Brown also points out that the details of the intensity distribution are contained in the wings of the Fourier transform of the distribution (which is actually what is measured) and the signal-to-noise ratio in this part essentially precludes obtaining the information. One can of course simply *assume* a certain distribution across the disk, go through the transform, and see if the result is discernably different from the standard form.

Klinglesmith: Both the Lyman lines and carbon continuous opacity must be included. Also, small overabundances of metals will increase the opacity effects.

Jordahl: Certainly. I intended to mention that although the Si and Mg continua seem to drop out around 10000°, the carbon continua should remain to a higher T_e and should be included. The point here is that these are *not minor* opacity sources, *not* that my models are complete.

Kraft: Suppose one maintained the position that all slowly rotating A's are in fact peculiar A's, so that if one sees a sharp-line normal A, it must be a rapid rotator seen pole-on. Vega falls in this class. Yet Vega is fundamental in all basic calibrations of T_e, R, etc. for early A's. Would this not constitute an embarrassment?

Jordahl: It would not bother the use of Vega as a spectrophotometric standard. But, yes, it would be poor to use such a star in the calibration of T_e vs. $(B - V)$ or T_e vs. spectral type. However, (as Dr. Collins pointed out), this assumption very severely limits the possible inclination of Vega, since it has such a low $v \sin i$ ($\leqslant 5$ km/sec) that no broadening is detectable. Vega does have an interferometric radius $\frac{3}{4}$ that of Sirius, but this is more easily explained as an evolutionary effect, than as a result of a star similar to Sirius rotating at or near break-up. To my knowledge, none of several analyses of Vega at very high dispersion have yielded any peculiarity in its spectrum that might be attributable to rotation.

ROTATION AND MACROTURBULENCE IN SUPERGIANT STARS

JEFFREY D. ROSENDHAL

Washburn Observatory, University of Wisconsin, Madison, Wis., U.S.A.

Abstract. A statistical investigation has been made of rotation and macroturbulence in early type Ia and Iab supergiants. The principal results are: (1) At all spectral types in the range B0–A5 both rotation and macroturbulence contribute to the observed line broadening. In the early B stars rotation is as important a broadening agent as large-scale mass motion. In the middle B and A stars turbulence dominates but there still is an appreciable contribution from rotation. (2) In spite of the fact that the stars observed seem to be losing mass, there is no strong evidence for significant angular momentum loss.

1. Introduction

Recent observational studies (Oke and Greenstein, 1954; Sandage, 1955; Abt, 1957; Faber and Danziger, 1969) of the compatibility of the observed axial rotation of stars and currently accepted theories of stellar evolution have suggested that for stars of luminosity classes IV, III and II earlier than spectral type G0 angular momentum is conserved during evolution. In the later spectral types of all luminosity classes the rotation is very low, and it is tempting to correlate this fact with the development of extensive outer convective zones and subsequent loss of angular momentum by stellar winds (Kraft, 1966, 1967).

In the higher luminosity domains of the Hertzsprung-Russell diagram the situation is considerably less clear. In particular, in the case of the stars of highest luminosity, the Ia and Iab supergiants, there is no *a priori* reason to expect angular momentum to be conserved, for many lines of evidence (Deutsch, 1956; Sargent, 1961; Abt, 1963; Morton, 1967; Hutchings, 1968) suggest more or less continuous mass loss and hence probable angular momentum loss throughout their evolution.

One can then ask the question whether a study of the change in rotation along an evolutionary track can be used to estimate the changes, if any, in the total angular momentum and from this infer the mass loss rate. The present investigation was undertaken in an attempt to answer this question.

The problem is not straightforward, however, and is complicated by the fact that in these stars the exact nature of the line broadening is not settled. Several good arguments against a strictly rotational interpretation of the line broadening and in favor of broadening due to some form of large-scale mass motion can be given (Huang and Struve, 1953, 1954; Slettebak, 1956; Abt, 1958). In particular, the absence of sharp-lined stars suggests that at least part of the observed line width is due to motion with a spherical rather than an axial form of symmetry.

Therefore the objectives of the present study have been (1) to determine the nature of the line broadening, (2) to see if broadening due to two macroscopic broadening agents (namely axial rotation and large scale turbulence) can be separated, and (3) if the separation is possible, to use the recovered rotational component to study the angular momentum problem.

A. Slettebak (ed.), Stellar Rotation, 122–129. All Rights Reserved
Copyright © 1970 by D. Reidel Publishing Company, Dordrecht-Holland

2. The Observational Program

The observational data consisted of coudé spectrograms of 64 B and A type Ia and Iab supergiants which were obtained with the 84-inch reflector of the Kitt Peak National Observatory. All stars were observed in the blue region at a dispersion of 13.5 Å/mm. Approximately 90% of all known supergiants observable in the Northern Hemisphere down to a limiting magnitude of $B = 8.0$ were observed.

A summary of the observational program is shown in Figure 1. A detailed list of the program stars is contained in Rosendhal (1970). Both field stars and cluster members were observed. We note that the B2-A0 stars are roughly evenly distributed between the two groups whereas the earliest spectral types are predominantly from the general field and the later A stars are almost all cluster members.

It was our original intent to observe only stars which were members of clusters with known distance moduli. In principle each has a known absolute magnitude and an evolutionary track can be associated with each star in order to determine its main sequence predecessor and to estimate the change in radius and moment of inertia which it has undergone during evolution (cf., Sandage, 1955; Abt, 1957, 1958). However, it was felt that there were not enough objects in this group for a meaningful statistical discussion and so the list was supplemented by a selection of field stars.

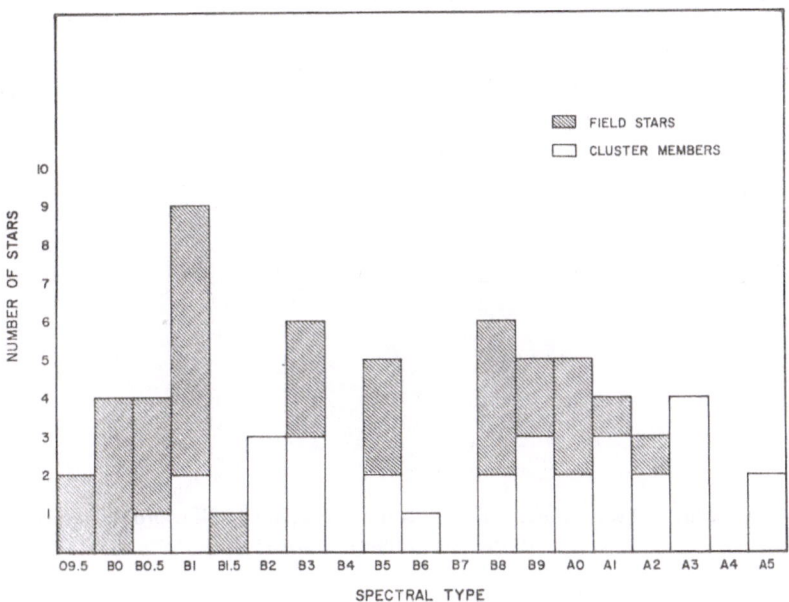

Fig. 1. Distribution of stars in the observational program. There is an even distribution with spectral type and the total sample is almost evenly divided between field stars and cluster members.

3. Determination of Rotational Velocities

The rotational velocities were obtained from a comparison of the observed line profiles

124   JEFFREY D. ROSENDHAL

with theoretically computed ones. The lines Mg II 4481, He I 4471 and Si IV 4089 were
the spectral features which were employed for this purpose. The details of the computations, the reasons for the choice of these lines and a comparison of the results for
different lines have been discussed by Rosendhal (1968, 1970).

A comparison of the observed and theoretically computed line profiles for several
typical cases is shown in Figure 2. In each case the solid line is the raw data, the dashed
line is the adopted smoothed profile (drawn before any theoretical profiles had been
computed) and the circles are the theoretical points.

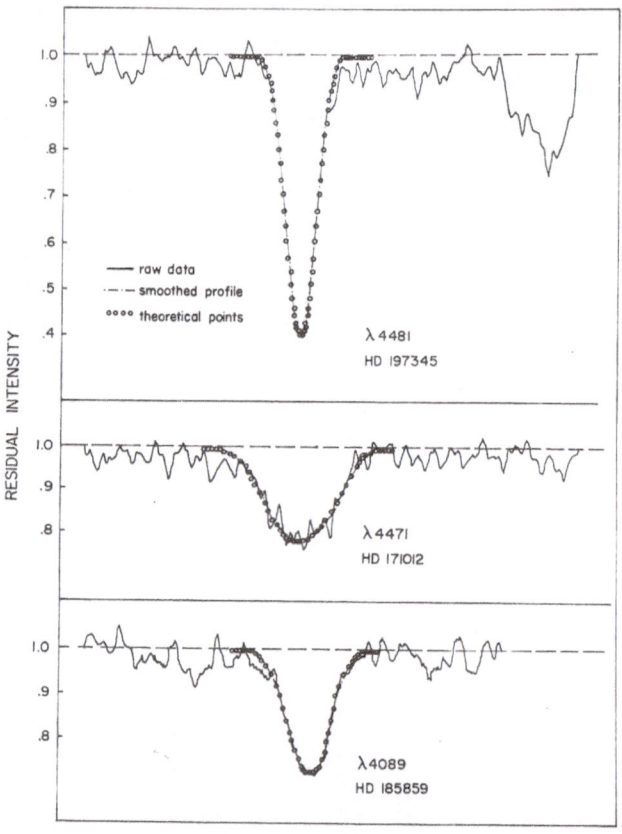

Fig. 2. A comparison of some observed and theoretically computed line profiles (reprinted courtesy
of the *Astrophysical Journal*).

4. Apparent Rotational Velocities

The results of the line profile computations are summarized in Figure 3 which shows
the behavior of the apparent line broadening expressed as rotation as a function of
spectral type. The most important features of this diagram are the absence of any
stars showing small apparent rotational velocities, the appreciable spread in broaden-

ing for the early spectral types, the decreasing spread toward the later spectral types and the relatively smooth decrease exhibited by the lower envelope of the points. The mean apparent rotational velocities for stars in selected small intervals of spectral type are given in the third column of Table I.

TABLE I

Statistical analysis of supergiant rotation

Group	Expansion factor	Apparent rotation (km sec^{-1})	Predicted rotation (km sec^{-1})	Recovered rotation (km sec^{-1})
A0-A5	15–23	≈ 40	8 ± 4	13 ± 6
B3-B6	7–10	≈ 60	18 ± 8	14 ± 9
O9.5-B1	< 5–6	≈ 90	26 ± 10	40 ± 20

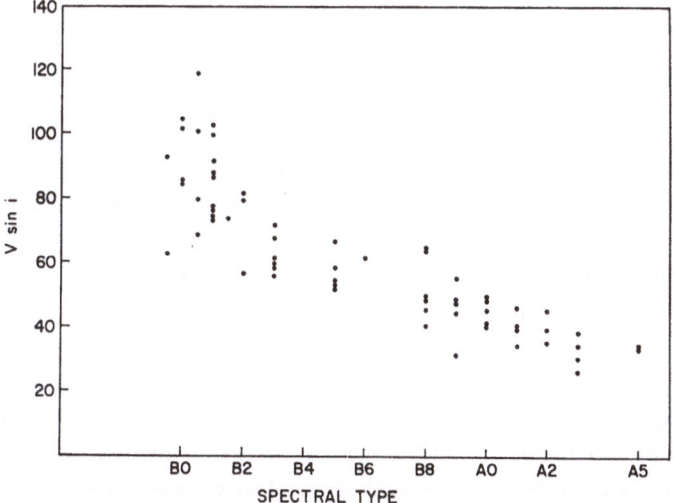

Fig. 3.   Line broadening expressed as rotation as a function of spectral type. If two lines have been used in the determination of rotation, the averaged results are shown (reprinted courtesy of the *Astrophysical Journal*).

5. Stellar Rotation and Stellar Evolution

The rotational velocities expected on the basis of evolution away from the main sequence with angular momentum conserved have been calculated with the aid of theoretical evolutionary tracks published by Hayashi and Cameron (1962), Iben (1966) and Stothers (1966). Details of the procedure have been published elsewhere (Rosendhal, 1970).

The two well-known limiting cases (Oke and Greenstein, 1954) of (1) no radial exchange of angular momentum (angular momentum conserved in spherical shells), and (2) free radial exchange of angular momentum (rigid body rotation at all times) have been considered. Detailed calculations have been made for case (1). Because of the absence of detailed published information on the density distributions of the appropriate models, only estimates have been made for case (2). In making these estimates, we have used the result (Oke and Greenstein, 1954; Kraft, 1968) that, for a given initial velocity, the final velocity for case (2) always exceeds that for case (1) by about a factor of 2. Recent computations by Faber and Danziger (1969) based on unpublished results of Iben indicate that this factor of two may be incorrect for the A stars. However their computations also indicate that this estimate is reliable for the B stars and it is felt that our basic conclusions will not be affected by these new results.

The expansion factors and rotational velocities expected on the basis of stellar evolution with angular momentum conserved in spherical shells are given in the second and fourth columns of Table I. The error estimates on the predicted velocities reflect uncertainties in effective temperature scales, bolometric corrections and predecessor rotational velocities and have been obtained by propagating possible limiting values through the computations.

6. Separation of Rotation and Macroturbulence

In order to separate the contributions to the observed line broadening arising from macroturbulence and axial rotation, we have made use of the fact that turbulence and rotation can be expected to exhibit different symmetries. As many supergiants as possible have been observed and the macroscopic line broadening initially computed as if it were all due to rotation. If results are grouped for stars with similar physical characteristics, in a large enough statistical sample one would expect to see a few stars which are either intrinsic slow rotators or are viewed at small inclination angles. For each group of stars we then attribute the minimum apparent rotation to broadening by large scale turbulence and shift the zero point of the distribution of rotational velocities in order to remove the contribution of large scale turbulence.

In adopting this procedure we have assumed that: (1) in any small box in the Hertzsprung-Russell diagram there is some large scale turbulent velocity which is characteristic of *all* stars in that box (cf., Meadows, 1961) and (2) the broadening functions for large-scale mass motion and rotation are linearly additive. The validity of these assumptions has been discussed elsewhere (Rosendhal, 1968, 1970). In the actual separation it has been assumed that the observed minimum is due to an aspect effect and a correction has been made (Rosendhal, 1968) to take account of the fact that in a small statistical sample the probability of seeing a star exactly pole-on is very small and that at least part of the observed minimum line broadening may be due to projected rotation.

The recovered rotational component of the observed line broadening for several groups of stars is given in the fifth column of Table I. The error estimates reflect the effect of photometric uncertainties on the observed minimum line broadening and also uncertainties in the geometrical orientation of the sharpest-lined star.

7. Discussion

Consideration of the third, fourth and fifth columns of Table I suggests the following conclusions:

(1) Purely on evolutionary grounds, it appears that axial rotation cannot be the sole source of line broadening in the Ia supergiants. Unless the final velocities from case (2) can be expected to exceed those from case (1) by a factor of 3 or more, the observed broadening expressed as rotation is too high in all cases to be consistent with any possible model of angular momentum conservation. We note that while the recent calculations of Faber and Danziger (1969) may raise some question as to the validity of this statement in the case of the A stars, they do not affect this conclusion for the case of the early B stars. In any event, the absence of sharp lined stars still remains a powerful argument for the presence of some sort of large-scale mass motion in addition to rotation.

(2) At all spectral types in the range B0–A5 both rotation and macroturbulence contribute to the observed line broadening. In particular, in the early B stars rotation is as important a broadening agent as large-scale mass motion. In the middle B and A stars turbulence dominates but there still is an appreciable contribution from rotation.

(3) In spite of the fact that the stars observed seem to be losing mass, we find no strong evidence for significant angular momentum loss. This conclusion is similar to that reached by Abt (1958) in his study of the Ib supergiants. In the later spectral types angular momentum appears to be conserved in spherical shells. Among the early B stars, on the other hand, the fact that the observed rotation is appreciably larger than that predicted from the evolutionary tracks may indicate that these stars are rotating as rigid bodies. We note that the members of this group have not expanded by very large factors. Abt's (1958) suggestion that during initial expansion (to a factor of about 4) a star rotates as a rigid body but that subsequently it commences to rotate differentially may also be true for the stars under consideration in this investigation.

The conclusion that angular momentum appears to be conserved is rather surprising. In fact this investigation was initiated with the expectation that the opposite conclusion would be reached.

We conclude by offering two suggestions to explain this result. The first possibility is that mass loss is not an efficient means of removing angular momentum from these stars. This, in turn, might imply something about the presence or absence of magnetic fields in these stars. In a different context Schatzman (1959, 1962) has shown that an outflow of ionized matter in the presence of a magnetic field exerts appreciable torques on a star and can be a very efficient way of braking a star's rotation. A second possibility is that emission and hence mass loss is a recent phenomenon in the history of these stars and the evolutionary changes occur on such a rapid time scale that there has not, as yet, been an opportunity for a significant amount of angular momentum to have been transported away.

A more extensive account of this work has appeared in the January 1970 *Astrophysical Journal* under the title 'Evolutionary Effects in the Rotation of Supergiants'.

References

Abt, H. A.: 1957, *Astrophys. J.* **126**, 503.
Abt, H. A.: 1958, *Astrophys. J.* **127**, 658.
Abt, H. A.: 1963, quoted by R. Weymann in *Ann. Rev. Astron. Astrophys.* **1**, 113.
Deutsch, A. J.: 1956, *Astrophys. J.* **123**, 210.
Faber, S. M. and Danziger, I. J.: 1969, this Volume, p. 39.
Hayashi, C. and Cameron, R. C.: 1962, *Astrophys. J.* **136**, 166.
Huang, S.-S. and Struve, O.: 1953, *Astrophys. J.* **118**, 463.
Huang, S.-S. and Struve, O.: 1954, *Ann. Astrophys.* **17**, 85.
Hutchings, J. B.: 1968, *Monthly Notices Roy. Astron. Soc.* **141**, 329.
Iben, I.: 1966, *Astrophys. J.* **143**, 516.
Kraft, R. P.: 1966, *Astrophys. J.* **144**, 1008.
Kraft, R. P.: 1967, *Astrophys. J.* **150**, 551.
Kraft, R. P.: 1968, in *Otto Struve Memorial Volume* (ed. by G. H. Herbig), in preparation.
Meadows, A. J.: 1961, *Monthly Notices Roy. Astron. Soc.* **123**, 81.
Morton, D. C.: 1967, *Astrophys. J.* **150**, 535.
Oke, J. B. and Greenstein, J. L.: 1954, *Astrophys. J.* **120**, 384.
Rosendhal, J. D.: 1968, unpublished Ph.D. thesis, Yale University.
Rosendhal, J. D.: 1970, *Astrophys. J.* **159**, 107.
Sandage, A. R.: 1955, *Astrophys. J.* **122**, 263.
Sargent, W. L. W.: 1961, *Astrophys. J.* **134**, 142.
Schatzman, E.: 1959, in *I.A.U. Symposium No. 10: The Hertzsprung-Russell Diagram* (ed. by J. L. Greenstein), (*Ann. Astrophys. Suppl. No.8*), p. 129.
Schatzman, E.: 1962, *Ann. Astrophys.* **25**, 18.
Slettebak, A.: 1956, *Astrophys. J.* **124**, 173.
Stothers, R.: 1966, *Astrophys. J.* **143**, 91.

Discussion

Faber: I would like to point out that your recovered velocities depend in some cases rather strongly on your assumption that turbulent broadening and rotational broadening add linearly. Assuming they add Gaussianly would give considerably higher rotational velocities. From mathematical experiments with test profiles, that is, by rotating profiles already rotated slightly, I find that two forms of macroscopic broadening appear to add much more like the convolution of two Gaussian profiles. If this is true, your subtraction scheme would be a substantial over-correction.

Rosendhal: I agree that there is some question as to the validity of the assumption of linearity. However, I am not sure that your numerical experiments are completely relevant to the problem. What you are doing is changing the unbroadened line profile on each small element of the star's disk and then broadening this new intrinsic line profile. It is not clear to me that this is the same as adding two macroscopic motions. It depends on the relative sizes of the actual macroscopic mass motions and the geometrical elements on the disk and I don't think it is possible to say anything about this. Also, if you look at the group of early B stars for which the factor of two between the evolutionary predictions for rigid body and differential rotation is approximately correct and assume that the broadening adds as the sum of the squares you then find that the recovered rotation is higher than the predicted rotation for both limiting cases. This is obviously unreasonable. Therefore, I think that treating the addition as the convolution of two Gaussians may be a substantial under-correction. The truth probably lies somewhere in between.

Hardorp: What type of line broadening did you use for HeI 4471?

Rosendhal: I assumed a Voigt profile, including both thermal and microturbulent Doppler broadening with a small value of the damping constant.

Hardorp: The helium lines are mainly broadened by Stark effect. In your case you may be able to squeeze this into a Voigt profile since the forbidden component is rather less important than on the main sequence. However, a detailed broadening theory for this line has been published by Griem in the *Astrophysical Journal*, December 1968.

Rosendhal: First, I see no evidence for the presence of the forbidden component on any of my plates. Second, at the electron pressures to be expected in supergiant atmospheres (as estimated crudely from counting Balmer lines, for example), unpublished calculations by Peterson indicate that the half-width of the Stark pattern is much smaller than any of the other characteristic half-widths and a negligible error is made by neglecting the Stark effect in this case. In any event, the observed line profiles show no trace of characteristic wings. It should also be remarked that it would seem to make no sense to apply a detailed broadening theory to stellar atmospheres whose structures are as poorly understood as those of the early-type supergiants.

Buscombe: In dealing with such a deep line as Mg II 4481 in A stars, have you taken account of curve-of-growth effects?

Rosendhal: Yes. The microturbulent velocity as obtained from curve-of-growth studies has been included in the intrinsic line profile in order to appropriately unsaturate the line.

Buscombe: Have you evidence of changes with time in the parameters affecting line-broadening?

Rosendhal: Yes. In a paper currently in preparation by Wegner and myself we have examined a time sequence of spectra of several A supergiants and have found that the microturbulence changes with time. We are still examining the data to see if there is any evidence for changes in the macro-turbulence.

Buscombe: Was there not a discussion by Steffey (Ph.D. thesis, University of Arizona, 1966) that what you describe as 'field supergiants' have their parentage in associations of young stars?

Rosendhal: Not that I am aware of. In any event, there appear to be no differences in the results which I obtained for field stars and for cluster members and so the parentage of the stars under consideration does not seem to be an important issue.

Bottemiller: On the question of slight variability, one may recall a paper by Serkowski where he reported variable polarization in three early-type supergiants.

Jaschek: Did you look into a correlation between the rotational velocities you measured and the existence of emission lines and/or variability in the spectrum or radial velocity?

Rosendhal: All the plates I took were in the blue and with a few exceptions there were no emission features on my plates. Therefore I can say nothing about such a correlation. I can only comment that the best lines for rotational velocity studies are in the blue and most emission features are in the red region of the spectrum. Several studies including one currently underway by Abt and myself indicate that the strengths of the emission features in some stars change with time. Therefore in looking for such a correlation you would have to take plates in both regions of the spectrum at approximately the same time.

PART III

STELLAR ROTATION IN BINARIES, CLUSTERS,
AND SPECIAL OBJECTS.
STATISTICS OF STELLAR ROTATION

ROTATION IN CLOSE BINARIES

(Review Paper)

MIROSLAV PLAVEC

Astronomical Institute, Ondřejov, Czechoslovakia

(Paper read by E. P. J. van den Heuvel)

1. Introduction

When studying the axial rotation of the components of binary systems, we ask the following fundamental question: How much is axial rotation affected by the other star? In particular, is there any synchronism between the periods of axial rotation and orbital revolution?

Thus in fact we are more interested here in the angular than in the linear velocity of rotation. It was pointed out by McNally (1965) that the angular velocity of rotation reaches its maximum near A5 and drops off rather rapidly on both sides so that the G0V and O5V stars have approximately the same average period of rotation. The accompanying Table I is an adaptation of McNally's figures.

In the last column of Table I, the orbital period of a binary is given in which both

TABLE I
Average rotational velocities of main-sequence stars

Spectrum (class V)	m $(m_\odot)$	R $(R_\odot)$	V_r (km/sec)	ωr $(10^{-5}\,\mathrm{sec}^{-1})$	P (days)	P_c (days)
O5	39.5	17.2	190	1.5	4.85	4.0
B0	17	7.6	200	3.8	1.91	1.75
B5	7	4.0	210	7.6	0.96	1.07
A0	3.6	2.6	190	10	0.73	0.78
A5	2.2	1.7	160	13	0.56	0.53
F0	1.75	1.3	95	10	0.73	0.40
F5	1.4	1.2	25	3.0	2.42	0.39
G0	1.05	1.04	12	1.6	4.55	0.36

components have equal mass (corresponding to spectral type in the first column) and are in contact (more exactly, both fill the critical Roche lobe). This is in fact the shortest possible period in each case; if the companion is less massive and/or more distant, the period will be longer. It is interesting to note that up to the late A's, the average period of axial rotation of single stars and the 'contact' period of revolution given here are about equal. Only for the F dwarfs and later types can we have the orbital period much shorter than the period of rotation.

2. Observational Material

It is clear that the problem of synchronism appears when we deal with close binaries,

A. Slettebak (ed.), Stellar Rotation, 133–146. All Rights Reserved

since only in these can a real physical interaction between the components be expec ted. In visual binaries, the components can be expected to evolve as well as to rotate as single stars. Investigations by Struve and Franklin (1955) and by Slettebak (1963) confirm this expectation.

A statistical study of the problem of synchronism was published by Plaut (1959). His statistics are based on the well-known surveys by Huang, by Slettebak, and by Slettebak and Howard. These data, of course, refer to the linear velocities of rotation. Therefore it is necessary to know the star's radius in order to compute the corresponding synchronised linear velocity. Plaut derived the stellar radii from photometric elements for eclipsing binaries, and from absolute magnitudes and spectral types for spectroscopic binaries.

Plaut concludes that the periods of rotation and revolution seem to be equal on the average for orbital periods shorter than 10 days, while for longer periods the synchronisation fails and the stars tend to rotate rather like single stars. However, Plaut's diagram shows a very large scatter. This is certainly partly due to the unknown value of orbital inclination i, which enters as $\sin i$ into the observed velocity but does not enter into the computed value of the 'synchronised' velocity $v_s = 2\pi R/P$. Large observational errors affecting R are another source of scatter. Apart from this, we should realize that orbital period alone cannot characterize the binary uniquely. A pair of O stars revolving with a period of 10 days can easily be understood to have synchronized rotation because the two bodies are almost in contact; while the same period in the case of two G dwarfs means that their radii represent only 0.07 of their separation, and interaction is certainly much weaker.

Practically only well-observed eclipsing binaries can yield complete data for a more detailed investigation into the problem of synchronism. Here we can get not only the radii of the stars but also the orbital inclination if we take it for granted that the axes of rotation are perpendicular to the orbital plane. Unfortunately, strong selection effects restrict our considerations almost entirely to periods of the order of a few days, and no statistically meaningful information can be obtained at this time about the systems with period one or two orders of magnitude longer.

The present discussion will be based mainly on two papers, by Koch *et al.* (1965) containing the rotational velocities of 19 eclipsing pairs, and by Olson (1968) who added another 9. Figures 1 to 4 are similar to those displayed in the original papers. I have added three stars studied by other authors: α CrB, RZ Sct, and RY Per. All three are exceptions rather than ordinary cases because of their rapid rotation, and are marked specifically in the figures.

Another change introduced here is the distinction between detached binary systems (circles) and semidetached ones (squares). This is because these two categories of close binaries are very likely to have different life histories. In the detached systems, both stars lie on or near the main sequence, and they are significantly smaller than their corresponding critical Roche lobes. In the semi-detached systems, only the more massive primary is a main-sequence star, while the less massive and colder secondary is a subgiant or giant, i.e. apparently more evolved. This paradox is in my opinion

correctly explained as a consequence of a large-scale mass transfer in which the originally more massive star lost a considerable part of its mass and developed into the
subgiant secondary (cf. Plavec, 1968a, b).

3. Observational Evidence on Rotation in Close Binaries

In Figure 1, the linear velocities of rotation of the components of close binaries are
plotted against their respective spectral types. Primary components are marked by
full symbols, secondary components with open symbols. The broken full line represents the mean velocities of rotation for single class V stars after Slettebak (1963). In

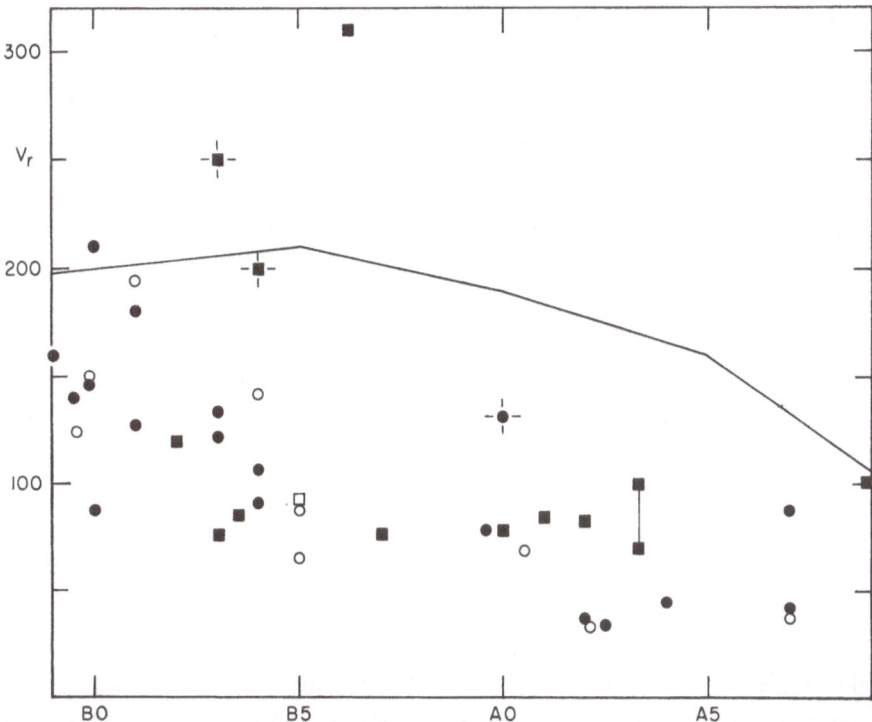

Fig. 1. Linear equatorial velocity of rotation V_r (km/sec) is plotted against spectral type. Circles:
detached systems, squares: semidetached systems. Open symbols: secondary (less massive) components, full symbols: primary components.

agreement with Koch *et al.*, and with Olson, we find a definite tendency of the components of close binaries to rotate more slowly than is the average for single stars of
the same spectral class. Thus the influence of the other component is certainly perceptible.

Figure 2 compares the actual velocity of rotation V_r with the calculated synchronized velocity V_s. In the case of perfect synchronism, the stars should lie on the 45°
line. Disregarding the very few striking deviations for the moment, we find that most

of the stars do scatter along this line, indicating that synchronization obtains in most cases either perfectly or approximately.

It is difficult to say how much of the scatter in Figure 2 is real. Certainly observational errors play an important role, as is shown by the following examples. Three

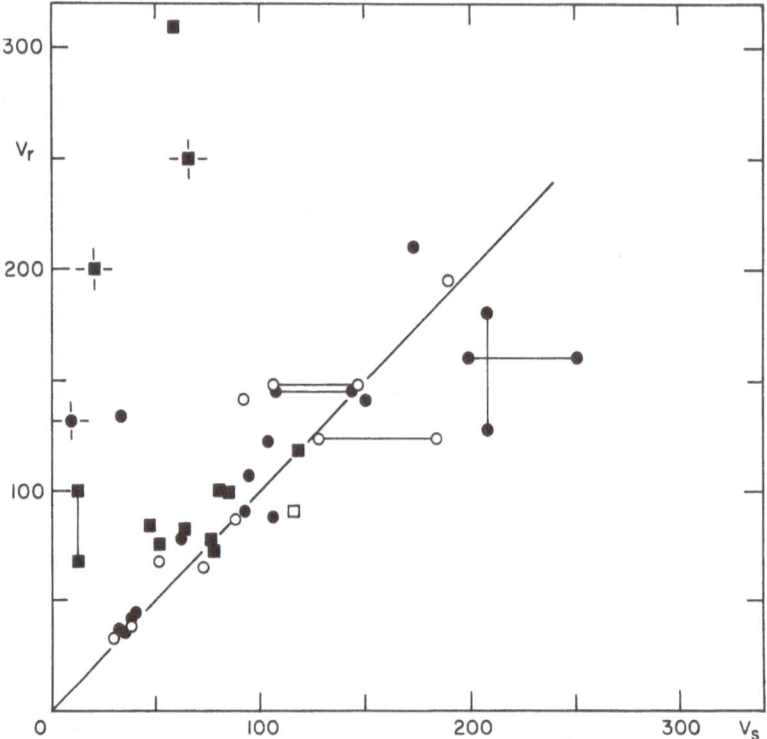

Fig. 2.   Observed linear velocity of rotation V_r is plotted against the 'synchronized' velocity V_s, both in km/sec.

stars lie apparently considerably below the line of synchronism: the primary of VV Ori, and both components of AO Cas. The synchronized velocity for VV Ori would be 208 km/sec, while Olson found the actual velocity to be 128 km/sec. A causal inspection of a few of the Victoria plates shows that the lines are very poorly defined. One would expect a higher velocity of rotation, and indeed Boyarchuk and Kopylov (1964) list a value of 180 km/sec. If something like this is more correct, then the discrepancy will be greatly diminished (cf. the point on the top of the vertical line in Figure 2). In AO Cas, the radii of both stars are quite uncertain, since the absolute dimensions are uncertain. In fact, Koch et al. found rather widely different values for the radii according to whether they computed them from the photometric elements or from the radiation law. The difference is shown by the two horizontal and parallel lines below the 45° line. In some cases the disagreement between the two solutions for the stellar radius is rather alarming. In AO Cas, the photometric solution gives much better agreement between V_r and V_s, but for Y Cyg (the two horizontal lines very close together above the $V_r = V_s$ line) just the opposite is true.

These examples show that not much weight can be placed on the accurate position of the points in Figure 2. Nevertheless, since the errors are probably at random, we may conclude the following: (1) No case is known of a component rotating considerably more slowly than required by synchronism. (2) As a rule, the rotation is synchronized with orbital motion, or may be somewhat faster. (3) A few stars rotate considerably more rapidly than they should if synchronism applied to them.

Among the non-conformists, there are two components of detached systems. The primary of α CrB should rotate at 9 km/sec if synchronism applied to it, but it rotates 14 times faster, at 132 km/sec which, however, is still considerably less than average for an A0V star. With a period of 17.4 days, and a companion of less than a solar mass, this deviation from synchronism is not surprising.

The other star is the primary of AR Cas which has a period of 6.1 days and a mass ratio probably even smaller than α CrB. Thus the influence of the companion will probably be weak, too. The expected synchronized velocity is 32 km/sec, while Olson observed 134 km/sec. Even this is considerably less than average for a B3V star. However, Batten (1961) reports that the rotational effect within eclipse indicates a velocity of 210 km/sec $\pm$ 50 km/sec; the upper half of this range would be a value rather typical for a single B3V star.

These two cases would indicate that synchronism breaks down at fairly short periods already. But the two other cases – admittedly less reliable – of δ Ori (5.7 days) and V 380 Cyg (12.4 days) indicate good synchronization. Figure 4 shows that in fact nothing statistically meaningful can be said about periods longer than 4 days!

As has already been said in connection with Plaut's diagram, orbital period alone is certainly not sufficient to determine the degree of interaction between components. One could expect that the synchronizing effect of the companion will increase with the fractional radius of the perturbed star; but Figure 3 yields no clear support to this contention.

4. Tidal Forces

In spite of the rather unsatisfactory supporting evidence from Figure 3, the classical tidal forces must be considered seriously in an attempt to explain the general trend towards synchronism. The problem was discussed recently by Huang (1966). He refers to the result obtained by Cowling (1938) that the time of adjustment of a star to an external gravitational field is of the order of the period of free adiabatic oscillations, which is as a rule much shorter than the orbital period. Thus the shape of the star is almost instantaneously adjusted to the perturbing gravitational field. Huang concludes: "As a result the two stars will have their longest axis along the line joining their centers, and the components, always facing each other, will rotate in synchronism with their orbital motion" (*ibid.* p. 50).

However, in the latter part of this sentence Huang was misled to a conclusion not warranted by Cowling's reasoning. It is true that if an external potential is applied to

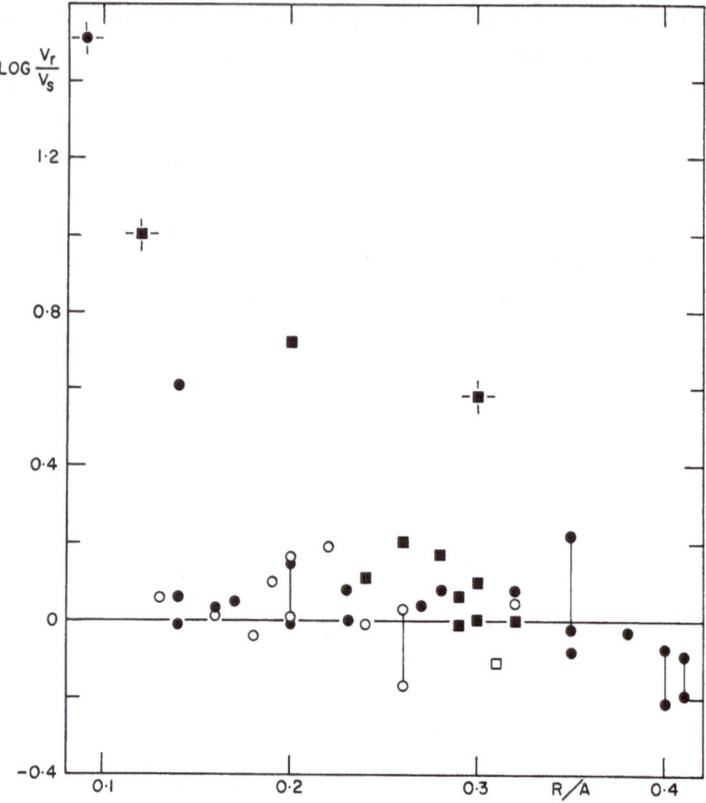

Fig. 3.   Departures from synchronism are plotted against fractional radii R/A. Zero line
corresponds to synchronism.

a star, it settles down to a hydrodynamical equilibrium in a very short time (of the
order of the free-fall time). Thus its shape is almost instantaneously adjusted to the
external field, but this does not imply that the rotation will adjust itself as well. If
there were no friction or dissipation of energy, the star would rotate indefinitely at
any given speed. A change in rotation requires a change of angular momentum, and
for this a torque is needed. There is, however, no torque unless there is some friction
inside the body of the star. The formula for the torque is

$$\Gamma = - \int \frac{\partial U}{\partial \varphi} \varrho \, dV \tag{1}$$

where U is the external potential, φ is the azimuthal angle, and ϱ is the local density
within an elementary volume dV. When ϱ has the same spatial symmetry as U, the
net torque is zero. But the potential is as a rule variable with time, and causes oscillations in the density distribution inside the star. These oscillations are not exactly in
phase with the perturbing potential, the phase lag being due to friction. It is just this
little phase lag that makes the net torque non-zero, and causes secular exchange of
rotational and orbital angular momenta. These dissipative tides are very much smaller

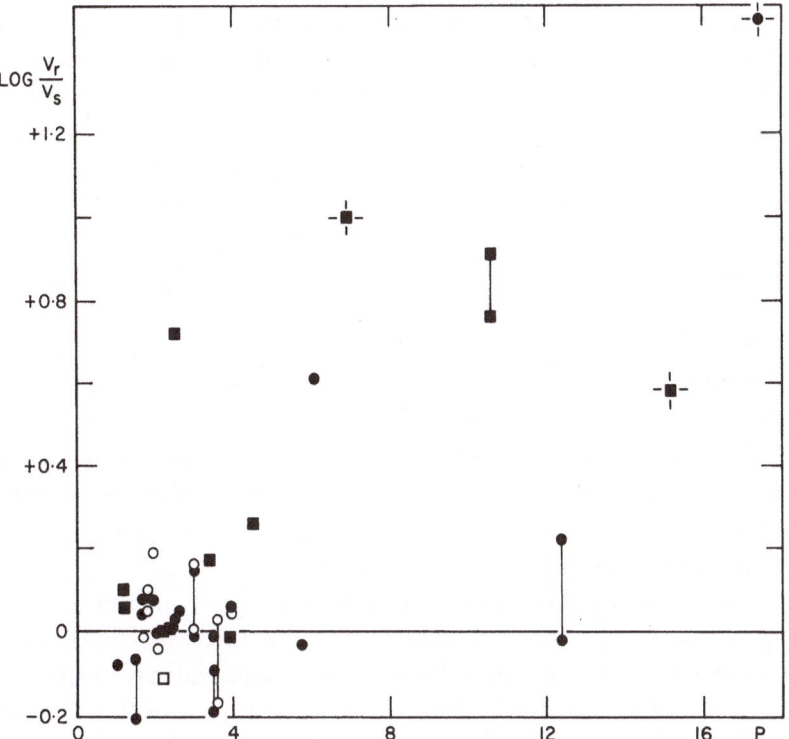

Fig. 4. Departures from synchronism are plotted against orbital period P (in days).

than the adiabatic tides that are exactly in phase with the perturbing potential and tend to adjust the shape of the star instantaneously to the external field of force. Therefore the adjustment of rotation to the condition of synchronism is a much slower process than the adjustment of the star's shape.

Recent years have witnessed a substantial revival of interest in the theory of tidal forces in the works by Kopal (1968), James (1967), Dziembowski (1967) and Zahn (1966). The following discussion is based mainly on the comprehensive work by Zahn.

The tidal effects are relatively large if the perturbed star has an extended outer convective zone, since the eddy viscosity is fairly large. Zahn finds that in this case the deviation of the actual angular velocity of rotation ω from the synchronized value ω_s decreases exponentially with time:

$$\omega - \omega_s = (\omega_0 - \omega_s) \exp(-t/\tau_c) \tag{2}$$

where the characteristic time constant τ_c can be taken as

$$\tau_c = Q C_c P^4 \tag{3}$$

with

$$Q = \tfrac{4}{9}(1 + q)^2, \tag{4}$$

where P is the orbital period in days, and $q = m_1/m_2$ is the mass ratio, m_1 being the perturbed star. C_c is a constant depending entirely upon the internal constitution of the perturbed star. For main-sequence stars with an extended outer convective zone, i.e. for late-type dwarfs less massive than the sun, Zahn finds a value of the order of 10^{-1} to 10^{+1} years. Among the spectroscopic binaries, red dwarfs may occur as secondaries in detached systems. Since Q will not be far from 1 in these cases, the characteristic time constant for a system with $P = 10$ days is of the order of 10^3 to 10^5 years, i.e. quite short compared with the main-sequence lifetime of these dwarfs. Thus in spectroscopic binaries of short or moderate period, the secondary components if less massive than the sun can always be expected to rotate in synchronism with orbital motion.

But this result does not solve the problem of our statistics, since there is not a single star of this kind in Figures 1 to 4. Typical stars in our sample are main-sequence stars more massive than the sun, possessing a small convective core and an outer radiative envelope. Now the radiative viscosity due to radiative damping is incomparably smaller than the turbulent viscosity. Therefore the radiative tides are extremely inefficient and their time scale is of the same order as the main-sequence lifetime of the star in question, if not longer.

In other words, it seems that the observed good degree of synchronism can actually not be explained as easily as one might think. If further investigations on more up-to-date stellar models confirm that the tidal forces do not provide an adequate explanation, I suggest that we could think of three possible ways out of this dilemma:

(1) We could perhaps preserve the tidal forces but assume that the most important evolutionary phase determining the angular momenta is the one preceding the main-sequence phase. During the initial contraction, a large-scale convection in stars of large size, and quite possibly also exchange of material, may fix the rotational velocities near the synchronised values. This would explain why synchronism is observed in such a young system as AG Per which appears to be a member of the II Persei association whose kinematic age is about 1.6×10^6 years. Moreover, it may well be that the very early stages of evolution of a binary determine the important circumstance that the axes of rotation are perpendicular to the orbital plane – if, of course, this is really a fact and not merely our oversimplified assumption!

If we make the assumption that the two components rotated in synchronism at a stage shortly preceding the main-sequence stage, we may perhaps argue that the final stages of (already rather slow) contraction lead to a slight increase in the velocity of rotation above the synchronized value, which appears to be the rule with main-sequence systems.

(2) Alternatively, perceptible changes of angular momenta on a relatively short time scale in the course of the main-sequence evolution of the components are perhaps possible if we admit that more powerful forces than the radiative tides may be at play. Dziembowski points out that the action of the Coriolis force on meridional circulation may be perceptible on a time scale considerably shorter than that of the radiative tides. What we need is a more detailed theoretical inquiry into these problems which should bear out whether there would be a trend towards synchronism as well as how efficient these forces are.

(3) As a third possible explanation of the trend towards synchronism, we should realize in any case that what we observe is the rotation of the outermost layers of the star only. If there occurs turbulence in these layers – and there appears to be more than one way of establishing it – the convective tides may create a kind of 'superficial synchronisation' masking the actual state of things underneath.

5. Mass Exchange and Rotation in Binaries

The last item leads rather naturally to another phenomenon which may affect rotation very perceptibly, and this is the mass exchange between components. I think a good argument in favor of this idea is to have a closer look at the most conspicuous deviations from synchronism in Figure 2. Omitting α CrB and AR Cas, we are left only with semidetached binaries, namely (from left to right) AW Peg, RY Per, U Cep, and RZ Sct; all of which are the primary components. Other primaries of this type, too, show a trend toward faster rotation than synchronism requires. Although in their case the deviations are not so striking, they seem to be well established; I have in mind in particular U Sge ($V_r/V_s = 1.5$), RS Vul (1.8), and RZ Cas (1.3).

Now it has already been mentioned that the best explanation for the existence of the semidetached systems is a large-scale mass exchange that starts when the more massive star reaches, in the course of its secular expansion, the limit of dynamical stability (the Roche limit). The material driven beyond this limit leaves the star very quickly and most of it is probably captured by the other component. The mass-losing star strives to adjust its structure, and in particular its radius, to this new condition of lower mass. The new equilibrium radius is as a rule smaller than the initial one, although this need not always be the case. But in practically all cases the equilibrium radius is larger than the new Roche radius, since the latter decreases more quickly with diminishing mass of the star in question and with increasing mass of the other component. As a consequence the mass-losing star cannot reach its equilibrium radius, since a smaller radius is forced upon it by external forces. Therefore the mass loss goes on and the process is in fact self-accelerating for some time as the star's deviation from thermal equilibrium increases.

Our model calculations were made on the assumption that the total mass and total orbital angular momentum of the binary system remain preserved – in other words, that the material is only being exchanged but does not escape from the system. If so, then it can easily be shown that the distance between components attains its minimum when both stars have equal mass, and approximately at the same time the Roche radius of the mass-losing star reaches a minimum, too. From now on, further mass loss is bound to increase the star's actual radius, because the dimensions of the system increase again. As a consequence, the deviation from thermal equilibrium diminishes steadily until the moment comes when the star's actual radius (as given by the instantaneous Roche radius) equals the equilibrium radius, so that thermal equilibrium is restored. From now on, the initial primary in fact continues its interrupted evolution on the nuclear time scale. Its mass has been reduced considerably (it is now the less massive component), and its internal constitution has been adjusted to this present condition. This does not mean that the star is similar to a main-sequence star of the same mass. What has changed is not only the total mass but also, for example, the mean molecular weight. As a consequence, the star is a subgiant overluminous for its mass. But its constitution has undergone no fundamental change: it still possesses a convective core in which hydrogen is being converted into helium. In fact, the hydrogen content within this core remains almost unchanged over the period of the rapid mass exchange described above, since the process is quite fast – its duration is of the order of 10^{-2} of the total main-sequence lifetime of a single star of the same initial mass.

As the hydrogen burning in the core continues, the star's envelope expands slowly. But the star filled its critical Roche lobe to begin with. Therefore the mass loss continues, but this time much more slowly and on a nuclear time scale. The mass-losing star is now the less massive one, and the absolute size of its critical Roche lobe now increases with decreasing mass, because with increasing disparity between the components their separation is bound to increase rapidly if the total angular momentum is to be conserved. Therefore the star balances always very near thermal equilibrium. This phase of slow mass loss usually lasts very long and may well have a duration comparable to the main-sequence lifetime of a single star with the same mass as the mass-losing star initially had. Since, moreover, the system can produce very deep eclipses if the orbital inclination is favorable, the probability is comparatively high that the binary will be observed just at this stage of evolution. We believe that most if not all Algol-like semidetached binaries are observed at the stage of slow mass loss. The chance of catching a system at the phase of rapid mass loss is low because of the short duration of that phase, but of course it is not negligible since the spectroscopic phenomena that may accompany this process may be quite conspicuous.

For our considerations about how this process can affect the axial rotation of the components it is necessary to have at least some quantitative picture of the process of mass exchange. No unique figures can be given since the rate of mass exchange depends on several parameters, in particular on the masses, mass ratio and on the degree of chemical inhomogeneity of the mass-losing star. For moderate masses and mass ratios, the case with the following initial conditions may be considered as rather representative: $m_1 = 5 \, m_\odot$, $m_2 = 3 \, m_\odot$, $X_{c1} = 0.25$ (Plavec et al., 1969). At the phase of rapid mass loss, the primary loses and the secondary gains $2.4 \, m_\odot$ within 7×10^5 yr, so that the average rate of mass loss is $3.4 \times 10^{-6} \, m_\odot/$yr. The process is, however, not uniform. The outflow is fastest shortly after the beginning, and attains a maximum rate of $2 \times 10^{-5} \, m_\odot/$yr. The star reaches minimum size and luminosity within only 1.2×10^5 yr after the beginning of mass loss. The subsequent phase of partial restoration of luminosity, when the star gradually returns to thermal equilibrium, takes 5.8×10^5 yr, and the average rate of mass loss is $2.4 \times 10^{-6} \, m_\odot/$yr. This is still very high compared to the slow phase of mass loss that follows, where the average rate is only $1 \times 10^{-8} \, m_\odot/$yr.

However, owing to the long duration of the latter phase – some 65×10^6 yr in the particular case considered here – the total amount of material transferred is certainly not negligible, amounting as it does to 0.6 $m_\odot$. The corresponding quantity for the fastest phase is 1 $m_\odot$, while the 'intermediate' phase is most efficient with 1.4 $m_\odot$ transferred. All this material is assumed to be captured by the other component. It is this mass-gaining component that is observed as the main-sequence, bright and massive primary in the Algol-like semidetached systems, and its velocity of rotation is usually measured. Now the transfer of material on it certainly affects this star's angular momentum profoundly – but very little can be said about this point at the present.

The truth is that the theory of evolution of close binaries through mass exchange is still very crude. The model computations performed by Kippenhahn, Weigert, Refsdal, Paczynski, by our Ondrejov group etc. have dealt so far exclusively with the mass-losing component. The assumption of conservation of the total orbital angular momentum is partly based on the well-known fact that in all binaries (except the W UMa systems) the angular momenta of axial rotation are small compared to the orbital angular momentum. This very argument when reversed means, however, that the axial angular momentum can be considerably changed by transfer of mass carrying with it a certain amount of orbital angular momentum.

The problem that must be solved can also be formulated as follows: Granted that the mass loss of the initially more massive component proceeds roughly as described above, is the other component capable of accommodating the incoming material on a time scale comparable to that of the mass loss? Observational evidence seems to answer this question in the affirmative, since we observe many Algol-like systems in which the primary behaves as a normal main-sequence star, and its rotation is nearly synchronized.

It can be argued, however, that many of the observed semidetached systems may be rather old, i.e. well advanced in the phase of slow mass loss which can be quite long. The present rate of mass exchange may already be quite low. In fact, in typical systems like U Sagittae, RW Tauri or U Coronae Borealis the process manifests itself only by rather weak spectroscopic phenomena (temporary appearance of emission and/or absorption satellite lines), and by changes of period. Indeed, the intermittent appearance of emissions and the erratic character of the period fluctuations makes one suspect that what we observe here may be atmospheric activity of the subgiant component rather than the continuous secular flow postulated by our theoretical evolutionary models. No secular increase of period has so far been established with certainty (Koch, 1969), which again is no argument against secular mass exchange since our observations cover too short an interval of time.

Whether the cause of the observed mass transfer is secular or superficial instability, there is little doubt that the material is streaming from the subgiant to the main-sequence component, partly falling directly on it and partly encircling it temporarily in the form of a ring. Although this picture is rather simple, we lack quantitative data about the rate of mass transfer. In fact, for our considerations about its effect on rotation we need more than this – namely we must know how the material flows and how it falls on the star. This is certainly a problem of hydrodynamics, and may easily involve magnetohydrodynamics as well if magnetic fields are present. This makes the whole problem almost formidably complex. No wonder, then, that hardly more than the crudest approximation has been attempted – namely the calculation of trajectories of discrete particles in a purely gravitational field.

The immediate aim of these calculations was in fact to explain the gaseous rings circulating around the more massive components in systems like RW Tau, U Sge or U Cep. In our work at Ondrejov (Plavec and Kříž, 1965) we did find certain trajectories that may contribute to a general circulatory pattern around the more massive star. They require a certain area of ejection from the surface of the subgiant, and fairly high initial velocities (over 100 km/sec); therefore some intrinsic force must be postulated in the atmospheres of the subgiants capable of expelling the particles.

If the ejected particles are not properly directed or have lower ejection velocities, they fall directly onto the more massive star. Presumably this is the fate of the major part of the expelled material, and it should fall on rather a limited area of the star's surface. Most of the trajectories tend to lie near the orbital plane, and centrifugal force lifts them ahead of the line joining the two stars. Thus they are in a sense 'focused' to an equatorial area centered around a point located, say, some 50° off the axis of symmetry of the system. The particles arrive with velocities of the order of 500 km/sec, but in general their direction is not normal to the surface, and there exists a finite positive component tangential to the surface and in the direction of axial rotation and orbital motion. The trend is therefore to accelerate the rotational velocity of the main-sequence component.

It may be argued that the discrete-particle approach is too crude. Prendergast (1960) attempted a hydrodynamical approach, although, in order to make the problem tractable, he again oversimplified the gravitational field. His solution describes not the streaming but a steady-state distribution of gas in the system, his main conclusion being that the streaming around the more massive component is essentially Keplerian motion around its center of gravity. The same conclusion was reached by Huang (1966). This means that if there is any interaction between the star and the surrounding gas, or if indeed parts of the gaseous ring mix with the superficial layers of the star, the result must inevitably be an increase of its velocity of rotation.

This result contradicts the conclusion by Van den Heuvel (1968) that the incoming material slows down the star's rotation, which statement he uses to support his contention that the Ap and Am stars (known as slow rotators) are members of close binaries in which mass exchange took place. Strictly speaking, Van den Heuvel did not consider the Algol-like binaries but systems of longer period – but even so I do not think his result is acceptable unless we admit that the motion of the gaseous masses is largely governed by a strong bipolar magnetic field. Of this we have no indication whatsoever in the Algol-like systems.

The root of the difference between Van den Heuvel's approach and ours is that he assumes that near the surface of the mass-gaining component, the gaseous particles move in roughly circular orbits with an angular velocity equal to the synchronized angular velocity of the system. In a field of force where gravity dominates this certainly cannot be so. It is much more appropriate to assume that the field there is determined by the attraction of the mass-gaining component only. Denoting the mean angular velocity in this field and at the surface of the star by ω_r, and the synchronized angular velocity by ω_s, we have for their ratio

$$\omega_r/\omega_s = (A/R)^{3/2}(1/1 + q)^{1/2}. \tag{5}$$

In a typical Algol system, the fractional radius of the more massive star $R/A \cong 0.2$, the mass ratio $q = m_2/m_1 \cong 0.3$, so that the ratio $\omega_r/\omega_s \cong 10$.

One should in fact rather wonder why only a small fraction of the observed main-sequence components of the semi-detached Algol-like systems rotates with velocities much in excess of the synchronized velocity. The reason probably is that in most cases the rate of mass exchange is already low, and the excess angular momentum brought to the more massive star by the falling material is small compared to its total angular momentum of axial rotation. Moreover, this excess is probably rather rapidly redistributed throughout the star. We could speculate that the rapid rotators like U Cephei or RZ Scuti are in an evolutionary stage where the rate of mass transfer is higher. For example, we could think that they are in the 'intermediate' phase of mass exchange where the rate is about two orders of magnitude higher. Or, because both represent relatively massive systems, that the rate of mass loss is generally higher in them.

If this idea is correct, then we should realize that in systems in which the rate of mass exchange is very high – in those which are in the 'rapid phase' – the effect of mass exchange on the rotation of the mass-gaining component must be even stronger, and leads perhaps to a formation of transient thick and partly or completely opaque disks or shells, and to rotational velocities that bring the star to the verge of stability. It has already been suggested (Plavec, 1968a) that β Lyrae and V 367 Cygni are at the stage of rapid mass exchange. But I should like to point out that this picture may be of great importance for a whole class of objects, namely for the shell stars – or at least for some among them. It is certainly possible that some of the shell stars create their extended atmosphere by internal forces of a single star. On the other hand the duplicity of 17 Leporis (Cowley, 1967) or AX Monocerotis (Cowley, 1964), and the strongly suspected duplicity of o Andromedae or φ Persei, indicate that duplicity may be essential for the phenomenon. If so, then both the rapid rotation of the star and its shell may be connected with mass exchange in the binary.

Acknowledgements

This paper was prepared while the author was a Visiting Scientist at the Dominion Astrophysical Observatory in Victoria, Canada. The author appreciates very much the hospitality offered to him by the director, K. O. Wright, and fruitful discussions with Dr. A. H. Batten. Very helpful was also correspondence with Dr. J.-P. Zahn.

References

Batten, A. H.: 1961, *J. Roy. Astron. Soc. Canada* **55**, 120.
Boyarchuk, A. A. and Kopylov, I. M.: 1964, *Izv. Krym. Astron. Obs.* **31**, 44.
Cowley, A. P.: 1964, *Astrophys. J.* **139**, 817.
Cowley, A. P.: 1967, *Astrophys. J.* **147**, 609.
Cowling, T. G.: 1938, *Monthly Notices Roy. Astron. Soc.* **98**, 734.
Dziembowski, W.: 1967, *Comm. Obs. Roy. Belg. Uccle* **B17**, 105.
Huang, S.-S.: 1966, *Ann. Rev. Astron. Astrophys.* **4**, 35.
James, R.: 1967, *Comm. Obs. Roy. Belg. Uccle* **B17**, 99.
Koch, R. H.: 1969, paper presented at the Copenhagen meeting of I.A.U. Commission 42.
Koch, R. H., Olson, E. C., and Yoss, K. M.: 1965, *Astrophys. J.* **141**, 955.
Kopal, Z.: 1968, *Astrophys. Space Sci.* **1**, 27, 284, 411.
McNally, D.: 1965, *Observatory* **85**, 166.
Olson, E. C.: 1968, *Publ. Astron. Soc. Pacif.* **80**, 185.
Plaut, L.: 1959, *Publ. Astron. Soc. Pacif.* **71**, 167.
Plavec. M.: 1968a, *Adv. Astron. Astrophys.* **6**, 201.
Plavec, M.: 1968b, *Astrophys. Space Sci.* **1**, 239.
Plavec, M, and Kříž, S.: 1965, *Bull. Astron. Csl.* **16**, 297.
Plavec, M., Kříž, S., and Horn, J.: 1969, *Bull. Astron. Csl.* **20**, 41.
Prendergast, K. H.: 1960, *Astrophys. J.* **132**, 162.
Slettebak, A.: 1963, *Astrophys. J.* **138**, 118.
Struve, O. and Franklin, K. L.: 1955, *Astrophys. J.* **121**, 337.
Van den Heuvel, E. P. J.: 1968, *Bull. Astron. Neth.* **19**, 449.
Zahn, J.-P.: 1966, *Ann. Astrophys.* **29**, 313, 389, 565.

Discussion

Deutsch: I think it hard to believe that only coincidence is responsible for Dr. Plavec's table, which shows that: $\langle P_{rot} \rangle$ and (P_{orb}) min are equal between types O and G along the main sequence.

Plavec: I suspect, too, that this is no coincidence – this is why I presented the table. Perhaps this relation has something to do with the formation of double stars.

Olson: I agree with Dr. Plavec that it is difficult to distinguish between binary components that are 'in synchronism' and 'near synchronism'. The radius must be known accurately, and this depends on the quality of the light and radial velocity solutions, which are often complicated by interaction effects. The rotational velocity determination must be as free of systematic error as possible. The systematic error does appear to be small for 68 Her, whose primary eclipse rotational disturbance has been carefully analysed from Kitt Peak coudé spectra by Dr. Koch and Dr. Sobieski (130th meeting of the American Astronomical Society). They found $v \sin i = 118$ km/sec, while from line widths I found $(116 \pm 4$ me) km/sec. My rotational velocity system shows no significant systematic differences from that of Dr. Slettebak for $v \sin i$ less than about 200 km/sec, which is the range of interest for most close binaries.

FISSION AND THE ORIGIN OF BINARY STARS

JEREMIAH P. OSTRIKER

Princeton University Observatory, U.S.A.

Abstract. Brief reviews of the classical 'angular momentum problem' and the statistics of upper-main-sequence binaries are presented as background for the suggestion that the close, early-type, binaries are produced by fission of rapidly rotating protostars.

Next, theoretical sequences of contracting, rotating stars are described. Recent work demonstrates that the zero-viscosity, polytropic sequences, have essentially the same properties as the McLaurin sequence. Thus, fission is possible for centrally condensed stars. Observations of close early-type binaries are compared with theoretical predictions for the minimum angular momentum in binary systems of given total mass; the agreement is excellent.

Finally, the existing theoretical objections to the fission hypothesis for the origin of binary stars are reviewed, and it is concluded that, although fission remains unproven, there are now no strong theoretical arguments against the process, and there is considerable observational support for its existence.

1. Background

A. THE ANGULAR MOMENTUM PROBLEM

It is generally believed that stars are formed from the diffuse gas and dust found in interstellar space (cf. Spitzer, 1968). Although the detailed processes of star formation are by no means well understood, a commonly accepted, over-all, model postulates conditions in the interstellar medium which would lead to large scale hydrodynamical or thermal instabilities occurring over extended regions; large masses ($M \gtrsim 10^4 M_\odot$) of gas and dust begin to collapse and then to fragment, the ultimate fragments being protostars. In a variant model suggested by McCrea (1961), fragmentation proceeds until the mass is reduced to quite small 'flocules'; these collide with one another, fuse during certain inelastic collisions, and ultimately approach proto-star proportions.

Detailed theories of star formation vary greatly in their predictions for the angular momentum content of protostars. However, all of the simpler (non-electromagnetic) theories lead to values of the angular momentum, J, much greater than observed in main-sequence stars. This point is easily made in rough quantitative fashion. Consider (a) the angular momentum of a spherical blob of gas, due simply to the fact that the local standard of rest rotates about the galaxy with a period of $\sim 10^8$ yrs; and (b) the angular momentum expected if the protostar cloud has rotational kinetic energy in equipartition with its translational kinetic energy ($|v_{tr}| \approx 10$ km/sec). In the two cases we find that a protostar with an original density of $\simeq 10^{-24}$ gm/cm^3 would have

$$\begin{aligned} J_a &\approx 10^{56} \, (M/M_\odot)^{5/3} \text{ gm cm}^2 \text{ sec}^{-1} \quad \text{(galactic rotation)}, \\ J_b &\approx 10^{58} \, (M/M_\odot)^{4/3} \text{ gm cm}^2 \text{ sec}^{-1} \quad \text{(equipartition)}. \end{aligned} \tag{1}$$

In comparison to this, even the most rapidly rotating main-sequence stars have

A. Slettebak (ed.), Stellar Rotation, 147–156. All Rights Reserved
Copyright © 1970 by D. Reidel Publishing Company, Dordrecht-Holland

$J \simeq 10^{51}-10^{52}$ gm cm^2 sec^{-1}. It is interesting to note parenthetically that, in McCrea's fusion model, the estimates given by Equation (1) are considerably reduced; the angular momentum due to galactic rotation appears primarily as the rotation of the resulting star cluster and the equipartition angular momentum is reduced by the square root of the number of 'flocules' comprising a single star. The values of J given by Equation (1) are in fact much larger than equilibrium stars would be able to contain with their observed masses and radii.* This is the angular momentum problem.

In order to 'solve' the problem various authors (cf. Spitzer, 1968, for references) have suggested that magnetic forces are adequate to transfer the 'excess' angular momentum to the surrounding medium. There are two principal objections to these mechanisms. First, the time scales for angular momentum transfer are longer than the relevant free fall times of the initial condensation if the prevailing field strength is $\simeq 3 \times 10^{-6}$ gauss. Second, there is much more angular momentum in binaries than in single stars; after finding an appropriate angular momentum loss mechanism one would then be required to turn it off – prevent angular momentum loss – when binaries are to be formed. Angular momentum transfer from protostars by magnetic fields may be astrophysically important. But for the reasons mentioned above, it may be useful to consider other approaches to the angular momentum problem.

In any case, it is easy to see that multiple star systems can store much more angular momentum in the form of orbital motions than the several stars could readily contain as spin. Thus the angular momentum problem would be greatly alleviated (if not solved) were there a simple mechanism for transforming the spin angular momentum of a single massive protostar into the orbital angular momenta of stars in a cluster. One can imagine the process proceeding in two stages. During the initial fragmentation and, perhaps, subsequent fusion stages of star formation, much of the angular momentum is fixed in the orbital motions of massive protostars. The subsequent gravitational interactions among the massive stars can, according to the detailed numerical calculations of Van Albada (1968b), lead to the formation, of wide (visual) binaries. Returning to the massive protostars, we see that, even in McCrea's model, each object is likely to have somewhat more angular momentum than is found in single main-sequence stars. In Sections 2 and 3 we shall suggest that, so long as the kinetic energy of the contracting star is large (compared to the gravitational energy) fission will occur, perhaps repeatedly, and that the resulting stars will have the observed properties with regard to surface velocity and dualism, of the upper main sequence.

B. BINARY STARS

In order for this solution to the angular momentum problem to be plausible, it is clearly required that most stars be found in gravitationally bound double (or triple,

---

* Of course a one-solar mass model *could* be constructed in equilibrium with, say, $J = 10^{57}$ gm cm^2 sec^{-1}; however, it would have a radius more nearly $R \simeq 10^{11} R_\odot$ than $1 R_\odot$!

etc.) systems. If we consider the upper* main-sequence stars we find our expectations confirmed (cf., for example, Blaauw, 1961). Approximately 75% of O-B stars are members of double or triple systems. An interesting, if tentative, additional piece of evidence is presented by Van Albada (1968a) and Blaauw and Van Albada (1969). Carefully examining the distribution of binary separations they find that the division of early-type binaries into close (spectroscopic) and wide (visual) pairs is probably real and not due to the obvious selection effects. For the upper main sequence, the data seem to indicate that, contrary to the earlier suggestions of Kuiper (1935), there is a real deficiency of binaries with separations in the range $10^0 \, \mathrm{AU} < a < 10^2 \, \mathrm{AU}$. This result, if accepted, implies that there are distinct processes for forming close and wide binary stars. Since many authors have suggested that widely separated binaries can be formed during the multiple encounters which occur in a young stellar association, it is attractive to examine the possibility that the close pairs are formed by fission.

2. Theoretical Sequences of Contracting, Rotating Stars

A. HISTORICAL BACKGROUND

The thoughts presented above are not new. They have provided part of the impetus behind centuries of elegant attacks on the problem of the equilibrium and stability of rotating, homogeneous, self-gravitating objects. Among others, McLaurin, Jacobi, Poincaré, Cartán, and Chandrasekhar have studied the subject (for a brief historical review cf. Chandrasekhar, 1967). Darwin (1916, see *Collected Works*, Vol. III) and Jeans (1929) in particular examined the question of fission. The problem, simply stated, is to enumerate the sequence of forms that an isolated object will pass through, if it is initially very large, nearly spherical, and slowly rotating, and then gradually and quasi-statically contracts. The classical problem of a *homogeneous* 'star' has been treated with considerable rigor; although real stars are not homogeneous, the classical tale bears retelling because (a) it is by now well understood, and (b) recent calculations show that similarly defined compressible sequences mimic in all essential respects the classical McLaurin sequence.

B. MCLAURIN SEQUENCE

Consider an axisymmetric star with fixed angular momentum J and mass M within which viscous and magnetic effects may be neglected. Define a model by the value of the density, ϱ, supposed uniform throughout the object. Define a limiting model, $\varrho \to 0$, to be uniformly rotating and require that the angular momentum of every

---

* Low mass ($M < 2 \, M_\odot$) stars must be considered separately, since they have lengthy contraction phases during which the transport of angular momentum by turbulent viscosity (cf. Von Weizsäcker, 1947) or stellar winds (cf. Schatzman, 1962) is likely to be important; the upper main sequence apparently passes directly from collapse to radiative contraction (Larson, 1968), skipping the convective Hayashi phase.

annular mass element in any member of the sequence have the same value as it had in the limiting model. Define a mean radius $\bar{r} \equiv (3M/4\pi\varrho)^{1/3}$. Then one can show that each member of the sequence (i) is an oblate spheroid with axes (a, a, c), $a > c$, such that the eccentricity $e \equiv (1 - a^2/c^2)^{1/2}$ is implicitly related to $\bar{r}$ as follows:

$$\bar{r} = \left(\frac{25}{18}\right)\left(\frac{J^2}{GM^3}\right) e^3 (1 - e^2)^{1/6} \left[(1 - \tfrac{2}{3} e^2) \sin^{-1} e - e(1 - e^2)^{1/2}\right]^{-1}, \quad (2)$$

(ii) rotates uniformly with angular velocity

$$\Omega = \left(\frac{162}{125}\right)\left(\frac{G^2 M^5}{J^3}\right) e^{-6} \left[(1 - \tfrac{2}{3} e^2) \sin^{-1} e - e(1 - e^2)^{1/2}\right]^2. \quad (3)$$

Note that, although the McLaurin sequence was defined (somewhat unconventionally) as a zero viscosity sequence and the angular velocity distribution was left to be determined by the dynamical constraints, it turned out that each member of the sequence was required to rotate as a solid body if the initial, nearly spherical member did. We may think of the McLaurin sequence as the sequence of an $n = 0$ polytrope [polytrope – star in which (pressure) $\propto$ (density)$^{(n + 1)/n}$]. Clearly there must exist analogous polytropic sequences having $n \neq 0$. It seems unlikely that the special property of the homogeneous McLaurin spheroids – uniform rotation – will be preserved in the more general case. Instead, even though the initial members are uniformly rotating, subsequent members of the generalized sequences are found to rotate differentially.

Returning to the $n = 0$ case we find the following behavior as ϱ varies from 0 to ∞ and $\bar{r}$ from ∞ to 0. The eccentricity varies monotonically from 0 to 1, c from ∞ to 0, a from ∞ to the asymptotic value $(25\pi/108) (J^2/GM^3)$, and Ω from 0 to $(27\pi/125) \times G^2 M^5/J^3$. The total energy varies monotonically from 0 to $-\infty$ and the ratio of $T/|W|$ of kinetic to absolute gravitational energy from 0 to $\tfrac{1}{2}$ monotonically.

The $n = 0$ sequence does *not* terminate at any point; there exists an equilibrium model for every $0 \leqslant \varrho < \infty$. However, a detailed analysis of the lowest incompressible modes by Lebovitz (1961) shows that the models are neutrally stable to a nonaxisymmetric mode at $e = 0.813$, $T/|W| = 0.138$ and become dynamically overstable at $e = 0.953$, $T/|W| = 0.273$. The neutral mode occurs at the classical 'point of bifurcation' where uniform, uniformly rotating, ellipsoids become dynamically possible; however, the bifurcation point has no significance in the present discussion, because no instabilities occur there so long as viscosity is neglected. The second point $(T/|W| = 0.273)$ leads to dynamical motions examined in the nonlinear regime by Rossner (1967) and Fujimoto (1968). The motion is quite complex but might very roughly be described as the end over end tumbling of a quite prolate $(a/c \approx 25)$ spheroid. An important next step would be to carry the numerical work further, relaxing the assumption, made by both Rossner and Fujimoto, that the bodies remain ellipsoidal. There are rather strong arguments (too detailed to be given here) for believing that, after the point of overstability is reached, fission will occur as the very thin, approximately spheroidal body breaks up into two or more objects orbiting about the common center of mass.

C. POLYTROPIC SEQUENCES

At this juncture the reader is no doubt wondering why the classical results have been summarized in such detail. Is it not known that these considerations are irrelevant to realistic centrally condensed stars? Has it not been shown (Jeans, 1919; James, 1964; Tassoul and Ostriker, 1970) that polytropes of index $n > 0.808$ 'fly apart' before any instabilities are reached? The answer is no. Rather, in the aforementioned papers it was shown that centrally condensed stars cannot store much kinetic energy and remain uniformly rotating. This is not surprising. We know without making any calculations that, for parts of a centrally condensed star within which the pressure forces are small compared to the inertial forces, the rotation law *must* approach Keplerian motion; that is $\Omega \propto \tilde{\omega}^{-1/2}$. If we restrict consideration to model stars forced to rotate uniformly, then we are restricting our attention to the interval $0 \leqslant T/|W| \ll 1$, within which interval we should not expect to find rotationally induced instabilities. What happens if we do not require uniform rotation?

The zero viscosity polytropic ($n \neq 0$) sequences defined earlier have been studied over the past few years by a group at Princeton University Observatory. The equilibria are numerically found using the SCF method (Ostriker and Mark, 1968) and the normal modes of oscillation studied using the virial techniques described by Tassoul and Ostriker (1968). Detailed results will be published elsewhere, but the important points are easily summarized by saying that *in all essential respects the $n \neq 0$ sequences resemble the classical $n = 0$ McLaurin sequence*. In particular, the sequences do not terminate. That is, for given (J, M) a model can apparently be constructed with central density ϱ_c, $0 \leqslant \varrho_c < \infty$. There is no 'rotational breakup'. For $n \neq 0$, neither the density nor the rotation are uniform nor, in general, are the equipotential surfaces spheroids. However, the dependence of a, c, Ω, E, $T/|W|$, and the normal modes on ϱ_c are topologically similar to the $n = 0$ case. In all cases examined, a point of bifurcation occurs at $T/|W| \fallingdotseq 0.14$ and overstability is reached at $T/|W| \fallingdotseq 0.26$. The numerical values of $T/|W|$ at the critical points are remarkably independent of n; to the level of accuracy presently available in the numerical work these critical values are constants. Thus it appears that real stars may, after all, undergo fission. The over-all picture is schematically reviewed in Figure 1. From a somewhat different point of view Roxburgh (1966) anticipated these conclusions without a detailed study of stability, and revived the fission hypotheses for the origin of close binaries among the low mass stars (W Ursae Majoris stars).

3. Applications of Theory

A. WHITE DWARFS

White dwarf stars obeying the Chandrasekhar equation of state are similar to polytropes, but the central densities are not arbitrarily adjustable. One can, however,

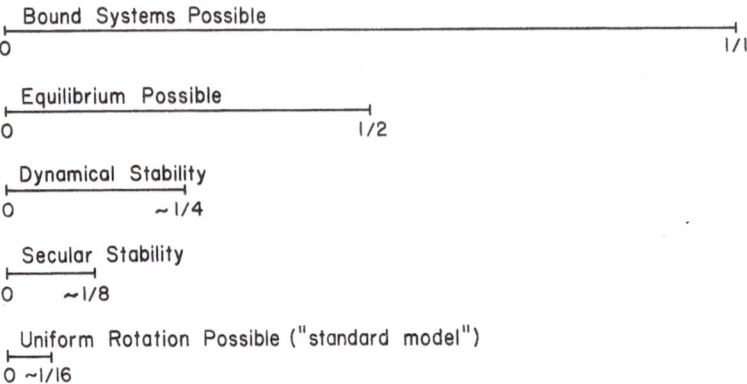

Fig. 1.   Schematic diagram showing the limits placed by equilibrium and stability requirements on the ratio $T/|W|$ between kinetic energy of rotation and gravitational energy. See text in Section 2.C.

consider a star of given mass and angular momentum distribution and construct the sequence of stars having larger and larger total angular momenta. The computations (Ostriker and Bodenheimer, 1968) again indicate that the sequences do not terminate. A maximum in Ω is found along the sequence but neither 'rotational mass loss' nor 'rotational breakup' ever occur. However, a stability analysis (Ostriker and Tassoul, 1969) indicates that for $T/|W| \doteq 0.26$ the models are overstable and probably subject to fission. Nevertheless, for a variety of reasons it is very unlikely that close pairs of white dwarfs can be formed by fission, and in fact no such pairs have yet been observed.

B. PRE-MAIN SEQUENCE EVOLUTION

A more promising line of research (with respect to the fission problem) has been followed by Bodenheimer (1969) and Bodenheimer and Ostriker (1970), who have examined the pre-main sequence evolution of massive stars. The calculations (for $3-12 M_\odot$ stars) begin at the start of the radiative contraction phase. For each star the mass, M, and total angular momentum, J, are assumed fixed. Since all the suggested processes for redistribution of angular momentum require longer than a Kelvin time, we further assume that the angular momentum of each annular element is preserved. Thus, the angular momentum distribution (chosen to be that of a uniformly rotating, uniform sphere) is an invariant. The study reached the following principal conclusions:

(1) For given M, the evolutionary track of a star with small J is like the nonrotating track, nearly horizontal in the $(\log L/L_\odot, \log T_{\text{eff}})$ plane. With increasing J the luminosity decreases and the tracks become more nearly vertical. Aside from the dependence of track shape and orientation on J, the high angular momentum models resemble slowly rotating models of lower mass with respect to contraction times and position (near the main sequence) after nuclear burning commences.

(2) For sufficiently high values of J (with given M) the stars become unstable to

nonaxisymmetric modes before they have contracted to the main sequence. It is interesting to note that if J is slightly less than the critical value $J_c(M)$, the surface velocity of the $12\,M_\odot$ model at the main sequence is ≈ 400 km/sec, approximately the maximum velocity observed * for that mass star. In all probability it is instability to fission that limits the angular momentum and hence the surface velocity of main sequence stars.

(3) Stars with angular momentum just greater than the critical value, $J_c(M)$, will presumably form close binary systems. We can compare the relation $J=J_c(M)$ with the observed relation between angular momentum and mass in the closest (lowest angular momentum) early-type binaries. The agreement is excellent, with the theoretical curve lying between the curves for binaries with mass ratio one $J=J_b(M, 1)$ and mass ratio two $J=J_b(M, 2)$. As is seen in Figure 2,

$$J_b(M, 1) > J_c(M) > J_b(M, 2). \tag{4}$$

A similar comparison for the low mass stars was made by Roxburgh (1966).

4. Summary

In conclusion, let us review the arguments that have been put forward *against* the fission theory for the origin of binary stars.

Early in this century mathematicians thought it possible for stars to proceed quasi-statically through a series of uniformly rotating configurations to a detached binary system; the evolution envisaged was first along the McLaurin sequence of spheroids, then, at the point of bifurcation, to the Jacobi ellipsoids, and then, at a second point of bifurcation, to the 'pear-shaped' configurations studied in detail by Darwin. At this stage James felt there was 'little doubt' that fission would occur and lead to two

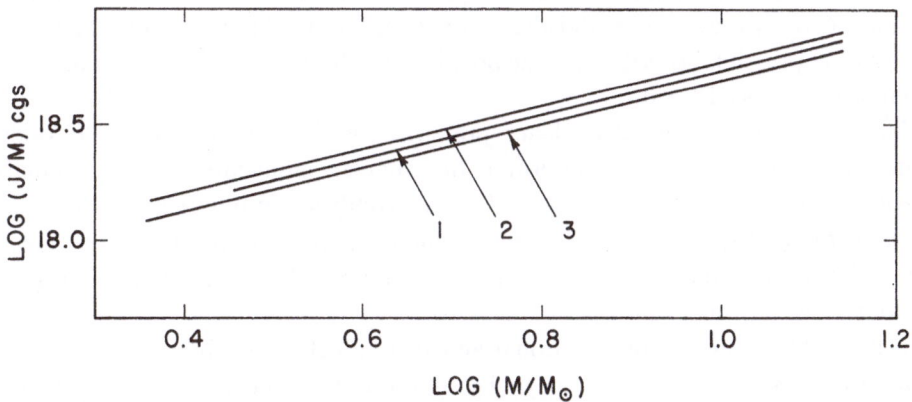

Fig. 2.   Estimated relations between mass and angular momentum per unit mass adapted from Bodenheimer and Ostriker (1970): (1) $J_c(M)$ – differentially rotating models which are suspected to undergo fission just prior to the onset of nuclear reactions; $J_b(M, 1)$, $J_b(M, 2)$ – main sequence contact binaries adapted from Kraft (1969), with the given total mass and mass ratios of 1 and 2 respectively.

* The 'observed' velocities for rapidly rotating stars perhaps underestimate $v_e \sin i$ by as much as a factor 1.4 according to Hardorp and Strittmatter's (1968) calculations of gravity darkened models.

detached, unequal, masses. However, Cartán (1922) proved unequivocally that the Jacobi ellipsoid is unstable at the point where the pear-shaped configurations branch off so that quasi-static evolution to a binary system was recognized as impossible. The possibility nevertheless remained that binary stars could form as a result of a dynamical instability. The argument for continuous quasi-static contraction fails on other grounds as well. It now seems unlikely that viscosity plays an important role during the early evolution of massive stars; in the absence of any dissipation a homogeneous star will not enter the Jacobi sequence at all but will follow the McLaurin sequence until $T/|W|=0.27$, at which point it is dynamically unstable. However, in the course of the subsequent dynamical evolution (described by Rossner, 1967) it may break into two or more pieces. Dynamically induced fission is probable here, but remains unproven.

A second objection is due to Lyttleton (1953). Since the dynamical evolution does not require friction, it must be time reversable. But, if we reverse the direction of time in an existing binary system it clearly does not revert to a single star but simply remains a binary system with the orbital directions reversed. Therefore a single system cannot evolve without dissipation into a binary. This argument is irreproachable as stated. However, other stages of stellar evolution proceed in a notably nonadiabatic way so it seems unreasonable to assume that the fission process alone should be energy conservative. Without the restriction to energy conservative systems, the argument of time reversability fails and Lyttleton's objection loses force.

As Roxburgh pointed out, a far more serious objection arose from the discovery of Jeans' (1929) and James' (1964) that centrally condensed ($n>0.808$) uniformly rotating objects could never reach the points of bifurcation or overstability. However, it is now clear that this result was an artifact of the constraint to uniform rotation. The self-consistent zero-viscosity, polytropic sequences ($n>1$ and $n<1$) described in Section 2.C do not terminate and they do become overstable at $T/|W|\eqsim0.27$. Thus, it seems compressible stars do have the possibility of becoming binaries by undergoing dynamically induced fission.

Finally, the observational evidence presented by Roxburgh (1966) and Bodenheimer and Ostriker (1970) concerning the angular momentum of close binaries provides experimental support for the fission hypothesis. While there is still a considerable amount of work to be done (cf. comments in Section 2.B) before one can say, with certainty, that the process is theoretically possible, there are now no strong theoretical arguments against it.

In the light of present observational and theoretical work, the fission hypothesis seems attractive, once again, as a partial solution to the angular momentum problem, and as a natural way of producing close binaries.

Acknowledgement

This work was supported by the Air Force Office of Scientific Research, U.S. Air Force, contract AF 49(638)-1555.

References

Blaauw, A.: 1961, *Bull. Astron. Inst. Neth.* **15**, 265.
Blaauw, A. and Van Albada, T. S.: 1969, *Bull. Astron. Inst. Neth.* (in press).
Bodenheimer, P.: 1969, paper presented at the 16th Liege Symposium.
Bodenheimer, P. and Ostriker, J. P.: 1970, *Astrophys. J.* (in press).
Cartán, E.: 1922, *Bull. Sci. Math.* 2e Ser. **46**, 316, 356.
Chandrasekhar, S.: 1967, *Commun. Pure Appl. Math.* **20**, 251.
Darwin, G. H.: 1916, *Collected Works*, Vol. III.
Fujimoto, M.: 1968, *Astrophys. J.* **152**, 523.
Hardorp, J. and Strittmatter, P. A.: 1968, *Astrophys. J.* **153**, 465.
James, R. A.: 1964, *Astrophys. J.* **140**, 552.
Jeans, J. H.: 1919, *Problems of Cosmogony and Stellar Dynamics*, Cambridge University Press, Cambridge.
Jeans, J. H.: 1929, *Astronomy and Cosmology* (reprinted by Dover Publications: New York, 1961), Chap. X.
Kraft, R. P.: 1969, *Otto Struve Memorial* (ed. by G .Herbig) (in press).
Kuiper, G. P.: 1935, *Publ. Astron. Soc. Pacific* **47**, 121.
Larson, R. B.: 1968, Thesis, Cal. Inst. of Technology.
Lebovitz, N. R.: 1961, *Astrophys. J.* **134**, 500.
Lyttleton, R.: 1953, *The Stability of Rotating Liquid Masses*, Cambridge Univ. Press., Cambridge.
McCrea, W. H.: 1961, *Proc. Roy. Soc. (London)* **A260**, 152.
Ostriker, J. P. and Bodenheimer, P.: 1968, *Astrophys. J.,* **151**, 1089.
Ostriker, J. P. and Mark, J. W-K.: 1968, *Astrophys. J.* **151**, 1075.
Ostriker, J. P. and Tassoul, J. L.: 1969, *Astrophys. J.* **155**, 987.
Rossner, L. F.: 1967, *Astrophys. J.* **149**, 145.
Roxburgh, I.: 1966, *Astrophys. J.* **143**, 111.
Schatzman, E.: 1962, *Ann. Astrophys.* **25**, 18.
Spitzer, L.: 1968, *Stars and Stellar Systems*, Vol. VII (ed. by B. M. Middlehurst and L. H. Aller), University of Chicago Press, Chicago, Chap. 1.
Tassoul, J. L. and Ostriker, J. P.: 1968, *Astrophys. J.* **154**, 613.
Tassoul, J. L. and Ostriker, J. P.: 1970, *Astron. Astrophys.* (in press).
Van Albada, T. S.: 1968a, *Bull. Astron. Inst. Neth.* **20**, 47.
Van Albada, T. S.: 1968b, *Bull. Astron. Inst. Neth.* **20**, 57.
Von Weizsäcker, C. F.: 1947, *Z. Astrophys.* **24**, 181.

Discussion

Jaschek: I would just like to comment that the investigation by Blaauw and Van Albada should be extended to later-type stars in which the statistics seem to behave very differently from what they found. In particular no break between close and wide binaries was found. This break is furthermore located at the very point where the incidence of selection is largest, so that its reality should not be accepted without further investigation.

Ostriker: I am not qualified to comment on the reliability of Blaauw and Van Albada's statistical conclusions. Van Albada's mechanism for producing wide (visual) binaries is in any case interesting and, I believe, plausible.

Fricke: I don't feel very satisfied with the neglect of angular momentum transport by meridional circulations during the pre-main-sequence evolution. The ratio of centrifugal force to gravity becomes of order 1 in your models and angular momentum transport within the Kelvin-Helmholtz time scale over appreciable distances of the star should be expected. I suggest that constant angular velocity might be the more reasonable constraint one should impose during the contraction phase.

Ostriker: For most of the proposed contraction phase the ratio of centrifugal to gravitational force is small. Even for the most rapidly rotating part of the star with the largest angular momentum considered and taking that star at its most rapidly rotating (most condensed) part of the contraction phase – even in this most extreme case, the ratio of centrifugal force to gravity is only 0.4. That is,

I believe that circulation currents should be included and intend to do so in future work, but I doubt the conclusions will be very much altered.

Furthermore, it is not obvious to me that circulation would lead the star towards constant angular velocity; in the absence of any detailed calculations constant angular momentum density seems, a priori, equally probable.

Collins: You have established that a sequence of models is in equilibrium. Have you established that any two neighboring models can be reached, i.e., can you move from one model to another without violating any physical laws?

Ostriker: The models are connected to one another in the same sense as models along the classical McClaurin Sequence. I have not investigated for either sequence what form (if any) the quasi-static contraction must take in order to transform one equilibrium model into the next one along the sequence.

Roxburgh: As I pointed out in 1966 it is to be expected that non-uniform rotation is developed during pre-main sequence contraction, and I used this in my papers on the fission theory. However, as I pointed out, during the adjustment period from fully convective phase to the radiative phase a high degree of differential rotation develops with an approximately radial variation.

The question I wish to put is whether your set of models, which with a misuse of the term are homologous, are in fact a sequence? That is whether you can in fact go from one model to the next in a sequential sense. You assume that in each model the angular velocity is constant on cylinders and that the angular momentum per unit mass is conserved. This may not be possible without at the same time destroying the angular momentum conservations for each individual fluid element that you must have in an inviscid fluid.

Ostriker: Your question was answered in part when I replied to Collins' question; the sequences are constructed so that every element conserves vorticity – and of course angular momentum as well. On your first point, we treated the contraction phase of high mass stars (5–12 $M_\odot$) specifically because these, according to recent calculations of Larson and Bodenheimer, may not have any Hayashi phase. The non-dynamical contraction phase is expected to be purely radiative for these stars and our calculations begin at the beginning of this quasi-static radiative contraction phase.

STRUCTURE OF CLOSE BINARIES. I: POLYTROPES

M. D. T. NAYLOR and S. P. S. ANAND

David Dunlap Observatory, University of Toronto, Canada

Abstract. An extension of the method devised by Monaghan and Roxburgh (1965) for rapidly rotating polytropes is used to study the structure of the primary component of a synchronous close binary system. Results are presented for polytropes of index 1.5, 2.0, 3.0, 4.0 and 4.9. We conclude that the extended Monaghan and Roxburgh method can be applied to real stars which are perturbed by both tidal and rotational forces.

1. Introduction

Recently there has been considerable work by a number of groups on the evolution of stars in binary systems (Plavec, 1968, and references therein). The results have been obtained by using the standard spherically symmetric equations of stellar structure with special boundary conditions (Plavec, 1968), and it was also explicitly assumed that the effects of the centrifugal and tidal distortions are small, and should not seriously change the overall picture of binary star evolution.

It is the purpose of the present investigation to examine this assumption by calculating detailed models which will include both distortion terms. In this our first communication, we will consider the problem of close binary polytropes as they should give an estimate of the accuracy which can be expected from the method we have chosen.

In a famous series of papers Chandrasekhar (1933a, b, c, d) developed a first-order perturbation analysis which he applied to the rotational problem, the tidal problem, and the binary star problem. Unfortunately when the separation between the components is only a few times the undisturbed radius of the primary (i.e. more massive) component, Chandrasekhar's method no longer gives accurate results near the surface of the polytrope unless modified (Monaghan, 1967; Martin, 1970). Since then there has been very limited progress in this direction (Takeda, 1934; Kuiper, 1941; Kopal, 1959; Orlov, 1961).

In the last five years several methods for finding exact solutions to the rotational problem have appeared (James, 1964; Stoeckly, 1965; Ostriker and Mark, 1968). However these methods are cumbersome, require large amounts of computer time, and are difficult to apply to real stars. For these reasons, Monaghan and Roxburgh (1965) (hereinafter referred to as MR) developed a simple perturbation method which gives fairly accurate results for rapid rotation but required much less computer time than the exact treatments. This technique was then used by Roxburgh *et al.* (1965), Faulkner *et al.* (1968), Sackmann (1968), and Sackmann and Anand (1970) to determine the structure of rapidly rotating stars.

The MR technique for rotating polytropes can easily be modified to study binary systems. This is accomplished by following the analysis of MR but including the tidal potential as well as the centrifugal potential. Results were obtained for polytropic models of the primary with indices 1.5, 2.0, 3.0, 4.0, and 4.9.

A. Slettebak (ed.), Stellar Rotation, 157–164. All Rights Reserved
Copyright © 1970 by D. Reidel Publishing Company, Dordrecht-Holland

2. Basic Equations and Method of Solution

Consider two gaseous masses in hydrostatic equilibrium which are revolving in circular orbits about their common centre of mass. Spherical polar coordinates (r, θ, ϕ) are introduced with origin at the centre of mass of the primary and the azimuthal angle ϕ is measured from the line joining the centres of mass of the two components. It is assumed that the rotation axis of the primary is perpendicular to the plane of the orbit and θ is taken to be the colatitude.

The equation of hydrostatic equilibrium for the primary is

$$\frac{\nabla P}{\varrho} = \nabla V + \nabla V_{\mathrm{T}} + \nabla \left[\tfrac{1}{3}\omega^2 r^2 \left(1 - P_2(\mu)\right) - \frac{M_2}{M_1 + M_2} R\omega^2 r P_1(\lambda) \right], \quad (1)$$

where V is the gravitational potential of the primary of mass M_1, V_{T} is the tidal potential of the secondary of mass M_2, ω is the uniform angular velocity of the primary about its rotation axis, and synchronism between rotation and revolution is assumed, R is the separation of the mass centres, and μ and λ are equal to $\cos\theta$ and $\sin\theta \cos\phi$ respectively. The term in square brackets is the centrifugal potential. The gravitational and tidal potentials must satisfy Poisson's and Laplace's equations, respectively. It is assumed that the primary can be represented by a polytrope of index n, and the following dimensionless variables are introduced,

$$\begin{aligned} &\varrho = \varrho_c \sigma^n, \quad \alpha = \omega^2/2\pi G\varrho_c, \\ &r = ax, \quad a^2 = K\varrho_c^{(1/n)-1}(n+1)/4\pi G, \end{aligned} \quad (2)$$

These variables can be substituted into Equation (1) and with the aid of Poisson's and Laplace's equations we obtain

$$\nabla^2_{x, \mu, \phi}\sigma = -\sigma^n + \alpha. \quad (3)$$

The form of the tidal potential to be used in this investigation has been derived by Chandrasekhar (1933b) to order $(\bar{a}/R)^5$ where $\bar{a}$ is the mean radius of the secondary,

$$V_{\mathrm{T}} = \frac{GM_2}{R} \sum_{j=1}^{4} \left(\frac{r}{R}\right)^j P_j(\lambda) + \text{constant}. \quad (4)$$

To obtain the solution we divide the polytrope into two regions. In the inner region σ is obtained by employing a first order perturbation analysis, whereas in the outer region it is assumed that the local gravitational potential is determined only by the mass contained in the inner region. The two solutions for σ are then matched on a spherical interface.

In the inner region the perturbing forces are small compared to the local gravitational force and following the method outlined by Chandrasekhar (1933a) and MR, σ is expanded in a power series and only first order terms are retained, i.e.

$$\sigma = \theta(x) + \alpha\Psi(x, \mu, \phi). \quad (5)$$

Furthermore we develop Ψ in terms of a series of surface harmonics,

$$\Psi = \psi_0(x) + \sum_{j=1}^{\infty} \psi_j(x) S_j(\mu, \phi), \tag{6}$$

where the $S_j(\mu, \phi)$ must satisfy

$$\frac{\partial}{\partial \mu}\left[(1 - \mu^2) \frac{\partial S_j}{\partial \mu}\right] + \frac{1}{(1 - \mu^2)} \frac{\partial^2 S_j}{\partial \phi^2} + j(j + 1) S_j = 0 \tag{7}$$

for each j.

In the outer region the technique is somewhat different. First the equation of hydrostatic equilibrium is integrated

$$K(n + 1) \varrho_c^{1/n} \sigma = V + \tfrac{1}{3}\omega^2 a^2 x^2 \left[1 - P_2(\mu)\right]$$

$$- \frac{M_2}{M_1 + M_2} R\omega^2 a x P_1(\lambda) + \frac{GM_2}{R} \sum_{j=1}^{4} \left(\frac{a}{R}\right)^j x^j P_j(\lambda) + \text{constant}. \tag{8}$$

The angular velocity is related to the masses and their separation by Kepler's law

$$\omega^2 = G \frac{(M_1 + M_2)}{R^3} (1 + \varepsilon), \tag{9}$$

where the quantity ε is of order $(\bar{a}/R)^5$ (Martin, 1970). Since ω^2 appears only in terms of order $(\bar{a}/R)^2$, the ε term will not enter the subsequent analysis which is consistent to order $(\bar{a}/R)^5$.

In this region we assume that the density is so low that its contribution to the total mass is negligible and hence the gravitational potential in the outer region is determined by the mass contained in the inner region. This approximation improves with increasing polytropic index. Thus the gravitational potential in the outer region is the Laplace solution.

$$\Phi = \frac{\beta_0}{x} + \frac{\alpha \beta_1}{x} + \alpha \sum_{j=1}^{\infty} \frac{S_j'(\mu, \phi)}{x^{j+1}}, \tag{10}$$

where β_0 and β_1 are constants, the $S_j'(\mu, \phi)$ satisfy the equation for surface harmonics and Φ equals $V/K(n+1)\rho_c^{1/n}$.

The constants and the forms of the surface harmonics are obtained by matching the two solutions on a spherical interface at radius $x = x_f$. The method used by MR to obtain the fitting point is to find that value of x for which the errors involved in the inner and outer solutions are of the same order of magnitude. In the final analysis their values of x_f were also influenced by the existing tables of the required functions. In any event the results should not be extremely sensitive to the position of the fitting radius, otherwise the method is unworkable. The validity of this assumption will be examined later. For convenience we have chosen the same values of x_f as MR.

The continuity conditions show that both $S_j(\mu, \phi)$ and $S_j'(\mu, \phi)$ involve only the spherical harmonics $P_2(\mu)$ and $P_j(\lambda)$ with coefficients A_2 and C_j say, for the $S_j(\mu, \phi)$ and similarly B_2 and D_j for the $S'_j(\mu, \phi)$. Our final solution for the inner region is

$$\sigma = \theta + \alpha\psi_0 + \alpha A_2\psi_2 P_2(\mu) + \alpha \sum_{j=2}^{4} C_j\psi_j P_j(\lambda), \tag{11}$$

and for the outer region

$$\sigma = \frac{\beta_0}{x} + \frac{\alpha\beta_1}{x} + \frac{\alpha\beta_2}{x^3} P_2(\mu) + \alpha \sum_{j=2}^{4} \frac{D_j P_j(\lambda)}{x^{j+1}} + \frac{1}{6}\alpha x^2 [1 - P_2(\mu)]$$

$$+ \frac{1}{2} \frac{M_2}{M_1 + M_2} \alpha \sum_{j=2}^{4} \left(\frac{a}{R}\right)^{j-2} x^j P_j(\lambda) + v_0 + \alpha v_1. \tag{12}$$

The constants are defined by:

$$\beta_0 = -\theta'x^2, \quad \beta_1 = \frac{x^3}{3} - x^2\psi_0',$$

$$v_0 = \theta + \theta'x, \quad v_1 = \psi_0 + x\psi_0' - \tfrac{1}{2}x^2,$$

$$A_2 = -\frac{5}{6}\frac{x^2}{(3\psi_2 + x\psi_2')}, \quad B_2 = \frac{x^5}{6}\left(\frac{x\psi_2' - 2\psi_2}{x\psi_2' + 3\psi_2}\right),$$

$$C_j = \frac{M_2}{M_1 + M_2}\left(\frac{a}{R}\right)^{j-2} c_j,$$

where

$$c_j = \frac{1}{2}\left[\frac{(2j+1)x^j}{(j+1)\psi_j + x\psi_j'}\right],$$

$$D_j = \frac{M_2}{M_1 + M_2}\left(\frac{a}{R}\right)^{j-2} d_j,$$

where

$$d_j = \frac{x^{2j+1}}{2}\left[\frac{j\psi_j - x\psi_j'}{(j+1)\psi_j + x\psi_j'}\right];$$

the prime denotes differentiation with respect to x, $j = 2$, 3, and 4, and the constants are evaluated at the fitting radius. New integrations were obtained for all functions, and their values, as well as those of the constants, are given in Table I.

For convenience a further simplification is introduced. It was shown by Chandrasekhar (1933c) that the parameters α, q, and η where $\eta = R/a$ and $q = M_2/M_1$ are not independent for synchronism but are related by

$$\alpha = \frac{2(1+q)x_1^2|\theta_1'|}{\eta^3}, \quad |\theta_1'| \equiv \left.\frac{d\theta}{dx}\right|_{x=x_1}$$

which is correct to first-order in α; the subscript 1 denotes the value at the Emden radius.

TABLE I
Function values and constants at the fitting radius

n	1.5	2.0	3.0	4.0	4.9
x_f	3.2	3.6	5.0	8.0	97.4
θ	1.0455 (−1)	1.1525 (−1)	1.1082 (−1)	1.0450 (−1)	7.6465 (−3)
θ'	−2.5875 (−1)	−1.8269 (−1)	−8.0126 (−2)	−2.7957 (−2)	−1.8179 (−4)
ψ_0	1.0423 (0)	1.2961 (0)	2.7202 (0)	8.1810 (0)	1.5688 (3)
ψ_0^1	4.9660 (−1)	6.7495 (−1)	1.2512 (0)	2.3563 (0)	3.2434 (1)
ψ_2	4.3134 (0)	4.5360 (0)	6.5108 (0)	1.3843 (1)	1.7738 (3)
ψ_2^1	1.0255 (0)	1.2617 (0)	1.9891 (0)	3.1887 (0)	3.6417 (1)
ψ_3	1.7411 (1)	2.2090 (1)	4.9772 (1)	1.8284 (2)	3.0095 (5)
ψ_3^1	1.1811 (1)	1.4617 (1)	2.7318 (1)	6.6699 (1)	9.2691 (3)
ψ_4	6.3560 (1)	9.3676 (1)	3.0845 (2)	1.8700 (3)	3.8377 (7)
ψ_4^1	6.7034 (1)	9.2461 (1)	2.3611 (2)	9.2231 (2)	1.5760 (6)
A_2	−5.2604 (−1)	−5.9504 (−1)	−7.0674 (−1)	−7.9557 (−1)	−8.9145 (−1)
B_2	−1.8427 (1)	−2.5151 (1)	−5.4356 (1)	−1.7732 (2)	−8.8761 (4)
β_0	2.6496 (0)	2.3676 (0)	2.0032 (0)	1.7889 (0)	1.7246 (0)
β_1	5.8375 (0)	6.8047 (0)	1.0384 (1)	1.4862 (1)	2.6347 (2)
ν_0	−7.2345 (−1)	−5.4242 (−1)	−2.8981 (−1)	−1.1910 (−1)	−1.0060 (−2)
ν_1	−2.4886 (0)	−2.7541 (0)	−3.5233 (0)	−4.4683 (0)	−1.5017 (1)
c_2	1.5781 (0)	1.7851 (0)	2.1202 (0)	2.3867 (0)	2.6743 (0)
c_3	1.0675 (0)	1.1583 (0)	1.3033 (0)	1.4167 (0)	1.5352 (0)
c_4	8.8644 (−1)	9.4333 (−1)	1.0329 (−1)	1.1018 (0)	1.1726 (0)
d_2	5.5282 (1)	7.5452 (1)	1.6307 (2)	5.3197 (2)	2.6628 (5)
d_3	2.3084 (2)	3.7936 (2)	1.4809 (3)	1.2384 (4)	7.3862 (8)
d_4	1.3130 (3)	2.6519 (3)	1.9105 (4)	4.0729 (5)	3.1498 (12)

The surface of the polytrope is specified by $\sigma = 0$. If $\sigma = 0$ and simultaneously $\partial\sigma/\partial x = 0$ at $\mu = 0$, $\lambda = 1$, then the effective surface gravity at the equator in the direction of the secondary is zero. The result is the so-called critical configuration.

The smallest equipotential surface (in volume) for which the gradient of the total potential is zero on the line joining the centres of mass is known as the inner Lagrangian surface or contact surface. Hence it is obvious that the primary lobe of the contact surface and the critical configuration are identical in our theory.

3. Discussion of Results

The parameters (x_c, η_c) appropriate to the critical configuration can be obtained for various q by solving $\sigma = 0$ and $\partial\sigma/\partial x = 0$ at $\mu = 0$, $\lambda = 1$ simultaneously. When q is zero our equations reduce to those of the rotational problem and in Table II the critical values of x_e and α_c are given, and, for comparison, the values of M R and James (1964) are also listed. The values of α_c differ markedly for the case $n = 1.5$ and the reason for this behaviour can perhaps be understood because the errors involved in the two-zone approximation for this polytropic index are the largest of all five cases. Moreover the difference between the two values obtained by the same method can be explained by noting that the constants a_2 (of M R) and B_2 of the present paper are not equal although they should be. It appears that the value of $\psi_2'(x_f)$ used by M R is incorrect. A similar

TABLE II

Comparison of the critical models of Monaghan and Roxburgh (MR), James (J) and the present investigation

n	1.5	2.0	3.0	4.0	4.9
x_e	5.3687	6.3748	10.1370	22.3408	257.2197
x_e (MR)	5.24	6.33	10.12	22.26	–
x_e (J)	5.3585	6.307	–	–	–
x_e/x_1	1.4694	1.4645	1.4698	1.4922	1.5004
α	3.7544 (-2)	1.9439 (-2)	3.9304 (-3)	3.2203 (-4)	2.0269 (-7)
α (MR)	4.10 (-2)	1.99 (-2)	3.95 (-3)	3.27 (-4)	–
α (J)	4.3624 (-2)	2.1604 (-2)	3.932 (-3)	–	–

situation exists for $n=4.0$ but the effect on the critical values of x_e and α_c is small because of the manner in which B_2 enters the equation.

The accuracy of the results improves with increasing polytropic index since the amount of mass being neglected becomes vanishingly small. A similar behaviour is expected in the binary star problem.

A sample of the binary results is given in Table III for $q=1.0$ and 0.1 where we list

TABLE III

Dimensions of the critical primary for different n and the corresponding Roche surface

$q = 1.0$						
n	1.5	2.0	3.0	4.0	4.9	Roche
x_c/η_c	0.5333	0.5317	0.5303	0.5296	0.5291	0.5000
y_c/η_c	0.3832	0.3941	0.3851	0.3851	0.3848	0.3742
z_c/η_c	0.3601	0.3622	0.3647	0.3650	0.3648	0.3562
w_c/η_c	0.4258	0.4247	0.4237	0.4231	0.4227	–

$q = 0.1$						
n	1.5	2.0	3.0	4.0	4.9	Roche
x_c/η_c	0.8305	0.8262	0.8217	0.8193	0.8183	0.7175
y_c/η_c	0.6550	0.6530	0.6508	0.6493	0.6485	0.5961
z_c/η_c	0.5598	0.5625	0.5651	0.5648	0.5642	0.5345
w_c/η_c	0.6917	0.6884	0.6848	0.6828	0.6820	–

the principal dimensions of the Roche primary and those obtained in the present work. For $n=4.9$ the largest differences, $\approx 6\%$ for $q=1.0$ and $\approx 14\%$ for $q=0.1$ occur in the value of x_c/η_c which appears to be quite sensitive to the relative importance of tidal and centrifugal forces, particularly for small q. The differences in the other radii are somewhat less. The fraction of these differences resulting from errors intrinsic to the method is difficult to estimate for all cases. However for $q=1.0$ and for all n, x_c/η_c should equal one-half exactly. It might appear that the difference in x_c/η_c between the $n=4.9$ model and the Roche model is a result of our assumption that the secondary can be represented by a mass point (i.e. we neglect terms of order $(\bar{a}/R)^6$), in addition to the use of the MR approximation scheme in the primary. However the largest part of this difference arises in the truncation of the expansion of $1/r'$ (Equation (4), Chandrasekhar 1933c) at $j=4$, because it is assumed that $\bar{a}/R$ is approximately the same size, or larger, than r/R. For the surface radii of the critical models this is not necessarily true, especially in the direction of the secondary. Reference to section III.3 of Kopal's monograph (Kopal, 1959) shows that r/R can be approximately 3 times larger than $\bar{a}/R$ for a mass ratio of 0.1. In the expansion

$$\frac{1}{r'} = \frac{1}{R} \sum_{j=0}^{\infty} \left(\frac{r}{R}\right)^j P_j(\lambda),$$

the summation should be continued until $(r/R)^j$ is of the order of $(\bar{a}/R)^6$. With this correction the dimensions of the $n=4.9$ polytrope and the Roche model agree to 4 significant figures, and in general, for $4.9 \geqslant n > 1.5$ the differences in radii between polytropes and the Roche model are less than 1%. A complete second-order theory has been derived by Martin (1970) which reduces these differences by still another factor of 2. However we feel that such accuracy is not necessary in the calculation of stellar models when the errors in the input data (e.g. opacities) can be much larger.

The effect of changes in the fitting radius (e.g. $\pm 10\%$ of the values previously quoted) has been investigated. In general it seems that an increase of the fitting radius introduces a smaller change in the results than does a decrease of x_f and this shows that the results are more sensitive to the mass approximation than to the first-order expansion. In more quantitative terms for the rotating case a 10% increase in x_f results in an increase of $\approx 0.6\%$ in the critical value of x_e for $n=1.5$ and this decreases to $\approx 0.2\%$ for $n=4.0$. The changes in α_c are larger and vary from $\approx 4\%$ for $n=1.5$ to $\approx 1\%$ for $n=4.0$. When x_f is decreased by 10% the changes are ≈ 3 times larger than those changes which occur when x_f is increased. As expected the changes decrease as n increases. In the binary case similar variations were obtained, with the percentage change in the separation being larger than the change in x_e.

In conclusion, we feel that small changes in the fitting radius will result in only minor variations of the critical parameters. It also appears that it is better to overestimate the fitting radius than to underestimate it.

4. Conclusion

Since our principal aim is the study of real binary stars rather than polytropes, the underlying purpose of the present investigation has been to determine whether the extended Monaghan-Roxburgh method is capable of treating such a problem. Hence our discussion of the detailed structure of close binary polytropes has deliberately been brief, but a complete and more accurate treatment will appear elsewhere (Martin, 1970).

Encouraged by the results of the present investigation we are extending this method to study the structure of real binary stars. These models will include detailed opacities and energy generation rates and where necessary we will allow for radiation pressure, partial electron degeneracy, and convective envelopes. The main sequence systems will be discussed in the second paper of this series.

Acknowledgements

We are grateful to Mr. P. G. Martin and Mr. G. G. Fahlman for their useful discussions. We are also indebted to the National Research Council of Canada and the Department of University Affairs of the Province of Ontario for their financial support.

References

Chandrasekhar, S.: 1933a, *Monthly Notices Roy. Astron. Soc.* **93**, 390.
Chandrasekhar, S.: 1933b, *Monthly Notices Roy. Astron. Soc.* **93**, 449.
Chandrasekhar, S.: 1933c, *Monthly Notices Roy. Astron. Soc.* **93**, 462.
Chandrasekhar, S.: 1933d, *Monthly Notices Roy. Astron. Soc.* **93**, 539.
Faulkner, J., Roxburgh, I. W., and Strittmatter, P. A.: 1968, *Astrophys. J.* **151**, 203.
James, R.: 1964, *Astrophys. J.* **140**, 552.
Kopal, Z.: 1959, *Close Binary Systems*, Chapman and Hall, London.
Kuiper, G. P.: 1941, *Astrophys. J.* **93**, 133.
Martin, P. G.: 1970, submitted to *Astrophys. Space Sci.*
Monaghan, J. J.: 1967, *Z. Astrophys.* **67**, 222.
Monaghan, J. J. and Roxburgh, I. W.: 1965, *Monthly Notices Roy. Astron. Soc.* **131**, 13.
Orlov, A. A.: 1961, *Soviet Astron.-A.J.* **4**, 845.
Ostriker, J. P. and Mark, J. W.-K.: 1968, *Astrophys. J.* **151**, 1075.
Plavec, M.: 1968, *Adv. Astron. Astrophys.* Vol. 6, Academic Press, New York, p. 201.
Roxburgh, I. W., Griffith, J. S., and Sweet, P. A.: 1965, *Z. Astrophys.* **61**, 203.
Sackmann, I. J.: 1968, Ph.D. Thesis, University of Toronto.
Sackmann, I. J. and Anand, S. P. S.: 1970, *Astrophys. J.* (in press).
Stoeckly, R.: 1965, *Astrophys. J.* **142**, 208.
Takeda, S.: 1934, *Kyoto Mem.*, *A.* **17**, 197.

Discussion

Roxburgh: When we did similar calculations several years ago I do not recollect any difference between the Monaghan and Roxburgh paper and the binary work by Durney and myself. The error for the $n = 4.9$ case seems enormous; you seem to be saying that the quadrupole moment of such a centrally condensed polytrope is very large – this I find difficult to accept. Since you can make the error in the approximation technique very small for this case you should be able to calculate the structure to a high degree of accuracy. The external potential can be expanded as a power series at the fitting point and keeping terms up to and including the fourth power should give an accuracy of much better than 1 % since the fitting point is so far inside the polytrope that $r/R \ll 1$ so that $(r/R)^6 \ll$ 0.01. All coefficients should be known to a high degree of accuracy, the quadrupole moment should be very small and you should reproduce the Roche model results.

Jackson: I have also calculated some models for close binary members using the same method. They were Cowling models, which include the opacity and nuclear energy generation as simple power laws. My results agree with those presented yesterday by Dr. Thomas. They show that the reduction in luminosity is due almost entirely to rotation and that the increase in volume is due mostly to rotation, with only a small contribution from the tidal force.

ROTATIONAL CORRELATION IN BINARY STARS

RAPHAEL STEINITZ and DIANE M. PYPER

Tel Aviv University, Tel Aviv, Israel

Abstract. A method for testing the possible correlation between axial rotations of pairs of stars is developed. The test is applied to a sample of visual binaries. It is concluded that some kind of coupling between the spins of components of visual binaries does in fact exist.

1. Introduction

The study of the distribution of orbital and spin angular momenta in binary and multiple star systems is vital to the development of a theory concerning the origin and evolution of such systems. Some studies of this type have recently been carried out. For instance, Huang and Wade (1966) conclude that the orbital angular momenta of binaries are randomly oriented in our galaxy. Another example is Slettebak's (1963) determination that no significant differences between the mean rotational velocities for components of visual binaries and single stars exist. However, these analyses, while they are of prime importance concerning possible theories of origin, are not concerned with coupling between the various angular momenta in a binary system.

In this paper, we develop a theory enabling us to determine the possible existence of coupling between the spin angular momenta in binaries. In Section 5 this theory is applied to a sample of visual binaries. A method for testing similar connections between orbital and spin angular momenta will be developed in a separate paper.

2. Mathematical Theory of Rotational Correlation

Let the observed rotational velocities of the primary and secondary stars in a binary system be u_1 and u_2. We wish to express the bivariate distribution function, $F(u_1, u_2)$, of the observed rotational velocities as a function of the true distribution of rotational velocities (v_1 and v_2) and the distribution of the angle (θ) between the spin axes of the two stars, i.e., as a functional of the distribution function $G(v_1, v_2, \theta)$.

To this end, we define the following quantities:

The primary and secondary stars have spins that are characterized by unit vectors S_1 and S_2, respectively. We choose the coordinate system with the z-axis aligned along S_1 and the (x, z) plane containing S_2. As stated above, θ is the angle between S_1 and S_2. The line of sight is defined by the unit vector A; the angles between S_1 and A and between S_2 and A are i_1 and i_2, respectively (Figure 1). Now, clearly,

$$u_1 = v_1 \sin i_1$$

and

$$u_2 = v_2 \sin i_2.$$

Finally, let ϕ be the angle between the (x, z) plane and the plane containing S_1 and A.

A. Slettebak (ed.), Stellar Rotation, 165–177. All Rights Reserved
Copyright © 1970 by D. Reidel Publishing Company, Dordrecht-Holland

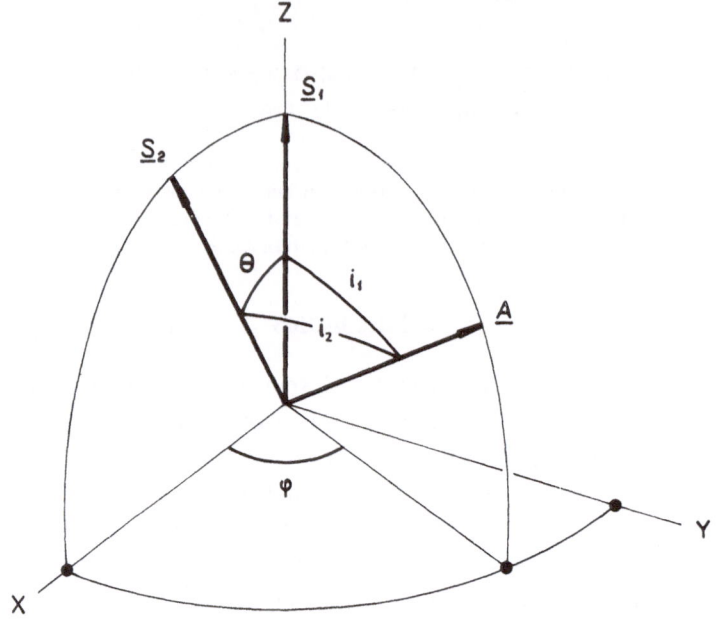

Fig. 1.   Geometry of the problem.

For the sake of brevity, we use the notation $E[q_1, q_2, \ldots q_r]$ to denote the event that $q_j \in [q_j, q_j + dq_j]$, for $j = 1, 2, \ldots, r$; and the notation $P(E[q_1, q_2, \ldots, q_r]) \, dq_1 \, dq_2 \ldots dq_r$ for the probability of E. Now, the probability of $E[v_1, v_2, \theta]$ is given by $G(v_1, v_2, \theta) \sin\theta \, dv_1 \, dv_2 \, d\theta$. The following particular cases should be considered:

(a) The angular distribution does not depend on θ. In that case,

$$G(v_1, v_2, \theta) = H(v_1, v_2).$$

(b) There is a dependence on θ, but that distribution is independent of the distribution of velocities. Then,

$$G(v_1, v_2, \theta) = H(v_1, v_2) \, g(\theta).$$

(c) The distribution of v_1 and v_2 are independent of each other, so that

$$G(v_1, v_2, \theta) = f_1(v_1) \, f_2(v_2) \, g(\theta).$$

Let the probability of $E[v_1, v_2, \theta; i_1, i_2]$ be given by

$$D(v_1, v_2, \theta; i_1, i_2) \, dv_1 \, dv_2 \, d\theta \, di_1 \, di_2.$$

This probability cannot be readily expressed in terms of θ, i_1 and i_2, since they are not wholly independent; the specification of the values of two of them will, in fact, limit the variation of the third. In order to determine $D(v_1, v_2, \theta; i_1, i_2)$, we shall first consider $E[v_1, v_2, \theta; i_1, \phi]$, since i_1, θ and ϕ are independent variables. Its probability is given by

$$P(E[v_1, v_2, \theta; i_1, \phi]) \, dv_1 \, dv_2 \, d\theta \, di_1 \, d\phi$$
$$= NG(v_1, v_2, \theta) \sin\theta \sin i_1 \, dv_1 \, dv_2 \, d\theta \, di_1 d\phi$$

where N is a normalizing factor.

In general (e.g., see Trumpler and Weaver, 1953), the transformation of probability densities from one set of variables $(q_1, ..., q_r)$ to a second set $(Q_1, ..., Q_r)$ is given by

$$P(E[Q_1, ..., Q_r]) = P(E[q_1, ..., q_r]) \cdot \left| J\left(\frac{q_1, ..., q_r}{Q_1, ..., Q_r}\right) \right|,$$

where

$$J\left(\frac{q_1, ..., q_r}{Q_1, ..., Q_r}\right)$$

is the determinant of the Jacobian matrix of the partial derivatives of $(q_1, ..., q_r)$ with respect to $(Q_1, ..., Q_r)$. Therefore,

$$P(E[v_1, v_2, \theta; i_1, i_2]) = P(E[v_1, v_2, \theta; i_1, \varphi]) \cdot \left| J\left(\frac{v_1, v_2, \theta, i_1, \varphi}{v_1, v_2, \theta, i_1, i_2}\right) \right|. \tag{1}$$

Now, as can be easily seen,

$$J\left(\frac{v_1, v_2, \theta, i_1, \varphi}{v_1, v_2, \theta, i_1, i_2}\right) = J\left(\frac{v_1, v_2}{v_1, v_2}\right) J\left(\frac{\theta, i_1, \varphi}{\theta, i_1, i_2}\right)$$

and

$$J\left(\frac{\theta, i_1, \varphi}{\theta, i_1, i_2}\right) = \frac{\partial \varphi}{\partial i_2}.$$

From the geometry (Figure 1), it follows that

$$\cos i_2 = \cos \theta \cos i_1 + \sin \theta \sin i_1 \cos \phi, \tag{2}$$

so that

$$\frac{\partial \varphi}{\partial i_2} = \frac{\sin i_2}{(Q(\theta, i_1, i_2))^{1/2}}, \tag{3}$$

where, in general,

$$Q(\alpha, \beta, \gamma) = 1 - \cos^2 \alpha - \cos^2 \beta - \cos^2 \gamma + 2 \cos \alpha \cos \beta \cos \gamma. \tag{4}$$

Therefore, from (1) and (3),

$$P(E[v_1, v_2, \theta; i_1, i_2]) = N \cdot G(v_1, v_2, \theta) \frac{\sin \theta \sin i_1 \sin i_2}{|(Q(\theta, i_1, i_2))^{1/2}|}. \tag{5}$$

In passing, it should be noted that except for the dependence on θ through $G(v_1, v_2, \theta)$ the expression for $P(E[v_1, v_2, \theta; i_1, i_2])$ is completely symmetric in θ, i_1 and i_2, as it should be.

We now express $P(E[v_1, v_2, \theta; u_1, u_2])$ in terms of $E[v_1, v_2, \theta; i_1, i_2]$:

$$P(E[v_1, v_2, \theta; u_1, u_2]) = P(E[v_1, v_2, \theta; i_1, i_2]) \cdot \left| J\left(\frac{v_1, v_2, \theta, i_1, i_2}{v_1, v_2, \theta, u_1, u_2}\right) \right|.$$

When evaluated, it is found that

$$J\left(\frac{v_1, v_2, \theta, i_1, i_2}{v_1, v_2, \theta, u_1, u_2}\right) = \frac{1}{[(v_1^2 - u_1^2)(v_2^2 - u_2^2)]^{1/2}}. \tag{7}$$

In order to obtain $F(u_1, u_2)$, it is necessary to integrate (6) over v_1, v_2 and θ. Clearly, the limits of integration for v_1 and v_2 are from u_1 to ∞ and from u_2 to ∞, respectively. It can readily be verified that $|i_1 - i_2| \leqslant \theta \leqslant i_1 + i_2$ for all values of i_1 and i_2 (Figure 1). Since we have transformed the set of variables $(v_1, v_2, \theta; i_1, i_2) \to (v_1, v_2, \theta; u_1, u_2)$, i_1 and i_2 should be expressed in terms of the new variables. The limits of integration over θ are therefore given by

$$\begin{aligned} \Theta_- &= |\arcsin(u_1/v_1) - \arcsin(u_2/v_2)| \\ \Theta_+ &= \arcsin(u_1/v_1) + \arcsin(u_2/v_2). \end{aligned} \tag{8}$$

Due to the above transformation, it is also necessary to rewrite $Q(\theta, i_1, i_2)$ as follows:

$$Q(\theta, i_1, i_2) = R\left(\theta, \frac{u_1}{v_1}, \frac{u_2}{v_2}\right) \equiv \sin^2\theta + \left(\frac{u_1}{v_1}\right)^2$$
$$+ \left(\frac{u_2}{v_2}\right)^2 + 2\left\{\cos\theta\left(\left[1 - \left(\frac{u_1}{v_1}\right)^2\right]\left[1 - \left(\frac{u_2}{v_2}\right)^2\right]\right)^{1/2} - 1\right\}. \tag{9}$$

After we make the appropriate substitutions, we finally obtain:

$$\frac{F(u_1, u_2)}{u_1 u_2} = N \int_{u_1}^{\infty} \int_{u_2}^{\infty} \int_{\Theta_-}^{\Theta_+} \frac{G(v_1, v_2, \theta)}{v_1 v_2}$$
$$\times \frac{\sin\theta \, d\theta \, dv_1 \, dv_2}{\left[R\left(\theta, \frac{u_1}{v_1}, \frac{u_2}{v_2}\right)\right]^{1/2} [(v_1^2 - u_1^2)(v_2^2 - u_2^2)]^{1/2}}. \tag{10}$$

When we consider case (b) above, (10) becomes:

$$\frac{F(u_1, u_2)}{u_1 u_2} = N \int_{u_1}^{\infty} \int_{u_2}^{\infty} \frac{H(v_1, v_2)}{v_1 v_2}$$
$$\times \left[\int_{\Theta_-}^{\Theta_+} \frac{g(\theta)\sin\theta \, d\theta}{\left[R\left(\theta, \frac{u_1}{v_1}, \frac{u_2}{v_2}\right)\right]^{1/2}}\right] \frac{dv_1 \, dv_2}{[(v_1^2 - u_1^2)(v_2^2 - u_2^2)]^{1/2}}; \tag{11}$$

while in case (c) we have:

$$\frac{F(u_1, u_2)}{u_1 u_2} = N \int_{u_1}^{\infty} \int_{u_2}^{\infty} \frac{f_1(v_1)}{v_1 (v_1^2 - u_1^2)^{1/2}} \cdot \frac{f_2(v_2)}{v_2 (v_2^2 - u_2^2)^{1/2}}$$
$$\times \left[\int_{\Theta_-}^{\Theta_+} \frac{g(\theta)\sin\theta \, d\theta}{\left[R\left(\theta, \frac{u_1}{v_1}, \frac{u_2}{v_2}\right)\right]^{1/2}}\right] dv_1 \, dv_2; \tag{12}$$

and under the special circumstance that $g(\theta) \equiv 1$ (case (a) above), we see that

$$\int_{\Theta_-}^{\Theta_+} \frac{\sin\theta \, d\theta}{\left[R\left(\theta, \frac{u_1}{v_1}, \frac{u_2}{v_2}\right)\right]^{1/2}} = \left[- \arcsin\left\{\frac{v_1 v_2}{u_1 u_2}\right.\right.$$

$$\times \left[\cos\theta - \left(1 - \left(\frac{u_1}{v_1}\right)^2\right)^{1/2} \left(1 - \left(\frac{u_1}{v_2}\right)^2\right)^{1/2}\right]\right\}\right]_{\Theta_-}^{\Theta_+} = \pi. \quad (13)$$

It is therefore evident that $N = \pi$.

Suppose we consider pairs of stars formed at random from a sample of single stars; then

$$\frac{F(u_1, u_2)}{u_1 u_2} = \int_{u_1}^{\infty} \frac{f_1(v_1) \, dv_1}{v_1 (v_1^2 - u_1^2)^{1/2}} \cdot \int_{u_2}^{\infty} \frac{f_2(v_2) \, dv_2}{v_2 (v_2^2 - u_2^2)^{1/2}}, \quad (14)$$

or,

$$F(u_1, u_2) = F_1(u_1) \cdot F_2(u_2), \quad (15)$$

as is to be expected; i.e., the observed bivariate distribution function is the product of the observed univariate distributions.

If members of binary systems were rotationally uncorrelated, it would be found that $F(u_1, u_2)$ would be represented by (15). However, if this is not found to be the case, it is, at least *a priori*, impossible to distinguish between the most general case $G(v_1, v_2, \theta)$ and case (b), where $G(v_1, v_2, \theta) = H(v_1, v_2) g(\theta)$.

3. Choice of a Relevant Sample

The choice of a relevant sample for testing the idea of rotational correlation will be determined by physical factors as well as considerations of observational selection effects.

Some very short-period spectroscopic binaries are interpreted as having synchronous rotation (see Struve, 1950), i.e. the rotational period of the components is equal to their orbital period. This is explained in terms of tidal forces, which are supposed to have brought about the synchronization. In some other close binaries, the period of the axial rotation of at least the primary is undoubtedly shorter than the orbital period of the pair. It has been suggested that this could be due to mass transfer between the components (e.g., see Batten, 1967). In either case, it would seem that the initial conditions which existed during the first phases of the evolution of the system no longer prevail.

If, on the other hand, visual binaries are not the result of the disruption of close systems (Chandrasekhar, 1944; Ambartsumian, 1937), and both components are on the main sequence, we cannot see any reason why such systems should not reflect now the conditions which prevailed during their formation period.

Observational selection effects should also be considered. Spectra of both components of a spectroscopic pair are visible only if their magnitude difference, Δm, is less than about one magnitude (Struve, 1950). This rather severe restriction does not apply in the case of visual binaries. As can be seen from Table I (Bečvář, 1964) the range in Δm for visual binaries is certainly larger than in spectroscopic binaries.

TABLE I

Magnitude difference and separation in visual binaries

Star	m_1	m_2	Sep ($''$)
66 Ari	6.11	12.2	1.0
β^1 Tuc	4.52	14.0	2.2

The actual value of Δm allowing one to recognize a visual binary as such depends in general on the telescope aperture and the apparent separation of the components. In a survey of rotational velocities, however, Δm is only limited by the instrumentation and by seeing effects, not by the physical characteristics of the pair. Additional selection effects are present when spectroscopic binaries are considered. Accurate values of rotational velocities cannot be measured in double-lined systems if the lines of the two components are too badly blended. There results a bias in favor of systems having large velocity amplitudes; i.e., large masses, short periods, or both. Moreover, this effect becomes larger as the line width increases.

Due to the arguments presented above, it is concluded that a sample of visual binaries would be the most meaningful one as a means of testing for rotational correlation in binaries.

4. The Data

The only large amount of data available to us concerning rotational velocities in visual binaries is a study of 116 pairs by Slettebak (1963).

Since the lower limit for the values of $v \sin i$ in this investigation is 25 km/sec, a meaningful discussion of the slowly-rotating late-type stars is impossible. We therefore have excluded all pairs having components of spectral type F or later. Giants and supergiants were rejected also on the grounds that they presumably no longer possess their original rotational velocities, due to their increase in radius and to possible mass loss.

From the list, therefore, all pairs which satisfy the following conditions were selected:

(a) The spectral types of both components are earlier than A9.

(b) Both components are on the main sequence.

When these criteria are applied, a sample of 50 pairs is obtained. The total range of spectral types is from O8 to A9. Visual binaries having components which are also single-lined spectroscopic binaries are included in the sample, as well as Ap and Am stars.

5. Statistical Analysis

A. OBSERVED VELOCITY CORRELATION

Table II shows the relative frequency distribution function, $F(u_1, u_2)$, of observed rotational velocities of the sample discussed in Section 4. The observed rotational velocities of the primary and secondary are u_1 and u_2, respectively. Due to the smallness of the sample, the velocity range has been divided into 50 km/sec sub-intervals.

TABLE II

Bivariate distribution of rotational velocities in visual binaries, $F(u_1, u_2)$

u_1 \ u_2	50	100	150	200	250	300		$\phi_2(u_2)$	$\bar{u}_1(u_2)$
0	0.12	0.02	0.02	0.02			0.02	0.20	85
50	0.04	0.10	0.06	0.04				0.24	96
100	0.04	0.04	0.04	0.04	0.02	0.02	0.02	0.22	148
150	0.02	0.04	0.02	0.04				0.12	108
200	0.02	0.02	0.02	0.02	0.02	0.02	0.02	0.14	175
250				0.04			0.02	0.06	225
300							0.02	0.02	325
$\phi_1(u_1)$	0.24	0.22	0.16	0.20	0.04	0.04	0.10		
$\bar{u}_2(u_1)$	79	111	112	155	175	175	195		

The marginal distribution, $\phi_1(u_1)$, gives the distribution of observed rotational velocities of primaries independent of the secondaries. The corresponding marginal distribution for the secondaries is $\phi_2(u_2)$. From these, another array (not shown), $\Phi(u_1, u_2) \equiv \phi_1(u_1) \cdot \phi_2(u_2)$ is derived. This array represents a bivariate distribution of observed velocities of pairs of stars which are formed by matching to each primary a secondary at random, out of the sample of selected binaries. The array $\Phi(u_1, u_2)$ represents, therefore, a distribution of two independent variables. Finally, the array of differences between the true distribution and the artificial one, $\Delta(u_1, u_2) \equiv F(u_1, u_2) - \Phi(u_1, u_2)$ is given in Table III. It should be noted that there is a marked tendency for positive differences to be found along the main diagonal.

According to Trumpler and Weaver (1953) there are various ways of testing correlation between variables in a bivariate distribution. As a first step, the correlation coefficient, r, was computed for $F(u_1, u_2)$. It was found that $r = 0.46$, while the correlation coefficient for the artificial distribution $\Phi(u_1, u_2)$ gives $r = 0.001$, as would be expected.

As a second step, the regression curves $\bar{u}_1(u_2)$ and $\bar{u}_2(u_1)$ were computed. These

TABLE III

Differences: $\Delta(u_1, u_2) = F(u_1, u_2) - \phi_1(u_1) \cdot \phi_2(u_2)$

u_1 \ u_2		50	100	150	200	250	300
	0.072	−0.024	−0.012	−0.020	−0.008	−0.008	0.000
50	−0.018	0.047	0.022	−0.008	−0.010	−0.010	−0.024
100	−0.013	−0.008	0.005	−0.004	0.011	0.011	−0.002
150	−0.009	0.014	0.001	0.016	−0.005	−0.005	−0.012
200	−0.014	−0.011	−0.002	−0.008	0.014	0.014	0.006
250	−0.014	−0.013	−0.010	0.028	−0.002	−0.002	0.014
300	−0.005	−0.004	−0.003	−0.004	−0.001	−0.001	0.018

curves give the variation of the conditional mean velocity of the primaries as a function of the velocity interval of the secondary and vice-versa. In terms of the velocity intervals, $u_1(i)$ and $u_2(j)$, the regression curves are defined to be:

$$\bar{u}_1\left(u_2(j)\right) = \sum_{i=1}^{7} u_1(i)\, F\left(u_1(i), u_2(j)\right) \Big/ \sum_{i=1}^{7} F\left(u_1(i), u_2(j)\right)$$

$$\bar{u}_2\left(u_1(i)\right) = \sum_{j=1}^{7} u_2(j)\, F\left(u_1(i), u_2(j)\right) \Big/ \sum_{j=1}^{7} F\left(u_1(i), u_2(j)\right). \tag{16}$$

The regression curves are graphically represented in Figures 2 and 3.

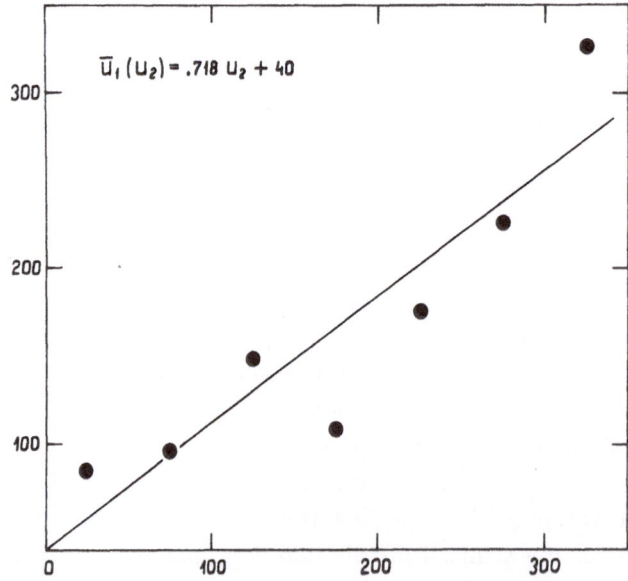

$$\bar{u}_1(u_2) = .718\, u_2 + 40$$

Fig. 2.   Regression of the mean rotational velocities of primaries (*ordinate*) as a function of the rotational velocity interval of secondaries (*abscissa*).

It can be seen that definite trends are present, in the sense that the mean observed rotational velocity of one of the components is an increasing function of the rotational velocity of the other. In fact, it appears that the relationship may be linear. A least squares solution for a straight-line fit to each regression curve is also shown.

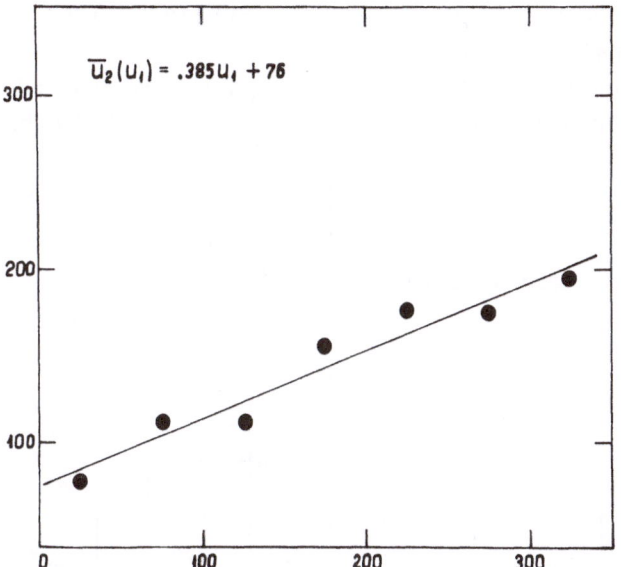

Fig. 3. Regression of the mean rotational velocities of secondaries (*ordinate*) as a function of the rotational velocity interval of primaries (*abscissa*).

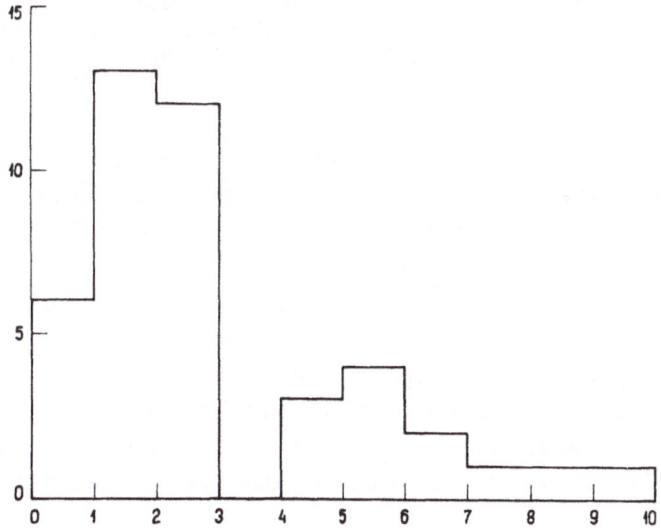

Fig. 4. Distribution of spectral type intervals (differences in subclasses) in the sample of binaries.

B. A SELECTION EFFECT

It appears that the sample selected for the statistical analysis is heavily biased in favor of pairs having similar spectral types. A histogram (Figure 4) of the distribution of spectral type intervals (the difference between subclasses) indicates that most intervals are less than or equal to three subclasses. It may be suspected, therefore, that the correlations discussed in the previous section are due to this fact and have nothing to do with rotational correlation as such.

Let $F(u)$ denote the distribution of observed rotational velocities of stars in the sample (each component considered as a single star) and let $P(|\Delta u|)$ be the probability

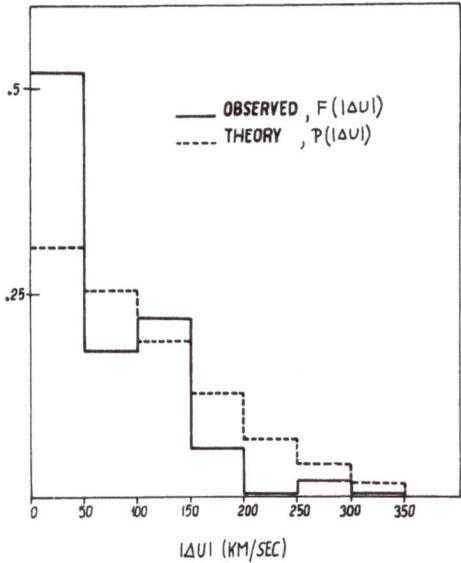
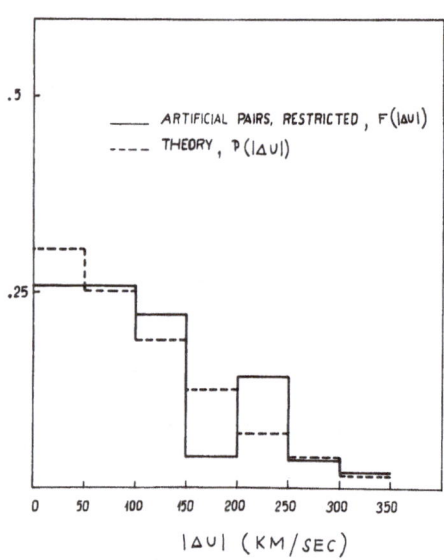

Fig. 5. Comparison of the distribution of rotational velocity differences in real pairs to the theoretical distribution (autocorrelation function).

Fig. 6. Comparison of the distribution of rotational velocity differences in restricted artificial pairs to the theoretical distribution.

that two stars chosen at random have an observed rotational velocity difference $\Delta u = u_1 - u_2$. This probability is given by the autocorrelation function of $F(u)$.

Figure 5 shows $P(|\Delta u|)$ as well as the distribution of the differences $|\Delta u|$ in the real pairs. It is evident that the latter distribution is more sharply peaked towards small values of Δu, compared to the theoretical distribution $P(|\Delta u|)$.

From the sample of binaries, new pairs were matched at random but with the restriction that their difference in spectral subclasses is $\leqslant 2$ (i.e., a restriction more severe than on the real pairs). The distribution of the differences $|\Delta u|$ in these pairs is compared to the theoretical distribution $P(|\Delta u|)$ in Figure 6. As can be seen, there is an appreciably closer correspondence between these last two distributions than those in Figure 5.

Following Slettebak (1963), the mean velocities of rotation of various subclasses of

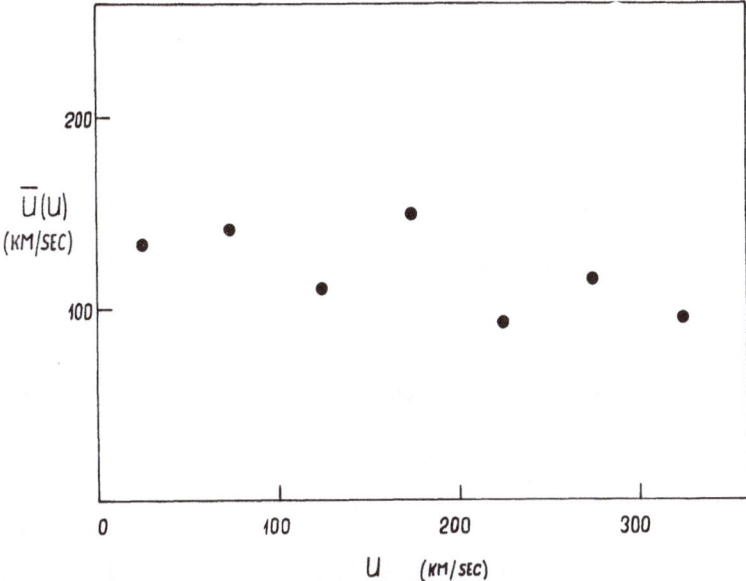

Fig. 7. Regression of the mean rotational velocities of components in artificial restricted pairs (*ordinate*) as a function of the rotational velocity interval of the other component (*symmetrized distribution*).

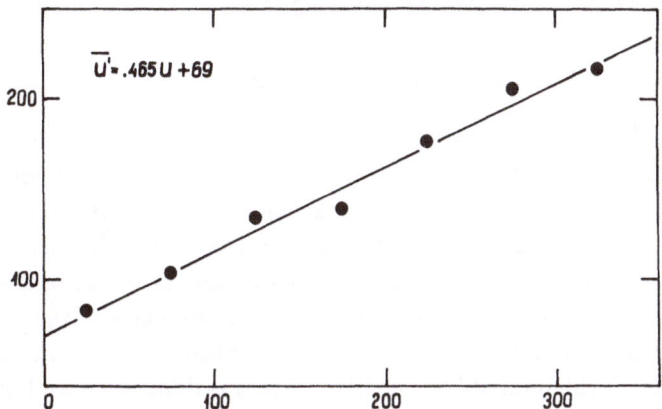

Fig. 8. Regression of the mean rotational velocities of components of true binaries (*ordinate*) as a function of the rotational velocity interval of the other component (*symmetrized distribution*).

the sample were derived (Table IV). From the small range in $\bar{u}$ (the mean velocity), it seems clear that a regression of one component of a binary on the other, i.e., the conditional mean velocities, cannot vary over a large range either, irrespective of the fact that their spectral types are similar. In order to check this, an analysis similar to the one carried out on the true binaries (see Subsection 5A) was performed on the artificial pairs restricted as to their difference in spectral type.

In order to obtain a more homogeneous sample, the distribution $F(u_1, u_2)$ was symmetrized, so that it is immaterial which star is the primary and which the secondary.

TABLE IV

Average velocity for various spectral
types in the sample of visual binaries

Spectral type	No. stars	$\bar{u}$
B0–B3	15	158
B5–B7	8	148
B8–A2	47	158
A3–A7	14	146

For comparison purposes, the same symmetrization was applied to the sample of artificial binaries.

The results are:

(1) The correlation coefficient is $r = -0.11$ compared to $r = 0.45$ for the real distribution.

(2) The regression $\bar{u}(u)$ of the artificial sample and that of the real sample are represented in Figures 7 and 8, respectively. The difference is evident.

In view of the results obtained in this section, we conclude that the regressions shown in Figures 2, 3, and 8 cannot be only the result of the fact that the visual binaries in our sample are heavily biased in favor of pairs having similar spectral types.

6. Concluding Remarks

If fission (Jeans, 1929; Roxburgh, 1966) is to account for the formation of binary systems, then one would in fact expect a high degree of rotational correlation to be present in binaries. In that sense our findings lend support to this theory.

However, it is impossible at this stage of the development of the theory of the origin of binaries to use our findings as a test case to support or disprove the capture and separate nuclei theories (see, e.g., Batten, 1967 for a summary of these theories).

It has been claimed that in some systems synchronization between axial and orbital rotations is present (Shajn and Struve, 1929), but it is difficult to see how this can be ascertained without knowing *a priori* that spin axes of members of binaries are in fact perpendicular to the orbital plane. Progress in this direction could be achieved in the following way:

Let $f_s(v_{sb})$ be the distribution of *true* rotational velocities of members of spectroscopic binaries, regarded as single stars, and let $F_{eb}(u)$ be the *observed* rotational velocities of members of *eclipsing binaries*.

If it is true that the spin axes are parallel to the orbital angular momenta, we should find that

$$F_{eb}(u) = f_s(v_{sb}).$$ (18)

As usual, we emphasize the need for more observational data. This would permit a more extended analysis to be carried out.

As a final remark, it should be pointed out that the idea of correlation in binary systems can be generalized to any property p measured in a quantitative way by testing whether or not

$$F(p_1, p_2) = f_1(p_1) \cdot f_2(p_2) \tag{19}$$

where $F(p_1, p_2)$ is the bivariate distribution function and $f_1(p_1)$ and $f_2(p_2)$ are the respective distributions for the components regarded as single stars.

References

Ambartsumian, V. A.: 1937, *Russian Astron. J.* **14**, 207.

Batten, A. H.: 1967, *Ann. Rev. Astron. and Astrophys.* **5**, 25.

Bečvář, A.: 1964, *Atlas Coeli Katalog II*, Československé Akadamie Věd, Prague.

Chandrasekhar, S.: 1944, *Astrophys. J.* **99**, 54.

Huang, S.-S. and Wade, Clarence Jr.: 1966, *Astrophys. J.* **143**, 146.

Jeans, J. H.: 1929, *Astronomy and Cosmogony*, Cambridge University Press, Cambridge.

Roxburgh, I. W.: 1966, *Astrophys. J.* **143**, 111.

Shajn, G. and Struve, O.: 1929, *Monthly Notices Roy. Astron. Soc.* **89**, 222.

Slettebak, A.: 1963, *Astrophys. J.* **138**, 118.

Struve, O.: 1950, *Stellar Evolution*, Princeton University Press, Princeton, N.J.

Trumpler, R. J. and Weaver, H. F.: 1953, *Statistical Astronomy*, University of California Press, Berkeley, Calif. U.S.A.

THE ROTATION OF THE MAIN-SEQUENCE COMPONENTS OF ALGOL-TYPE SEMI-DETACHED ECLIPSING BINARIES*

E. P. J. VAN DEN HEUVEL[†]

Lick Observatory, University of California, Santa Cruz, Calif., U.S.A.

Abstract. Newly determined rotational velocities of the main-sequence components of 14 Algol-type semi-detached systems and of 2 detached systems are presented. Combination of these data with the existing data on the rotation of the components of semi-detached systems shows that (i) in systems with primaries of spectral type B8 or later and with $P<5$ days, deviations from synchronism between rotation and revolution are small in 14 out of 15 cases. The average rotational velocity of the primaries in such systems is 75 km/sec, viz. only 40% of the average rotational velocity of single main-sequence stars in the same spectral region: (ii) primaries of spectral type earlier than B8 in systems with short as well as long periods tend to rotate more than twice as fast as one would expect from synchronism. A tentative explanation for these results is presented.

In close binaries with periods less than about 4 days and with both components on the main sequence the axial rotation of the components is generally observed to be synchronised with the orbital revolution (Plaut, 1959; Olson, 1968). However, in semi-detached Algol-type systems with periods of less than 4 days such a synchronism is not always observed; for instance in the systems of U Cephei (Struve, 1944, 1963) and RZ Scuti (Hansen and McNamara, 1959) the main-sequence components rotate much faster than would be expected in case of synchronism.

Algol-type semi-detached (s-d) systems are expected to be the products of extensive mass exchange in evolving close binaries. During such a mass exchange both the binary period and the rotational angular momentum of the components may change. In the computations of the evolution of close binaries one generally assumes for computational reasons that no exchange between orbital angular momentum of the exchanged matter and rotational angular momentum of the components occurs. However, the above mentioned observations of non-synchronism, as well as computations of particle trajectories in close binary systems (cf. Kruszewski, 1966) show that such an exchange of angular momentum probably takes place in nature. One therefore does not a priori expect the components of s-d systems to rotate synchronously with the orbital motion, especially not during or just after the stage of mass exchange. One may then wonder whether the deviations from synchronism will last for a long time after the mass exchange is terminated, or whether tidal (re-)synchronisation will occur in a fairly short period of time. One may try to answer such a question by examining the rotation of primaries of a random sample of s-d systems, and see how large the percentage of synchronous rotators is. Since the random sample probably will show a random distribution of ages (following the mass exchange stage) the fraction of synchronised primaries in the random sample is then expected to be equal

---

* Contributions from the Lick Observatory, No. 305.
† On leave of absence from Sterrewacht Sonnenborgh, Utrecht, The Netherlands, July 1, 1968 – September 1, 1969.

A. Slettebak (ed.), Stellar Rotation, 178–186. All Rights Reserved
Copyright © 1970 by D. Reidel Publishing Company, Dordrecht-Holland

TABLE Ia

The 16 Algol-type systems for which the rotational velocities of the primaries were obtained in this investigation [a]

No. in SB Catalogue	Name	Spectrum	m_v	Period (days)	v (km/sec)	v_{syn}	$f(m)$ or m_2/m_1
8	TV Cas	A0	7.0 – 7.9	1.8126	70	60	0.128
73	RY Per	B4 + gF5	8.5 –10.3	6.8636	280	64	0.015
–	IZ Per	B5 (?)	8.0 – 9.2	3.6877	185	37	–
139	RZ Eri	Am (?) + G8	8.0 – 9.2	39.2807	45	1.8	0.06
426	TW Dra	A6 + G5	7.5 – 9.8	2.8067	50	34	0.0744
453	W UMi	A3	8.5 – 9.6	1.7011	75	61	0.11
563	Z Vul	B3, 4 V + A2, 3 IV	7.9 – 9.3	2.4549	195	103	m_2/m_1 = 0.42
578	V505 Sgt	A1 V + F6	6.5 – 7.7	1.1829	102	91	0.137
589[b]	V477 Cyg	A3 + F5	8.5 – 9.4	2.3470	40	44	m_2/m_1 = 0.7
625	W Del	A0e + gG5	9.5 –12.5	4.8060	30	22	0.013
637	S Equ	B8 V	8.2 –10.1	3.4360	52	36	0.0043
664	EK Cep	A0	8.0 – 9.4	4.4277	< 20	24	0.160
671	AW Peg	A5e + F5 IV	7.5 – 8.2	10.6225	120	9.7	0.16
677[b]	CM Lac	A2	8.1 – 9.1	1.6047	55	44	m_2/m_1 = 0.75
722	Y Psc	A3 + K0	9.0 –12.0	3.7659	37	27	0.019
725	XX Cep	A8	8.5 – 9.6	2.3373	47	34	0.005

[a] The first column gives the number of the system in the 'Sixth Catalogue of Orbital Elements of Spectroscopic Binaries' (Batten, 1967). The last column lists the mass ratio or the mass function.
[b] Detached.

E. P. J. VAN DEN HEUVEL

TABLE Ib

Semi-detached systems for which rotational velocities of primaries were found in the literature [a]

No. in SB catalogue	Name	Spectrum	m_v	Period (days)	v (km/sec)	v_{syn}	$f(m)$ or m_2/m_1
28	U Cep	B8	7.0 – 9.8	2.4929	200–300	100	0.16
75	RZ Cas	A2	6.0 – 7.7	1.1953	82	63	0.023
84	β Per	B8	2.0 – 3.3	2.8673	65	50	0.0255
84	β Per	B8	2.0 – 3.3	2.8673	50 [b]	50	0.0255
111	λ Tau	B3 V + A4 IV	3.4 – 3.9	3.9529	75	77	0.075
409	δ Lib	A0 V + sgG	4.9 – 5.9	2.3273	78	76	0.115
465	AI Dra	A0 + G2	7.0 – 8.1	1.1988	99	86	0.099
513	RZ Scuti	B2	7.5 – 8.7	15.1902	250	18	0.0766
513	RZ Scuti	B2	7.5 – 8.7	15.1902	240 [b]	18	0.0766
557	RS Vul	B5 + G	7.0 – 8.0	4.4773	90	46	$m_2/m_1 = 0.31$
559	U Sge	B9n + gG2	6.5 –10.1	3.3862	76	88	0.119

[a] The first column gives the number of the system in the 'Sixth Catalogue of Orbital Elements of Spectroscopic Binaries' (Batten, 1967). The last column lists the mass ratio or the mass function.
[b] As determined in this investigation

to the fraction of the lifetime of an Algol-type s-d system (after the mass exchange stage) during which its primary rotates synchronously.

Inspection of the literature showed us that for only nine s-d systems are accurate rotational velocities, determined from spectra with dispersions of about 20 Å/mm or better, available. The data for these systems are given in Table Ib. The rotation of Algol was determined by Slettebak (1954); the values for U Cephei and RZ Scuti were taken from the above mentioned sources: the data for the six other stars were given by Olson (1968). Since this sample is rather small it was decided to study the rotation of the 16 systems in Table Ia. Apart from the systems already studied by others, these are practically all Algol-type systems in the 'Sixth Catalogue of Orbital Elements of Spectroscopic Binaries' (Batten, 1967) brighter than 9^m5 at maximum and accessible at Lick Observatory during the late spring and early summer. It was later found that the systems of CM Lac and V477 Cyg are not s-d, hence 14 s-d systems in our sample remain.

Of each system, as well as of Algol and RZ Scuti, at least two spectra were obtained. We used the 20″ Schmidt camera of the coudé spectrograph of the 120″ telescope, with a dispersion of 16 Å/mm at 4300 Å. The widening used in general was 0.5 mm, although some plates were taken with 0.35 mm widening. In addition spectra of 20 bright A and B stars with known rotational velocities (Slettebak, 1949, 1954, 1955) were taken, which served as standards. All spectra were taken on baked Kodak IIa-O plates. Direct intensity tracing was made of all plates. For the determinations of the rotational velocities the line MgII 4481 Å was used for spectral types later than B8. For spectral type B8 and earlier the line HeI 4026 Å was used. For spectral type B8 both lines were used. The rotational velocities were determined in a way somewhat similar to the one used by Abt and Jewsbury (1969), viz. curves of $v \sin i$ vs. halfwidth and central depth of the standard lines were constructed for the standard stars. The calibration of the halfwidth of MgII 4481 against Slettebak's rotational velocities indicates a p.e. per star of about 10% in $v \sin i$. The central depth gives a less well defined relation, with a p.e. about twice as large. Measurements of the half-width and central depth of the standard lines in the spectra of the Algol-type systems then give 2 values of $v \sin i$. Thirdly, direct visual comparisons were made between the line profiles for the standard stars and the program stars, from which another value of $v \sin i$ was obtained. A similar procedure was followed with respect to the HeI 4026 line. Here the p.e. of the $v \sin i$ values derived from the halfwidth relation is about 15%, and of the central depth relation about 20%.

The finally adopted rotational velocity for each star is the average of the 6 values obtained from the 2 plates per star; in determining this average, the values obtained from the central depth were given half weight only.

From the scatter in the 6 values of $v \sin i$ for each star we estimate the internal p.e. in the final average to be about $\pm 10\%$. Table I lists the finally adopted values of the rotational velocity v; the inclinations i were taken from the best available orbital studies. The table furthermore lists the velocity v_{syn} expected in the case of synchronism. For the computation of this velocity we used the best available determination

TABLE II

The spectral type vs. radius relation adopted
for systems for which no accurate radius
determinations were available

Spectral type	Radius (in $R_\odot$)
A5	1.87
A3	1.92
A0	2.00
B8	2.50
B5–6	3.3
B4–5	4.0
B3	4.7
B2	5.2

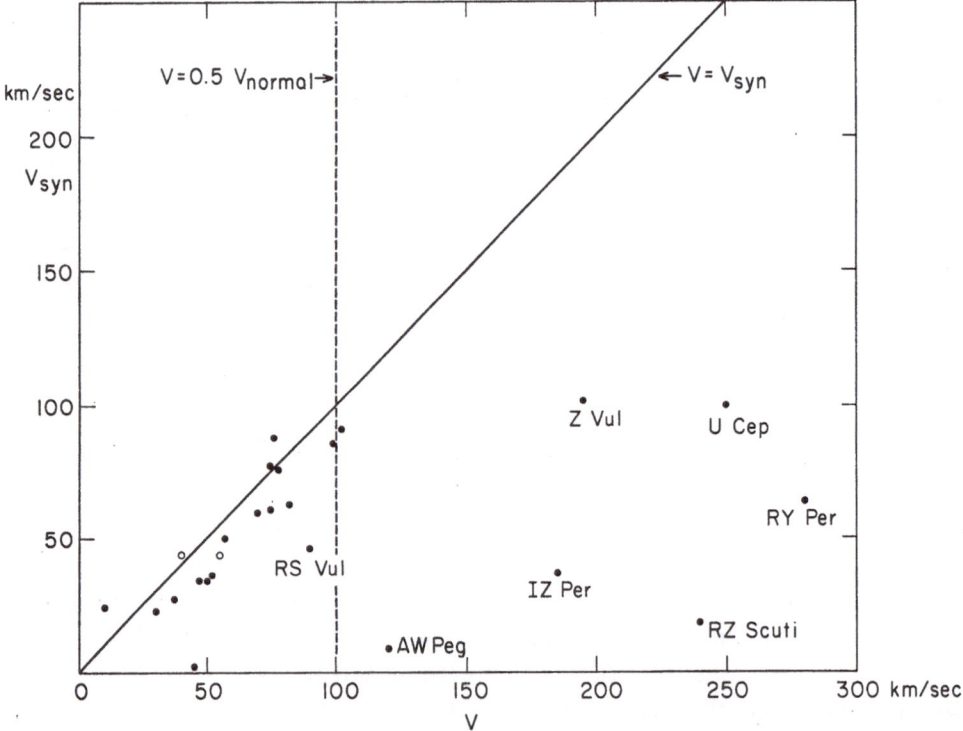

Fig. 1. Plot of the rotational velocities v of primaries of the Algol-type systems from Table I against the velocities v_{syn} expected in case of synchronism between rotation and revolution. Filled circles represent semi-detached systems; open circles indicate the 2 detached systems.

of the radius from photometric and spectroscopic investigations. In cases in which no value for the radius was available we used the main-sequence relation for radius vs. spectral type given by Harris *et al.* (1963) as specified in Table II.

In Figure 1 v is plotted vs. v_{syn}. The non-semi-detached systems are indicated by open circles. The figure and the table show that within the observational uncertainties, in about 67% of all cases rotation and revolution in s-d systems are synchronised. There seems to be a slight tendency to rotate about 10 km/sec faster than synchronous, although it is not certain whether this is a real effect. It is interesting to notice that all non-synchronous primaries in systems with $P < 5$ days have spectral type B8 or earlier. Systems with $P < 5$ days and primaries of spectral type B8 or later show synchronous rotation in 14 out of 15 cases. The systems RZ Eri and AW Peg, which both have $P > 10$ days, have primaries which still rotate much more slowly than single A-or B-type main-sequence stars. The table shows that the mean rotational velocity of primaries of type B8 or later in s-d systems is about 75 km/sec, which is about 40% of the average value for single late B and early A stars.

Contrary to what is observed for spectral type B8 and later, the primaries of type earlier than B8 rotate on the average more than twice as fast as expected in case of synchronism. λ Tauri is the only synchronous rotator in this group of 6.

These results seem to indicate that in systems with primaries of spectral type B8 or later, and with periods less than 5 days, synchronism is restored quite rapidly after the mass exchange stage, whereas in systems with earlier type primaries resynchronisation might take more than three quarters of the lifetime of the system, after the stage of mass exchange.

The interpretation of this difference in rotational behaviour may perhaps be found in the way in which these systems were produced. For the types of evolution of close binary systems we will use here the notation by Kippenhahn and Weigert (1967), viz.: case A for evolution in which the primary already fills its Roche lobe before the end of core hydrogen burning, case B for the evolution in which the primary fills the Roche lobe after the end of core hydrogen burning but before the onset of helium burning. From the work of Kippenhahn et al. (1967) and of Refsdal and Weigert (1969) it is known that Algol-type s-d systems can be produced either by case B evolution with initial primary mass smaller than about 2.8 $M_\odot$, or by case A evolution for any value of the initial primary mass. The primary masses of the Algol-type systems produced in case B are not expected to be much larger than 4 $M_\odot$; for an average initial mass ratio of primary and secondary of 2.0, and an initial primary mass of 2.8 $M_\odot$, the mass of the new primary after the mass exchange is expected to be about 3.8$M_\odot$ (cf. Kippenhahn et al., 1967), which is about the mass of a B8 main-sequence star. For this reason, Algol-type systems produced by evolution according to case B are expected to have in general only primaries of spectral type B8 or later.

On the other hand, although case A can in principle occur for any spectral type, it is most common among stars earlier than about B8, as Table III shows. The left-hand side of this table shows the distribution of periods of unevolved B-type spectroscopic binaries as given in the Sixth Catalogue (Batten, 1967); the right-hand side of the table shows the limiting periods $P_{max\ H}$ and $P_{max\ He}$ defined as follows. In systems with $P < P_{max\ H}$ the primary already fills the Roche lobe before the end of core hydrogen burning; in systems with $P < P_{max\ He}$ it fills the Roche lobe before the onset of core helium burning. Consequently, systems with $P < P_{max\ H}$ are in case A, and systems with $P_{max\ H} \leqslant P < P_{max\ He}$ are in case B. Comparison of the left- and right-hand side of

TABLE III

Left: the distribution of periods of unevolved B-type spectroscopic binaries in the 'Sixth Catalogue of Orbital Elements of Spectroscopic Binaries' (Batten 1967). Right: the periods $P_{\max H}$ and $P_{\max He}$ (defined in the text) for three stellar masses, for an original mass ratio of secondary and primary of 0.5.

P (days)	B0-B5	B6-B9.5	M	$P_{\max H}$	$P_{\max He}$
<1	0.9%	2.6%			
$1-1.5$	7.9	7.9	$3M_\odot\ (\sim B8)$	$0^d.99$	$35^d.4$
$1.5-2$	9.6	7.9			
$2-3$	21.1	15.8			
$3-9$	28.9	31.6	$5M_\odot\ (\sim B5)$	$1^d.36$	$92^d.2$
$9-27$	17.6	21.0			
$27-81$	7.0	7.9	$9M_\odot\ (\sim B2)$	$2^d.00$	$406^d.4$
>81	7.0	5.3			

Table III shows that case A occurs among about 18% of the early B stars, but among less than 10% of the later B systems. In view of the decrease of $P_{\max H}$ with stellar mass, it occurs among only a few per cent of the A stars (which have about the same period distribution for spectroscopic binaries as the later B stars). For this reason, case A is quite common among unevolved spectroscopic binaries of spectral type earlier than about B8, but is quite rare at later types. For this reason one expects most of the s-d systems with primaries of spectral type B8 or later to be the products of case B evolution, whereas systems with primaries of spectral type earlier than about B8 are expected to be mostly the products of case A.

Why then do the products of case A evolution rotate much faster than the products of case B evolution? The reason might be that in case B most of the mass-exchange takes place in a relatively short time, and there is in general no long lasting stage of slow but considerable mass exchange (Kippenhahn et al., 1967; Refsdal and Weigert, 1969). In this case, therefore, once the mass exchange is over, tidal forces can start their synchronising work. On the other hand, in case A only a relatively small amount of mass is transferred rapidly, followed by a long-lasting stage of slow mass exchange (Plavec et al., 1968). Hence, in this case the surface layers of the primary always consist of newly accreted material, presumably of high angular momentum, which did not yet have time to adjust itself to synchronism. We suggest that this may be the reason for the different rotational behaviour of systems with primaries of spectral type earlier and later than about B8.

One may wonder what will happen to the systems with primaries of spectral type later than about B8, after the subgiant component has evolved into a white dwarf. According to the work of Kippenhahn et al. (1967) and of Refsdal and Weigert (1969) the subgiant has a lifetime of only about 15 to 20% of the remaining lifetime of the primary. One therefore expects this main-sequence primary to have a white dwarf companion during 80% or more of its lifetime after the mass exchange. From the fact that, after due allowance for inclination effects, some 2.5% of the A and B stars are primaries in Algol-type s-d systems (Van den Heuvel, 1969a) one expects some 10–12.5% of the A and B stars to have been primaries in s-d systems, and to have a white dwarf companion now. The fact that for spectral type later than about B8 the primaries of Algol-type s-d systems are slow rotators indicates then, that one expects the 10–12.5% of the late B and A stars which have white dwarf companions, to be relatively slow rotators. This might be the explanation for the observed occurrence of an overabundance of slow rotators among the early A stars (Conti, 1965; Deutsch 1967). Similarly, one expects those blue stragglers in galactic clusters that are evolved spectroscopic binaries to have rotational velocities comparable to those of the stars in Table I. Indeed the A-type blue stragglers in M67 studied by Deutsch (1966) have rotational velocities not larger than about 100 km/sec and the blue straggler 40 Cancri in Praesepe is a very slow rotator (Conti *et al.*, 1965). (The position of the blue straggler 68 Tauri in the colour magnitude diagram of the Hyades can be explained by ordinary evolution of a single star (Van den Heuvel, 1969b).)

Acknowledgement

This research has been supported in part by the ONR under Contract Number N00014-69-C-0155.

References

Abt, H. and Jewsbury, C. P.: 1969, *Astrophys. J.* **156**, 983.
Batten, A.: 1967, 'Sixth Catalogue of Orbital Elements of Spectroscopic Binaries', *Publ. Dom. Astrophys. Obs.* **13**, 8.
Conti, P. S.: 1965, *Astrophys. J.* **142**, 1594.
Conti, P. S., Wallerstein, G., and Wing, R. F.: 1965, *Astrophys. J.* **142**, 999.
Deutsch, A. J.: 1966, *Astron. J.* **71**, 383.
Deutsch, A. J.: 1967, in *The Magnetic and Related Stars* (ed. by R. C. Cameron), Baltimore, p. 181.
Hansen, K. and McNamara, D. H.: 1959, *Astrophys. J.* **130**, 792.
Harris, D. L., Strand, K. Aa., and Worley, C. E.: 1963, in *Stars and Stellar Systems* (Chicago) **3**, 273.
Kippenhahn, R. and Weigert, A.: 1967, *Z. Astrophys.* **65**, 251.
Kippenhahn, R., Kohl, K., and Weigert, A.: 1967, *Z. Astrophys.* **66**, 58.
Kruszewski, A.: 1966, *Advances Astron. Astrophys.* **4**, 233.
Olson, E. C.: 1968, *Publ. Astron. Soc. Pacific* **80**, 185.
Plaut, L.: 1959, *Publ. Astron. Soc. Pacific.* **71**, 167.
Plavec, M.: 1968, *Advances Astron. Astrophys.* **6**, 201.
Plavec, M., Kriz, S., Harmanec, P., and Horn, J. 1968, *Bull. Astron. Inst. Czech.* **19**, 24.
Refsdal, S. and Weigert, A.: 1969, *Astron. Astrophys.* **1**, 167.
Slettebak, A.: 1949, *Astrophys. J.* **110**, 498.
Slettebak, A.: 1954, *Astrophys. J.* **119**, 146.
Slettebak, A.: 1955, *Astrophys. J.* **121**, 653.
Struve, O.: 1944, *Astrophys. J.* **99**, 222.
Struve, O.: 1963, *Publ. Astron. Soc. Pacific* **75**, 107.
Van den Heuvel, E. P. J.: 1969a, *Astron. J.* **74**, 1095.
Van den Heuvel, E. P. J.: 1969b, *Publ. Astron. Soc. Pacific* **81**, 815.

Discussion

Abt: What periods would you expect for the evolved binaries with white dwarf companions?

Van den Heuvel: About the same as or slightly larger than those of Algol-type semi-detached systems. There is a selection effect which favours detection of Algol-systems with short periods. Due to this effect the systems with a white dwarf component may well have larger average periods than the average period of the Algol-type systems in Table I. The expected percentage of the remnants of Algol-type systems given here is therefore also probably an underestimate.

I would further like to stress that the mass ratios of the systems with white dwarfs are very small, and the binary character of such systems may easily escape detection.

BINARY ORBIT PERTURBATIONS AND THE ROTATIONAL ANGULAR MOMENTUM OF STELLAR INTERIORS

STANLEY SOBIESKI

Goddard Space Flight Center, National Aeronautics and Space Administration, Greenbelt, Md., U.S.A·

Abstract. Calculations show that a significant variation in the minima of eclipsing binaries should arise for systems where axial precession exists. Several different angular velocity distributions are assumed in order to estimate the expected photometric variation as a function of the model parameters. It is found that the solid body rotation approximation is a reasonable representation unless interiors rotate more rapidly than present models predict.

Plavec (1960) has applied the detailed perturbation theory describing the interaction of close binary components as developed by Kopal (1959) to estimate the periodic variation of the orbital motion of twelve binaries typical of detached, semi-detached, and contact configurations. The perturbations are expressed in terms of apsidal motion, nodal regression, nutation and their cross products. He found that with the exception of the apsidal term, the perturbation periods are short, most less than one year, with amplitudes too small to be readily detected. The purpose of this work is to re-investigate the axial precession or nodal line regression by numerical evaluations of stellar models with the view of estimating the magnitude of observable photometric effects. The technique to be used was proposed originally by Luyten (1943).

For arbitrary axial orientations, the precession of the rotation axes of binary components must be accompanied by the precession of the orbital plane on the invariable plane of the system in order that the total angular momentum be conserved. In the case of an eclipsing binary, this perturbation causes the apparent inclination of the orbit to change with respect to the celestial sphere thereby changing the eclipse minima. The maximum differential inclination, $\Delta i_{\max}$, of the orbital plane to the invariable plane occurs when the angular momentum vectors of the two components are coplanar and at the same azimuthal angle. In this case,

$$\tan(\Delta i_{\max}) = \frac{J_1 \sin\theta_1 + J_2 \sin\theta_2}{L + J_1 \cos\theta_1 + J_2 \cos\theta_2}, \tag{1}$$

where L is the orbital angular momentum, J_1 and J_2 are the rotational angular momenta of the two components, and θ_1 and θ_2 are the inclination angles of the J_i vectors to the orbital normal. It has been shown by Kopal (1942) and by Hosokawa (1953) that the axial inclinations, θ_i, can be found from an analysis of the velocity curve during eclipse phases. During these phases the so-called rotational disturbance appears asymmetric in both phase and amplitude about conjunction if the eclipsed component rotates about an axis inclined to the orbital normal as projected on the plane of the sky. Since the effect is second order, high quality spectrographic data are required and the system must be free of 'complications'. Two such investigations have been reported. Kopal (1942) found a value of $\theta_1 = 15°$ for the B8 component of

A. Slettebak (ed.), Stellar Rotation, 187–192. *All Rights Reserved*

β Persei and Koch and Sobieski (1969) recently reported finding a θ_1 not less than $4°$ for the B2 component of 68 Herculis. A value of $\theta_i = 15°$ will be adopted for all calculations below.

The angular momentum, J, for a star of radius R in the case of radial symmetry is given by

$$J = \tfrac{8}{3}\pi\Omega_0 \int_0^R \varrho(r)\, w(r)\, r^4\, dr, \qquad (2)$$

where Ω_0 is the surface value of the angular velocity, $w(r)$ is the angular velocity distribution, and $\varrho(r)$ is the density distribution with distance from the stellar center. Equation (2) was integrated numerically using the density distributions for the 10 $M_\odot$ and 2.5 $M_\odot$ models of Schwarzschild (1958) combined with the angular velocity distributions published by Roxburgh (1964) and by Clement (1969). Their $w(r)$ distributions based on an electron scattering opacity were coupled with the 10 $M_\odot$ model density distribution while the $w(r)$ distribution for the models involving the Kramer opacity were used with the 2.5 $M_\odot$ density distribution. Since Clement's $w(r)$ distributions are neither axisymmetric nor radially symmetric, a distribution appropriate to an astrographic latitude of $30°$ was adopted to simplify the calculations. His radiative braked (R.B) and viscous braked (V.B.) cases are considered separately. Calculations were also made for the case of solid body rotation for reference purposes. A surface angular velocity for each mass model and a representative orbital angular momentum for each system investigated were read from a mean curve passed through the basic data given in Table I. For these tabulated systems, solid body rotation and a radius of gyration equal to 0.2753, appropriate to the Eddington standard model, were assumed to calculate the angular momenta listed in column 8. Absolute dimensions are from the Kopal and Shapley Catalogue (1956) and equatorial velocities are those determined by Koch *et al.* (1965). With the boundary condition thus derived the angular momenta for the different angular velocity distributions, calculated by using Equation (2), are listed in Table II. These results were applied, in turn, to the solution of Equation (1). The values of Δi_{max} thus found are tabulated in Table III.

Several conclusions apropos of these calculations can be drawn.

(1) For even moderate axial inclinations, the estimated values for the differential inclination are surprisingly large. A characteristic value and a maximum value for Δi_{max} can be taken as $0°.4$ and $1°.4$, respectively. These values will be used below to estimate characteristic photometric variations.

(2) Since the orbital angular momentum is significantly greater than the rotational angular momentum, the contributions by the two components are approximately additive. The differential inclination is not a strong function of mass. Although the rotational angular momentum is larger for the more massive stars, this increase is counterbalanced by the larger orbital angular momentum expected in close systems.

(3) The most significant result, however, is that with the exception of the radiative braked model the differential inclination is sensibly model independent. The result

TABLE I

Basic data for close binary systems

	Sp. C.	Mass $(M_\odot)$	V_{eq} (km/sec)	Radius $(R_\odot)$	$\Omega_0 \times 10^5$ (sec^{-1})	Period (days)	$\log (J/M)$	$\log (L/\Sigma M)$
σ Agl	B3	6.8	123	4.2	4.21	1.95	17.435	19.24
	B4	5.4	142	3.3	6.18		17.393	
WW Aur	A3	1.81	41	1.9	3.10	2.53	16.614	18.83
	A3	1.75	39	1.9	3.10		16.614	
U Cep	B8	2.9	310	2.4	18.56	2.49	17.594	18. 70
	G8	1.4	–	2.9			–	
AH Cep	B0.5	16.5	210	6.1	4.95	1.78	17.830	19.21
	B1	14.2	195	5.5	5.09		17.752	
Y Cyg	B0	17.4	146	5.9	3.56	3.00	17.658	19.25
	B0	17.2	148	5.9	3.56		17.658	
68 Her	B2	7.9	119	4.5	3.80	2.05	17.441	18.80
	B5	2.8	90	4.3	3.01		17.310	
RX Her	B9.5	2.75	78	2.1	5.34	1.78	16.937	18.67
	A0.5	2.33	68	1.8	5.43		16.810	
δ Lib	A0	2.6	85	3.5	3.49	2.33	17.196	18.64
	G2	1.1	–	3.5			–	
U Oph	B4	5.30	107	3.4	4.52	1.68	17.283	18.94
	B6	4.65	87	3.1	4.03		17.153	
β Per	B8	5.2	60	3.6	2.39	2.87	17.056	18.66
	K0	1.0	–	3.8			–	
ζ Phe	B6	6.1	100	3.4	4.23	1.67	17.254	18.84
	A0	3.0	–	2.0			–	
V Pup	B1	16.6	180	6.0	4.31	1.45	17.756	19.17
	B3.5	9.8	–	5.3			–	
U Sge	B7	6.7	76	4.1	2.66	3.38	17.215	18.78
	G2	2.0	–	5.4			–	
μ Sco	B1.5	18.0	225	4.8	6.73	1.45	17.755	19.12
	B3	9.3	–	5.3			–	
BH Vir	F9	0.87	90	1.1	11.76	0.82	16.718	17.92
	G2	0.90	90	1.2	10.78		16.756	
RS Vul	B4	4.6	80	3.9	2.95	4.48	17.217	18.78
	F9	1.4	–	5.3			–	

TABLE II

Rotational angular momenta

Mass ($M_{\odot}$)	$\log J$ (cgs units)			
	Solid	Rox.	V.B.	R.B.
2.5	50.60	50.66	50.68	51.36
10.0	51.73	51.92	51.79	52.30

TABLE III

Computed differential inclinations

M_1	M_2/M_1	θ_1	θ_2	$\log L$ (cgs)	Δi_{max}			
					Solid	Rox.	V.B.	R.B.
2.5	0.25	15°	0°	52.04	0°.27	0°.31	0°.32	1°.42
2.5	1.00	15°	0°	52.39	0°.12	0°.14	0°.14	0°.63
2.5	1.00	15°	15°	52.39	0°.24	0°.28	0°.29	1°.27
10.0	0.25	0°	15°	53.10	0°.02	0°.03	0.°03	0°.12
10.0	0.25	15°	0°	53.10	0°.32	0.°48	0°.36	1°.10
10.0	0.25	15°	15°	53.10	0°.34	0.°51	0°.39	1°.22
10.0	1.00	15°	0°	53.46	0°.14	0°.21	0°.16	0°.48
10.0	1.00	15°	15°	53.46	0°.27	0°.42	0°.31	0°.96

found for solid body rotation becomes a fair general approximation and one must expect that only for rapid core rotation will model delineation through observation be possible.

The photometric variation of the eclipse minima can be related directly to the differential inclination through the geometric eclipse depth, p_0. In the notation of Russell and Merrill (1952),

$$\cos i = r_g(1 + kp_0)$$

while the brightness at minimum is

$$l_0 = (1 - \alpha_0 L_s)$$

for occultations and

$$l_0 = (1 - \tau \alpha_0^{tr} L_g)$$

for transits. Since $\alpha_0 = \alpha_0(k; p_0)$, the variation of the eclipse minima, expressed in magnitudes is given by

$$\Delta m = - 2.5 \log \left\{ \frac{l_0(k, p_0 \pm \Delta p_0)}{l_0(k, p_0)} \right\}, \tag{3}$$

where

$$\Delta p_0 = - \frac{\sin i}{r_s} \Delta i.$$

Table IV lists line results found by applying Equation (3) to several representative model solutions as well as to the two 'observed' systems β Per and 68 Her. The last column in Table IV lists estimates of the nodal regression period calculated using the method and relevant constants given by Plavec.

TABLE IV

Photometric variations

Model	i	r_s/r_g	r_s	L_s	Δi	Δm^{oc}	Δm^{tr}	P_1/P_0
Assumed	84°	0.85	0.30	0.85	$+1°.4$	0.181		103
$M_1 = 4\ M_\odot$					$+1°.4$	0.052		
$M_2 = 1\ M_\odot$					$-0°.4$	-0.048		
					$-1°.4$	-0.171		
Assumed	84°	0.85	0.30	0.15	$1°.4$		0.087	12
$M_1 = 10\ M_\odot$					$0°.4$		0.020	
$M_2 = 2.5\ M_\odot$					$-0°.4$		-0.019	
					$-1°.4$		-0.082	
β Per	80°.5	0.74	0.20	0.72	$1°.4$	0.133		64
					$0°.4$	0.039		
					$-0°.4$	-0.039		
					$-1.°4$	-0.142		
68 Her	78°.6	0.80	0.28	0.19	$1°.4$		0.058	52
					$0°.4$		0.017	
					$-0°.4$		-0.017	
					$-1°.4$		-0.058	

Notes: Limb-darkening of 0.6 assumed in all cases.
$L_s + L_g \equiv 1$.
$P_1/P_0 =$ nodal regression period expressed in units of the orbital period for assumed synchronism of rotation.

One may conclude that the magnitude of the photometric variation is adequately large to be detected by photoelectric means. Since the expected precessional period can be quite short for semi-detached and contact systems, each series of eclipse observations must be treated separately. This applies both to the photometric data necessary for determining the depth of the eclipse minimum and to the velocity data required to define the instantaneous axial inclination.

References

Clement, M.: 1969, *Astrophys. J.* **156**, 1051.
Hosokawa, Y.: 1953, *Pub. Astron. Soc. Japan* **5**, 88.
Koch, R. H. and Sobieski, S.: 1969, Presentation at AAS Meeting, Albany, New York.
Koch, R. H., Olson, E. C., and Yoss, K. M.: 1965, *Astrophys. J.* **141**, 955.
Kopal, Z.: 1942, *Astrophys. J.* **96**, 399.
Kopal, Z.: 1959, *Close Binary Systems*, Wiley, New York.
Kopal, Z. and Shapley, M. B.: 1956, *Jodrell Bank Annals* **1**, 4.

Luyten, W. J.: 1943, *Astrophys. J.* **97**, 274.
McNally, D.: 1965, *Observatory* **85**, 166.
Plavec, M.: 1960, *Bull. Astron. Czech.* **11**, 197.
Roxburgh, I.: 1964, *Monthly Notices Roy. Astron. Soc.* **128**, 157.
Russell, H. N. and Merrill, J. E.: 1952, *Contr. Princeton Obs.*, No. 26.
Schwarzschild. M.: 1958, *Structure and Evolution of the Stars,* Princeton University Press, Princeton, N.J., p. 254–255.

STELLAR ROTATION IN OPEN CLUSTERS

(Review Paper)

HELMUT A. ABT

Kitt Peak National Observatory*, Tucson, Ariz., U.S.A.

Abstract. The characteristics of rotational velocities in fifteen open clusters and associations are summarized. The differences in mean rotational velocities observed between individual groups are found to be due primarily to three mechanisms: evolutionary expansion, tidal coupling of rotational and orbital motion in binaries, and magnetic braking in Ap stars.

1. Introduction: The Problem

Recent work on stellar rotation in open clusters and associations has been motivated primarily by the following questions:

(1) Are the stellar rotational velocities in various clusters the same or different for each group, and

(2) If different, what causes the differences?

This paper is confined entirely to a discussion of answers to these questions.

As was true for so many spectroscopic problems, the first results came from Struve (1945), who pointed out that the Pleiades and Hyades have very different mean rotational velocities as a function of spectral type than do field stars. The brighter stars in the Pleiades and Hyades have unusually high and low mean rotational velocities, respectively, although Anderson *et al.* (1966) and Kraft (1965) showed that the fainter stars in these clusters have the same mean rotational velocities as the field stars. An example of the differences between clusters is shown in Figure 1. It seems apparent that clusters often do differ in their mean projected rotational velocities, so the remainder of this paper will deal with possible explanations for such differences.

The question immediately arises whether the mean measured values of $V \sin i$ are unusual because of high or low equatorial rotational velocities, V, or because of preferential inclinations, i, of rotational axes. It seems very likely that the former case occurs for the following reasons, which are partly due to Kraft (1965):

(1) Clusters, such as the Hyades and Coma, that have many characteristics in common also share an unusual dependence of $\langle V \sin i \rangle$ on spectral type. Since the directions of these two clusters from the solar system are very different ($b^{II} = -24°$ and $+84°$, respectively), it would be surprising if their V's and i's combined to give the same values of $\langle V \sin i \rangle$.

(2) Investigation (Kraft, 1965) of the inclinations of orbital planes of visual binaries in these two clusters showed no preferential orientation, and one might expect the rotational and orbital axes to be roughly aligned.

---

* Operated by the Association of Universities for Research in Astronomy, Inc., under contract with the National Science Foundation.

(3) It would be surprising if the B stars in a cluster showed one preferential orientation and the A stars a different preferential orientation.

(4) The difference between the $\langle V \sin i \rangle$ for field stars and for the brightest stars in the α Persei cluster, for example, is larger than would occur for random orientation of axes (for the field stars) and complete alignment seen equator-on (for the cluster).

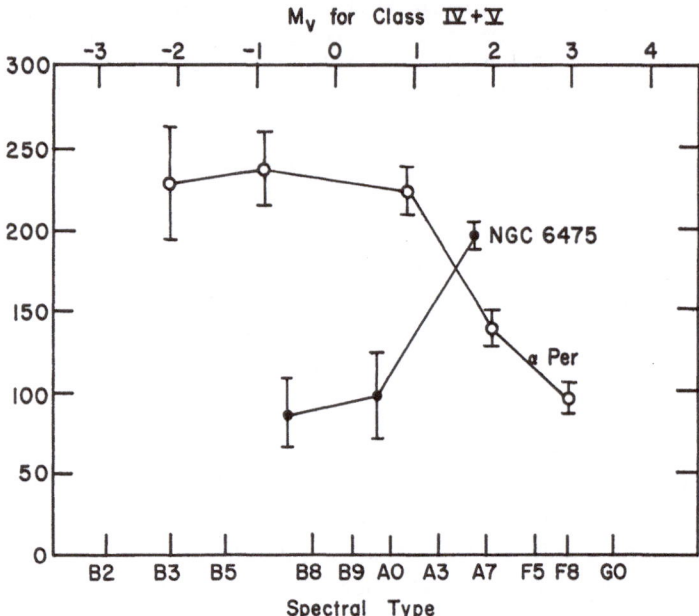

Fig. 1.   Mean rotational velocities in two open clusters. The stars have been grouped by absolute magnitude; the corresponding spectral types, using Blaauw's (1963) calibration, are given. Half the length of the vertical bars is the probable error in the mean value of $V \sin i$ for each group.

(5) Huang and Wade (1966) showed that the orbital planes of field eclipsing binaries show no preferential galactic orientation; in closely-spaced binary systems we would expect rough alignment of rotational and orbital axes, so the lack of a preferential galactic orientation should also apply to the rotational axes of field stars.

Therefore we shall assume in the following discussion that alignment of axes does not contribute appreciably to the unusual projected rotational velocities that are observed.

2. Cluster Data and Discussion

The first compilations of data from various clusters and associations were by Treanor (1960) and Abt and Hunter (1962), although most of the groups discussed by them have been studied more thoroughly and more carefully since then. Table I gives references to the most recent studies involving appreciable numbers of stars per cluster.

The results from these fifteen groups can be summarized as follows:

(1) Relative to field stars, most clusters show unusually large or unusually small

TABLE I

References to rotational velocity studies

Cluster or association	Reference	No. of stars
α Persei	Kraft (1967a)	73
Coma	Kraft (1965)	30
h and χ Persei	Slettebak (1968b)	83
Hyades	Kraft (1965)	78
IC 4665	Abt and Chaffee (1967)	27
M39	Meadows (1961)	18
NGC 2516	Abt *et al.* (1969)	30
NGC 6475	Abt and Jewsbury (1969)	27
Pleiades	Anderson *et al.* (1966)	57
Praesepe	Dickens *et al.* (1968)	56
Ursa Major	Geary (1969)	48
I Lacerta	Abt and Hunter (1962)	26
I Orion	McNamara and Larsson (1962), McNamara (1963)	83
II Perseus	Treanor (1960)	10
Sco-Cen	Slettebak (1968a)	82

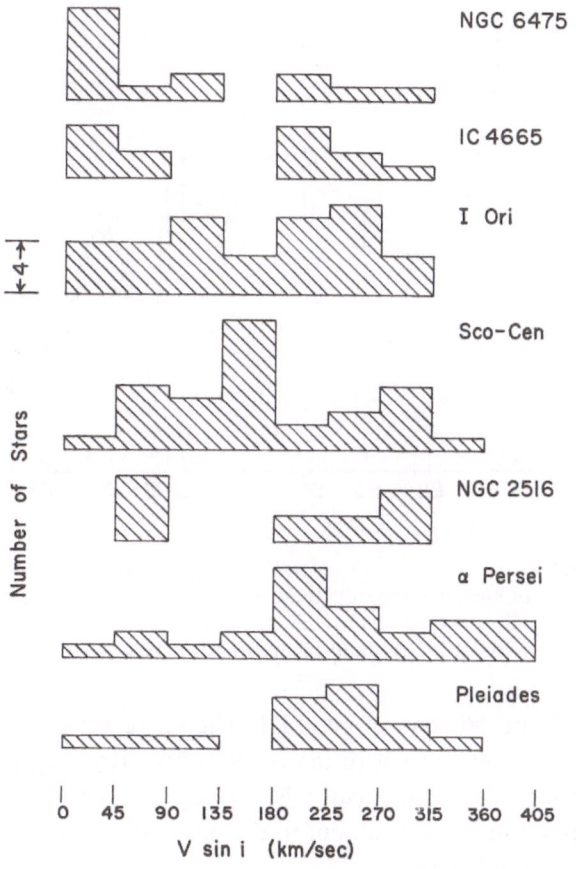

Fig. 2.   Frequencies of values of $V \sin i$ in seven clusters and associations. All Ap stars
have been excluded.

mean rotational velocities for their brighter stars, although their fainter stars tend to
conform to the field-star relation.

(2) In many clusters there is a bimodal distribution of values of $V \sin i$ such that
there are very few stars of intermediate ($100 < V \sin i < 200$ km/sec) line broadening
(Figure 2).

(3) In some clusters, such as the Pleiades (Struve, 1945), Sco-Cen (Slettebak, 1968a)
and NGC 2516 (Abt *et al.*, 1969), there is a tendency for the rapid rotators to concentrate to the projected center of the cluster, but this concentration is not apparent for
other groups, such as *h* and χ Persei (Slettebak, 1968b).

Before attempting to interpret some of these characteristics, let us summarize the
data on field stars. It is a remarkable tribute to our host that his initial papers (Slettebak, 1949, 1954, 1955; Slettebak and Howard, 1955) after 15–20 years still provide
the best fundamental data on stellar rotational velocities; very few of his results
require revision because of line doubling, photometric errors, or errors of analysis.
His distribution of $\langle V \sin i \rangle$ with spectral type is shown in Figure 3 for 279 stars of

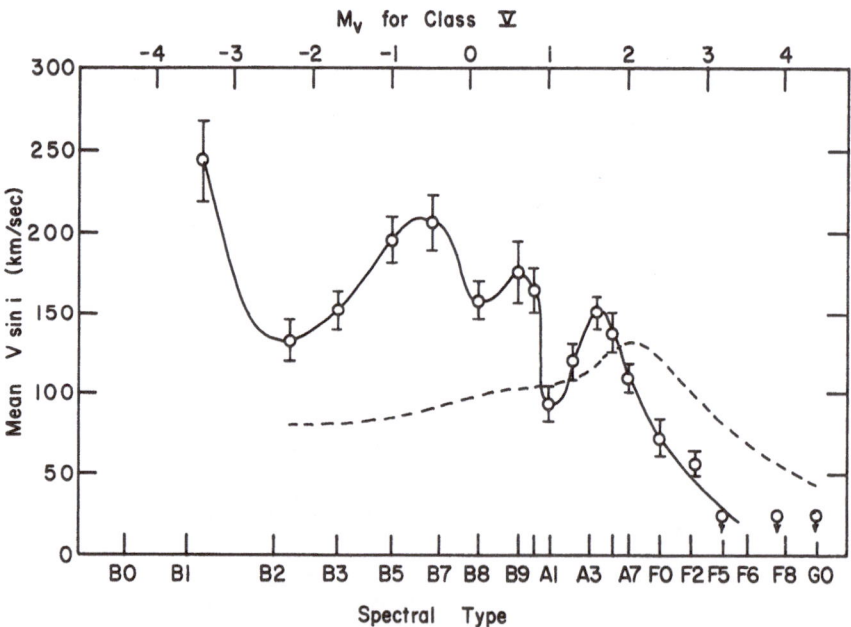

Fig. 3.   The distribution of $\langle V \sin i \rangle$ with spectral type for stars of luminosity class V (individual
means with probable error bars, and the solid line) and class III (dotted line), as derived by Slettebak
and Howard. The calibration of type vs. absolute magnitude is taken from Blaauw (1963). Note the
minima at B2, A1, and perhaps B8 for the class V stars.

luminosity class V and 56 stars of class III. The curve for class V stars shows the
marked decrease at about F2 where the outer convective zone commences, an uncertainty among the O-type and early B-type stars where turbulence contributes
appreciably but to an unknown amount to the line broadening, and various maxima
and minima among the B's and A's.

The maxima and minima, originally pointed out by Boyarchuk and Kopylov (1958) from less homogeneous material, have been discussed by Abt and Hunter (1962), who suggested that appreciable contributions to the sample of field stars by individual clusters and associations having abnormally high or low rotational velocities at some spectral types may be distorting the mean curve. For instance, removal of the 21 Cassiopeia-Taurus B1-B3 stars from the sample of 66 such field stars will partly eliminate the minimum at B2. Inclusion of the extensive material on A stars by Palmer *et al.* (1968) makes this explanation less likely (see the discussion at the end of this paper).

With regard to the causes of differences between clusters, we can think of three likely explanations, which will be discussed in the following three sections.

A. EVOLUTIONARY EXPANSION

As the brighter stars in a cluster leave the zero-age main sequence and expand, their rotational velocities decrease at a rate that is inversely proportional to the radius if the outer layers retain their original angular momenta, or at a slower rate if angular momentum is transferred within the stars such that they always rotate as rigid bodies.

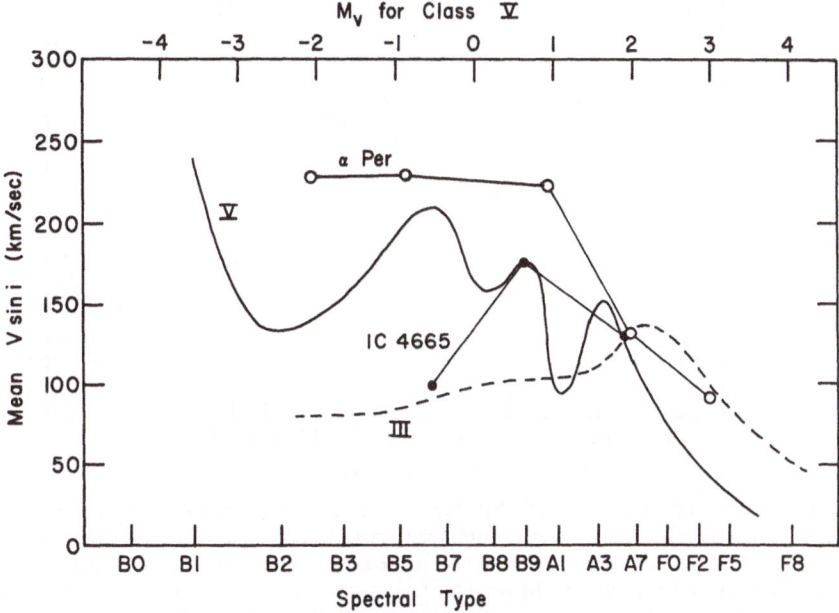

Fig. 4.   Dependence of mean rotational velocities on spectral type in the α Persei and IC 4665 clusters, compared with the relation for field stars of luminosity classes V and III.

A third possibility is that of differential rotation involving a rapidly rotating core and slowly rotating envelope (Dicke, 1964), or the inverse. Observationally, the rotational velocities of the brightest stars in clusters should approach those of class III field stars. In Figure 4 we see a cluster (IC 4665) in which such an evolutionary expansion would seem to explain the low rotational velocities of the brightest stars. Unfortunately we do not yet have good spectral classifications for the stars in that cluster and some others (NGC 6475) showing that dip, to see whether most or all of their brightest stars are of luminosity class III. However, the brightest stars in IC 4665 have a maximum radial expansion from their zero-age sizes (Abt and Chaffee, 1967) of only 26%, which is probably much too small to explain their 2-fold decrease in $V \sin i$ relative to field stars. What is even more convincing in demonstrating that evolutionary expansion is not the only cause of abnormal rotational velocities in clusters is the case of the α Persei cluster, where the evolved stars have larger, rather than smaller, mean rotational velocities than field stars.

A group in which the brighter stars have abnormally low rotational velocities and in which good MK types are available by Morgan (Morgan *et al.*, 1955), Slettebak and Howard (1955), or Sharpless (1952) is the I Orion association. In Figure 5 we show the

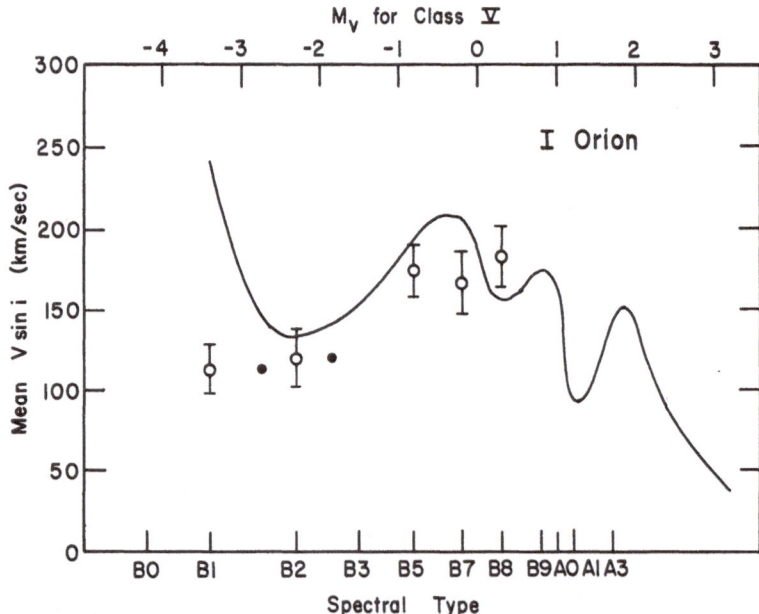

Fig. 5.   The mean rotational velocities (McNamara and Larsson, 1962) of 35 bright I Orion stars (left two circles) with MK types indicating luminosity class V, and 33 fainter stars (right three circles) assumed to be of class V (McNamara, 1963) are plotted according to their mean spectral types. The solid line is for class V field stars. If the brighter Orion stars are plotted according to their absolute magnitudes (black dots), the discrepancy with the field stars is much less.

mean rotational velocities for class V stars in I Orion relative to the class V field star relation. The fainter stars in the association are approximately normal (meaning like field stars), but the B0-B1.5 stars rotate too slowly. In this case the inverse evolutionary effect may be present: the class V association stars may be less evolved than the average class V field star. The black dots in Figure 5 show to where the left two means are shifted if those points are plotted by absolute magnitude, using an apparent distance modulus of 8.5 mag., rather than by spectral type. In this case the discrepancy with field stars is less marked.

We conclude that the evolutionary expansion and consequent effect on the rotational velocities of stars should be considered, but that this effect alone cannot explain many of the differences between individual clusters and field stars.

B. TIDAL COUPLING IN BINARIES

There are at least two ways in which the initial rotational velocities of stars may be gradually modified: by tidal interaction in closely-spaced binaries and by magnetic braking in magnetic stars. The latter effect will be discussed in the next section. If some clusters differ from each other or from field stars in their numbers of spectro-

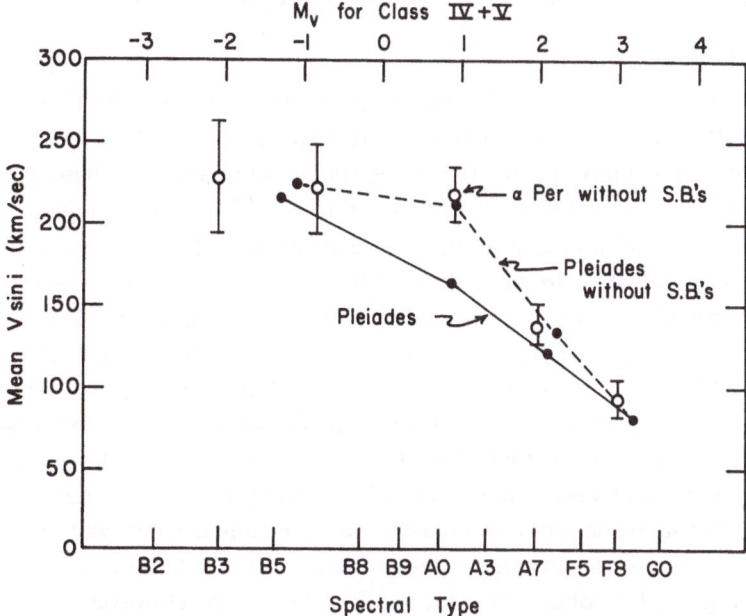

Fig. 6.   Mean rotational velocities for all the Pleiades stars (solid line) and excluding the spectroscopic binaries (dashed line) compared with similar quantities for the α Persei cluster (circles with error bars), excluding binaries.

scopic binaries, we might expect that their modified mean rotational velocities would differ.

An interesting test case is that of the Pleiades and α Persei clusters. Eggen (1965) suggested that since these two clusters, and several others, have common space motions and similar color-magnitude diagrams, they have a common origin. Kraft (1967a) inquired whether their rotational velocity distributions are similar. Those distributions are different if all their stars are included. However, although both clusters have few spectroscopic binaries, the Pleiades has more than the α Persei cluster among the early A-type stars, where the rotational velocity distributions depart significantly. After excluding all definite spectroscopic binaries (and the few Am and Ap stars) in both groups, the rotational velocity distributions were found to be similar (Figure 6).

It is now apparent that some clusters, such as the α Persei cluster (Heard and Petrie, 1967) and Pleiades (Abt *et al.*, 1965) have very few spectroscopic binaries while others, such as NGC 6475 (Abt *et al.*, 1970) have many. Also, where the frequency of binaries is low, the mean rotational velocities are high, and where the binaries are frequent, the mean rotational velocities are low.

We conclude that tidal interaction in closely-spaced binaries is effective in reducing rotational velocities, and that a large part of the differences between clusters in their mean rotational velocities is due to different binary frequencies. In a sense we have succeeded only in shifting the problem from trying to explain the various mean rotational velocities to trying to explain the different binary frequencies.

C. MAGNETIC BRAKING

The rotational velocities of Am (Slettebak, 1955; Abt, 1965) and Ap (Slettebak, 1955; Abt *et al.*, 1967) stars are consistently low relative to normal stars. For the Am stars the slow rotation seems to be due to tidal coupling in binaries, which are very frequent among such stars. However, it seems likely that binaries are not frequent among Ap stars (Babcock, 1958) and that another mechanism must be found for producing the low rotational velocities in these stars. After consideration of the extremely large amounts of angular momentum available in gaseous nebulae, we do not think that single stars are likely to reach the main sequence with low rotational velocities. It seems more likely that magnetic braking, either within a star rotating differentially or between the stellar and interstellar magnetic fields, will cause the slow rotation.

A test case for this idea is that of NGC 2516 that, according to Eggen (1965), has a similar space motion and color-magnitude diagram to the Pleiades and α Persei cluster. Kraft found the latter two clusters to have similar rotational velocities, after allowance for differences in binary frequencies, so we are somewhat more confident about finding similar rotational velocities in NGC 2516. However, that cluster has many Ap stars (Abt and Morgan, 1969): seven of the 23 brightest main-sequence stars are definite or probable Ap stars. The binary frequency for this cluster is not yet known. With the Ap stars included, the rotational velocities are different than for the Pleiades (Figure 7), both with binaries included, but with the Ap stars excluded, the rotational velocities are similar.

We conclude in this section that different frequencies of Ap stars in clusters can help cause differences in mean rotational velocities.

D. OTHER MECHANISMS

Other mechanisms that may produce differences between clusters by modifying stellar rotational velocities are (1) differential rotation, either in the surface layers or with depth, (2) age effects, such as those demonstrated by Kraft (1967b) (this was suggested

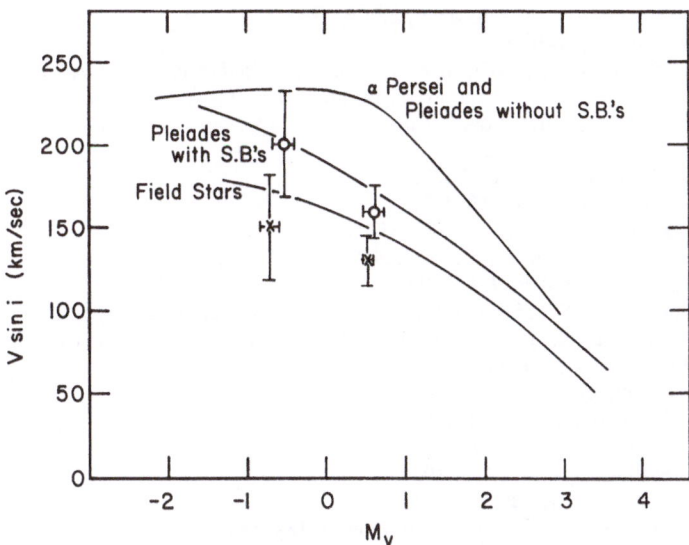

Fig. 7.   Mean rotational velocities (x's) and probable error bars for 30 stars in NGC 2516. If the seven definite or likely Ap stars in this cluster are excluded, the resulting mean rotational velocities (circles) agree with the Pleiades relation, including spectroscopic binaries in both clusters (reprinted courtesy of the *Astronomical Journal*).

by Ostriker in the Colloquium discussion), (3) magnetic field effects, (4) mass loss, and (5) possible effects of the initial total cluster mass upon the cluster characteristics. For most of these it is not clear why stars of the same mass and age but in different clusters should produce differing results.

The goal for coming research is to see whether, after allowance for all the obvious effects (evolution, differing duplicity, differing frequencies of peculiar stars, etc.), the known clusters have a common dependence of mean rotational velocity on spectral type.

References

Abt, H. A.: 1965, *Astrophys. J. Suppl.* **11**, 429 (No. 102).
Abt, H. A. and Chaffee, F. H.: 1967, *Astrophys. J.* **148**, 459.
Abt, H. A. and Hunter, J. H., Jr.: 1962, *Astrophys. J.* **136**, 381.
Abt, H. A. and Jewsbury, C. P.: 1969, *Astrophys. J.* **156**, 983.
Abt, H. A. and Morgan, W. W.: 1969, *Astron. J.* **74**, 813.
Abt, H. A., Barnes, R. C., Biggs, E. S., and Osmer, P. S.: 1965, *Astrophys. J.* **142**, 1604.
Abt, H. A., Chaffee, F. H., and Suffolk, G.: 1967, *Astron. J.* **72**, 783.
Abt, H. A., Clements, A. E., Doose, L. R., and Harris, D. H.: 1969, *Astron. J.* **74**, 1153.
Abt, H. A., Levy, S. G., Baylor, L. A., Hayward, R. R, Jewsbury, C. P., and Snell, C. M.: 1970, *Astron. J.* **75**, in press.
Anderson, C. M., Stoeckly, R., and Kraft, R. P.: 1966, *Astrophys. J.* **143**, 299.
Babcock, H. W.: 1958, *Astrophys. J.* **128**, 228.
Blaauw, A.: 1963, *Basic Astronomical Data* (ed. by K. Aa. Strand), University of Chicago Press, Chicage, Ill., p. 383.
Boyarchuk, A. A. and Kopylov, I. M.: 1958, *Astron. Zh.* **35**, 804; trans.: 1959, *Soviet Astron.* **2**, 752.
Dicke, R. H.: 1964, *Nature* **202**, 432.
Dickens, R. J., Kraft, R. P., and Krzeminski, W.: 1968, *Astron. J.* **73**, 6.

Eggen, O. J.: 1965, *Ann. Rev. Astron. Astrophys.* **3**, 235.
Geary, J. C.: 1969, in preparation.
Heard, J. F. and Petrie, R. M.: 1967, in *Determination of Radial Velocities and Their Applications, I.A.U. Symposium No. 30*, Academic Press, London, p. 179.
Huang, S.-S. and Wade, C., Jr.: 1966, *Astrophys. J.* **143**, 146.
Kraft, R. P.: 1965, *Astrophys. J.* **142**, 681.
Kraft, R. P.: 1967a, *Astrophys. J.* **148**, 129.
Kraft, R. P.: 1967b, *Astrophys. J.* **150**, 551.
McNamara, D. H.: 1963, *Astrophys. J.* **137**, 316.
McNamara, D. H. and Larsson, H. J.: 1962, *Astrophys. J.* **135**, 748.
Meadows, A. J.: 1961, *Astrophys. J.* **133**, 907.
Morgan, W. W., Code, A. D., and Whitford, A. E.: 1955, *Astrophys. J. Suppl.* **2**, 41.
Palmer, D. R., Walker, E. N., Jones, D. H. P., and Wallis, R. E.: 1968, *Roy. Obs. Bull.*, No. 135.
Sharpless, S.: 1952, *Astrophys. J.* **116**, 251.
Slettebak, A.: 1949, *Astrophys. J.* **110**, 498.
Slettebak, A.: 1954, *Astrophys. J.* **119**, 146.
Slettebak, A.: 1955, *Astrophys. J.* **121**, 653.
Slettebak, A.: 1968a, *Astrophys. J.* **151**, 1043.
Slettebak, A.: 1968b, *Astrophys. J.* **154**, 933.
Slettebak, A. and Howard, R. F.: 1955, *Astrophys. J.* **121**, 102.
Struve, O.: 1945, *Pop. Astron.* **53**, 259.
Treanor, P. J., S.J.: 1960, *Monthly Notices Roy. Astron. Soc.* **121**, 503.

Discussion

Jaschek: Are the periods of all Ap stars which are spectroscopic binaries equal to the periods of their magnetic variations? If not, would that not throw doubt upon the possibility that magnetism influences rotation?

Abt: The only Ap star in a binary with identical magnetic and orbital periods is HD 98088, which has identical binary, rotational, and magnetic periods. In long-period systems there is probably a tendency toward synchronism, but it need not be achieved.

Preston: The spectroscopic binary HR 710 seems to be an exception. The magnetic field does not vary periodically in the binary period of 3 days.

Rosendhal: If you add together the results for the individual clusters do you recover something which looks like the field star distribution?

Abt: Yes, very closely.

Ostriker: Isn't it likely that, for the later spectral types, a fourth explanation (suggested by Kraft) must be considered? An age effect may exist with the older stars (at about the same spectral type) having the lower surface velocities.

Abt: Yes, this may explain why the $\langle v \sin i \rangle$ for the later-type stars in a young cluster like the Pleiades is larger than in an older cluster like the Hyades and Coma.

Kraft: NGC 2516 is very remarkable because of the large number of Ap's. Even if the rectified rotation function (omitting Ap's) is the same as that of the Pleiades and α Per cluster, it does not support Eggen's suggestion of a common origin of these three clusters because one cannot explain why NGC 2516 has so many Ap's and the other two clusters have almost none.

Abt: Yes. Perhaps all we can do is shift the problem from trying to explain different rotational velocities to trying to explain different frequencies of Ap stars and binaries. On the other hand, we do not know how much the subclusters in an association differ from each other. We should look at all the characteristics (rotational velocities, frequencies of binaries, Ap and Am stars) of subclusters in a larger complex like the I Orion association; there is some evidence for such differences in the Sco-Cen association.

Slettebak: Your comment to the effect that clusters with large total mass tend to show larger $v \sin i$'s than clusters with small total mass is extremely interesting. Do you find a fairly strict correlation among the clusters thus far investigated?

Abt: No, data are lacking on the lengths of main sequences in many clusters and associations, so we probably cannot tell at present.

Collins: Do the data relating to $\langle v \sin i \rangle$ vs. spectral type and indicating the dip at B2 include Be stars? If not, would their inclusion remove the dip at B2?

Abt: That sounds like a very reasonable suggestion. There seems to be no reason why the Be stars, which occur primarily in the vicinity of the B2 dip, should be excluded unlike, for example, the Am stars, whose rotational velocities may represent later modifications. On the other hand, there are nearly as many Be stars with types of B3 and B5 as of B2, so that including the Be stars does not remove the dip, unless the types of Be stars are often too late for their masses.

Van den Heuvel: Dr. Conti has shown that among the early A-stars there are stars similar to Am stars which, however, can only be recognized at high dispersion, and which therefore will mostly be classified as 'normal' at the dispersions which are employed in the work you referred to. Should not these be left out, or did you leave them out? The dip near A1 in the $\langle v \sin i \rangle$ curve for field stars might disappear in such a case.

Abt: If one wishes to recover the original rotational velocities of stars, we should probably exclude the Sirius type or hot Am stars because their rotational velocities represent later modifications. But such stars have not been identified in the complete sample of Slettebak stars.

Deutsch: I think Conti has found that nearly enough of the sharp-line stars near A0 are abnormal, either in the sense of showing the Ap or 'hot Am' characteristic, to account for the excess number of sharp-line stars. These stars would also depress the value of $\langle v \sin i \rangle$.

Buscombe: (1) From preliminary estimates of rotational velocity for members of the Scorpio Centaurus Association, I showed (*Irish Astron. J.* **7** (1965), 63) that the frequency distribution as a function of spectral type extends towards hotter spectral types the trend earlier established for II Per, which is a coeval association of the same space velocity stream. The very high incidence of duplicity among Sco-Cen stars must be taken into account. Moreno (*Santiago Publ.* **5** (1969), 72) has evidence of the correlation of narrow-band photometric indices on $v \sin i$ from Slettebak's more reliable individual determinations. The divergent ($\langle v \sin i \rangle$, spectral class) distribution for members of NGC 6475 confirms my earlier assertions that this cluster is not a subgroup of the Sco-Cen association; NGC 6475 is more evolved and has a different space motion.

(2) Hyland (unpublished Ph.D. dissertation, Canberra, 1967) while investigating model atmospheres for the nine brightest stars (including Si-rich B peculiars) in each of two southern clusters, established that the relative frequency of fast rotators is high in IC 2602, but low in IC 2391.

(3) There is growing evidence that several sets of classification for B stars based on the precepts of Morgan and Keenan, show considerable influence of the use of slow or fast rotators among the standards.

Abt: Regarding the third point, the originators of the MK system cannot be responsible for results obtained when their recommendations are not followed, especially with regard to the value of low dispersions ($\simeq 125$ Å/mm) and low resolution (25–30 μ). Collins has shown that under those conditions the rotational effects on spectral classifications should be trivial, except for stars near the critical velocities.

CALIBRATION OF INTERMEDIATE-BAND PHOTOMETRIC
PARAMETERS AND V Sin i EFFECTS

DAVID L. CRAWFORD

Kitt Peak National Observatory*, Tucson, Ariz., U.S.A.

Calibration of four-color and Hβ photometry, in terms of intrinsic color and absolute magnitude, for the B-, A- and F-type stars is about finished, though still subject to small changes. The data for the bright stars in the northern and southern hemispheres and for a number of the brighter open clusters were used in obtaining these calibrations. Some of the data have been published; much more of it will be going to press shortly. The calibration is interesting not only for itself and its applications, but because deviations from it may be due in part to rotational velocity effects.

For the A stars (β between $2^m.880$ and $2^m.700$) the calibration for intrinsic color may be expressed simply as $(b-y)_0 = 2.943 - 1.0\beta - 0.1\ \delta c_1 - 0.1\ \delta m_1$. δc_1 and δm_1 are determined as the difference between the observed value and the value of c_1 and m_1 for a given β. The standard relations are given in condensed form in Table I. All bright stars observed within the limits of β and having δc_1 less than $0^m.280$ were used in obtaining the calibrations. In other words, about 500 stars. The 25 largest residuals were all positive (in the sense that positive indicates reddening) and, if these are

TABLE I
Standard relations, A and F Stars
(Preliminary)

β	$(b-y)_0$ $\delta c_1 = 0$ $\delta m_1 = 0$	m_1	c_1 (Z.A.) June '69	M_v (Z.A.) June '69
$2^m.880$	$0^m.063$	$0^m.207$	$0^m.93$	$2^m.3$
$2^m.860$	$0^m.083$	$0^m.207$	$0^m.89$	$2^m.4$
$2^m.840$	$0^m.103$	$0^m.208$	$0^m.85$	$2^m.5$
$2^m.820$	$0^m.123$	$0^m.206$	$0^m.82$	$2^m.6$
$2^m.800$	$0^m.143$	$0^m.203$	$0^m.78$	$2^m.7$
$2^m.780$	$0^m.163$	$0^m.196$	$0^m.74$	$2^m.8$
$2^m.760$	$0^m.183$	$0^m.188$	$0^m.70$	$2^m.8$
$2^m.740$	$0^m.203$	$0^m.181$	$0^m.66$	$2^m.9$
$2^m.720$	$0^m.223$	$0^m.176$	$0^m.60$	$3^m.1$
$2^m.700$	$0^m.243$	$0^m.172$	$0^m.54$	$3^m.3$
$2^m.680$	$0^m.265$	$0^m.170$	$0^m.48$	$3^m.5$
$2^m.660$	$0^m.294$	$0^m.171$	$0^m.43$	$3^m.7$
$2^m.640$	$0^m.324$	$0^m.178$	$0^m.38$	$3^m.9$
$2^m.620$	$0^m.356$	$0^m.193$	$0^m.34$	$4^m.3$
$2^m.600$	$0^m.390$	$0^m.216$	$0^m.30$	$4^m.6$

* Operated by the Association of Universities for Research in Astronomy, Inc., under contract with the National Science Foundation.

A. Slettebak (ed.), Stellar Rotation, 204–206. All Rights Reserved
Copyright © 1970 by D. Reidel Publishing Company, Dordrecht-Holland

omitted, the rms scatter for one star is $\pm 0\overset{m}{.}013$. By fitting a normal curve to the negative residual tail, a dispersion of $\pm 0\overset{m}{.}011$ results (for one star). I interpret the slightly higher red tail as due to small values of reddening. Therefore the rms scatter for one star (which is equal then to the cosmic scatter since all stars were included) is $\pm 0\overset{m}{.}011$. This cosmic scatter includes the effects of spectroscopic binaries, metallic line stars, effects of rotation, etc. The calibration fits the observed unreddened clusters to $0\overset{m}{.}003$. There is little or no rotational velocity effect apparent in the calibration, and the relation of the $(b-y)_0$, β, and δc_1 to MK spectral types is quite good.

The absolute magnitude calibration may be expressed as $M_v = M_v$ ($\delta c_1 = 0$, in other words 'zero age main sequence') $- 8 \delta c_1$. This calibration was obtained by using the slopes of absolute magnitude vs. β relations for open clusters and fitting the zero point by trigonometric parallax stars. The factor 8 was determined by stars within clusters and from cluster to cluster. The observed values for cluster stars were corrected for this before the shape fitting was made. The Hyades were not used in the calibration. The rms scatter is estimated to be $0\overset{m}{.}3$ for one star.

For the F stars, these same types of equations hold, though the coefficients are slightly different. The rms scatter for one star is about $0\overset{m}{.}015$, larger than for the A stars, but still quite usable for determining intrinsic colors of individual stars. The range of β is $2\overset{m}{.}720$ to $2\overset{m}{.}600$. Near G0 the stars with values of δc_1 greater than $0\overset{m}{.}100$ were omitted. The coefficient of β becomes larger as β gets smaller and the coefficient of δm_1 becomes larger as β gets smaller. In the absolute magnitude calibration the factor times δc_1 is 11. This is determined both from data for trigonometric parallax stars and from observations in NGC 752. The rms scatter appears to be $0\overset{m}{.}3$.

For the B stars a simple relation for intrinsic color that fits the observations quite well is $(b-y)_0 = -0.116 + 0.097c_0$. ($c_0$ is the value of c_1 corrected for interstellar reddening.) The effects of reddening on c_1 are about 20% those on $(b-y)_0$ and so one iteration using the above equation accurately determines both the reddening on c_1 and on $(b-y)$. The rms scatter as estimated from nearly unreddened stars and a few galactic clusters that appear unreddened appears to be $0\overset{m}{.}01$ for one star. The effects of reddening on m_1 are 30% those on $(b-y)$ and on $(u-b)$ are 1.7 times those in $(b-y)$. The calibration appears to fit the clusters quite well and also appears to fit luminosity classes III and II within the above rms scatter.

The calibration for absolute magnitude is given in terms of a relation between β and absolute magnitude, and it appears that a small correction for spectral type or δc_0 or $\delta \beta$ is necessary. The observational errors in β of about $\pm 0\overset{m}{.}010$ would yield an error in absolute magnitude from this source alone ranging from between $0\overset{m}{.}2$ for the late B's to $0\overset{m}{.}7$ for the early B supergiants. As an example of the dispersion expected on the average, for the stars in h and χ Persei observed to the 15th magnitude, the average β is $2\overset{m}{.}610$ and the formal rms scatter for one star is $\pm 0\overset{m}{.}5$. The calibration itself was obtained by fitting the shape of the relation from the observations in clusters and then fitting the zero point by the 'known' distance moduli of the α Persei and Pleiades clusters. The relation between the observed values and the MK spectral types is very good.

Effects of $v \sin i$ must be reasonably small: of the order or less than $0^{m}.2$ in M_v. Other factors affect the dispersion equally or even larger, for example, unresolved double stars, peculiar stars, age differences, and the 'Hyades anomaly' in the F-type stars. Unfortunately I have been unable to thoroughly analyze the data for $v \sin i$ effects, but in the α Persei and Pleiades clusters, for the A stars where there should be little or no differential age effect, I have applied individual reddening corrections to the stars and find that the clearest correlation with $v \sin i$ is in the parameter δc_1. δc_1 when plotted against $v \sin i$ shows a linear slope of about $0^{m}.04$ in δc_1 for 100 km/sec difference in $v \sin i$. This is valid over a range from 0 to 250 km/sec. Small individual differences with spectral type and between the clusters exist but these have not yet been looked at in detail.

The calibrations will be published within the next year, and in the meantime are available on request as well as are the data on the clusters. I am especially open to suggestions of important things to look for in the data and of ways to analyze the data in light of theory and from the experience of spectroscopists.

Discussion

Faber: What was the spectral type of the reddest stars observed in the α Per cluster, and did you detect any change in the slope of the relation between δc_1 and $v \sin i$ dependent upon spectral type?

Crawford: G0 is the latest spectral type. The slope seems the same for different types, but there may be some small differences that I haven't seen yet.

Buscombe: I hope the *photometric* data will be published in the form of *photometric* parameters. It is unfortunate that a trend to convert by an unpublished transformation to a *spectroscopic* parameter (equivalent width) has begun to creep into the photometric literature.

Crawford: The observed data – $(b-y)$, m_1, c_1, V, β – will be published, of course.

Deutsch: Dr. Strömgren has found that δm_1 appears to correlate with $v \sin i$ in the A stars. Can you comment on this effect and its interpretation?

Crawford: The Am stars have negative δm_1's and low velocities. If these data for these stars are omitted, there appears to be little effect left, certainly not as large as in δc_1. The question is: what are the independent parameters?

MAXWELLIAN DISTRIBUTIONS FOR STELLAR ROTATIONS

ARMIN J. DEUTSCH

Hale Observatories,
Carnegie Institution of Washington and California Institute of Technology, Pasadena, Calif., U.S.A.

Abstract. For stars in the main-sequence band, the distributions of 'line-widths' $V \sin i$ are rediscussed, with the inclusion of recent measurements by Walker and Hodge for B stars, and by Preston for Ap stars. The Ap stars are counted with normal stars of the same color-class. Over the whole range from B2 to A2, the observed incidence of sharp-line stars far exceeds the incidence predicted on the hypothesis that the true rotational velocities have a Maxwellian distribution. The 'extra' slow rotators are probably close binary remnants, in the sense of Van den Heuvel. However, it remains unclear how to distinguish individual members of this group of stars.

1. Introduction

For the Greenbelt Symposium in 1965, an attempt was made to find whether the line-width measurements of Slettebak and Howard (1955) are conformable with a Maxwellian law for the true distribution function of rotational velocities among B and A stars in the main-sequence band (Deutsch, 1967). The Maxwellian law was found to be admissable for stars in the range B5–B9 and for those in the range A3 to F0. However, in the ranges B2–B5 and B9–A2 the incidence of sharp-line stars appeared to be excessively high, as Conti (1965) had first pointed out for stars near A0V. In these parts of the main-sequence band, the line-width statistics were found to be compatible with bimodal Maxwellian distributions. The conjecture was put forward that the excessively abundant sharp-line stars represent older populations, which consist of metamorphs of red giants, or Population I analogues of 'horizontal-branch stars'. However, no reason was given as to why the line-widths statistics failed to show similar excesses of sharp-line stars in the ranges of spectral types B5–B9 and A2–F0.

The peculiar A stars and the metallic line stars introduce serious ambiguities into the spectral classification of A-type stars. According to the photometry of Eggen (1963) and others, one can best circumvent this difficulty by grouping nearby A stars on a criterion of color-class. Accordingly, this was done for the A stars in the 1965 work. However, it was not done for the B stars. This resulted in the exclusion from the statistics of a number of Ap stars with the colors of B stars.

2. Simple Maxwellian Distributions

At the same time that we correct this inconsistency in the 1965 work, we are able to incorporate additional measurements of $V \sin i$ that have recently been made on the system of Slettebak and Howard. In advance of publication, Dr. G. W. Preston has kindly made available to me his determinations of $V \sin i$ for all known Ap stars brighter than $V=6.0$ and north of $-15°$ declination. These comprise 18 objects with (unreddened) colors in the range $-0.05 \leqslant (B-V)_0 \leqslant \pm 0.10$, and 43 objects in the range $-0.18 \leqslant (B-V)_0 \leqslant -0.03$.

In these ranges of $(B-V)_0$, which for normal stars correspond to spectral-type ranges B5 $-$ to B9 $+$ and B9 $+$ to A2 $+$, respectively, the line-widths of Slettebak are complete only to $V=5.0$. To combine the two sets of data, therefore, I have extrapolated the Slettebak measurements to $V=6.0$ by applying the factor 4 to the numbers of stars he found in each increment of $V\sin i$. The parentheses in Table I and in the figures are intended to denote that the corresponding numbers have been obtained by such an extrapolation. The same extrapolation has been made to obtain the 'observed' numbers of stars given in the respective boxes of the histograms.

In the range of spectral types B2–B5, inclusive, the measurements of Walker and Hodge (1966) give the line-widths $V\sin i$ for many more stars than were measured by Slettebak and Howard (1955). The bright limit of the main-sequence band was arbitrarily set at 1.5 magnitudes above the zero-age main sequence, and the line-widths were counted for all the stars that Walker and Hodge assigned to this strip in the Hertzsprung-Russell diagram.

Table I shows the numbers of stars for which $V\sin i$ was taken to be known,

TABLE I

Maxwellian distributions for $V\sin i$

Sp. Type		B2 to B5	B5 $-$ to B9 $+$	B9 $+$ to A2 $+$
$(B-V)_0$		-0.24 to -0.16	-0.18 to -0.03	-0.05 to $+0.10$
N (normal)		$\rbrace$ 197	(208)	(316)
N (pec.)			43	18
$1/j$ (km/sec)		185	164	151
$V\sin i < 50$ { Obsd.		81	53	90
$V\sin i < 50$ { Pred.		13	23	14

according to these precepts, in each of three intervals along the main-sequence band. In the interval B2–B5, of course, there is no subgroup of stars that are recognizable analogues of the Ap stars encountered between B5 and A2.

For the stars counted in each range of spectral type, Table I also gives the root-mean-square value of $V\sin i$. This is the parameter $1/j$; it is equal to the modal value of the true rotational velocities, if these have a Maxwellian distribution. With these values of $1/j$, and the formulae given in Deutsch (1967), one can then predict, for each sample of N stars, how many should have $V\sin i < 50$ km/sec. Evidently these predictions are grossly incorrect; each of the three samples of Table I contains many more sharp-line stars than would be found if the true distribution were Maxwellian. In particular, on including the Ap stars with color-classes in the range B5 $-$ to B9 $+$, we now find an abnormality in the statistics where none appeared in the 1967 work. The present position is that all along the main-sequence band from B2 to A2, slow rotators have far too large an incidence to comport with a Maxwellian law for the true distribution of rotational velocities.

This result is likely to be of greater significance than one might think. For, by using the methods of classical statistical mechanics, one can rigorously prove the following theorem relative to the angular velocities ω of the stars. If the distribution of ω is isotropic, and if the component ω_x in a system of Cartesian coordinates is distributed independently of ω_y and ω_z, then the distribution of the positive scalar ω is given by $f(j\omega)\,d(j\omega)$, where f is the Maxwellian law,

$$f(x) = \frac{4}{\sqrt{\pi}} x^2 \exp(-x^2);$$

or by the sum $\sum_i a_i f(j_i\omega)\,d(j_i\omega)$, with, $\sum_i a_i = 1$. The proof of this theorem is given in Appendix I.

Now, the angular-velocity vectors of stars are known to be directed at random, and one would suppose that little or no coupling should occur among $(\omega_x, \omega_y, \omega_z)$ for objects that rotate about one of their principal axes of inertia. Therefore our theorem ought to apply. However, the results in Table I have ruled out simple Maxwellian distributions for stellar rotations. Accordingly, we shall now try to represent the observed statistics with curves derived from bimodal Maxwellian laws, for these represent the next most simple distributions which our theorem will allow.

3. Bimodal Maxwellian Distributions

To obtain the parameters for the bimodal distributions, recourse was had to a computer program written by Mr. J. F. Bartlett, of the Booth Computing Center at the California Institute of Technology. Bartlett's program finds the best-fitting bimodal distributions on the basis of a maximum-likelihood calculation, starting from the histograms that have been observed. The smooth curves in the figures give the distributions of $V \sin i$ that would be produced by these optimum bimodal Maxwellian laws for V. Each curve is the weighted sum of two apparent distributions; in units of percent per km/sec, these apparent distributions are, respectively,

$$\varphi_y = 100\, j_y \varphi(j_y V \sin i)$$
$$\varphi_0 = 100\, j_0 \varphi(j_0 V \sin i),$$

where

$$\varphi(x) = 2x \exp(-x^2).$$

The numerical parameters given in Figure 1, then, have the following significance: 53% of the 197 stars represented belong to a 'young' population having a Maxwellian distribution of V with a mode $1/j_y = 248$ km/sec; and 47% belong to an 'old' population having a Maxwellian distribution of V with a mode $1/j_0 = 39$ km/sec. Because no account was taken of the distribution of $V \sin i$ within the first box ($V \sin i < 50$ km/sec), it is necessary to add that these calculations really determine only an upper limit to the parameter $1/j_0$, except possibly in the range B2–B5.

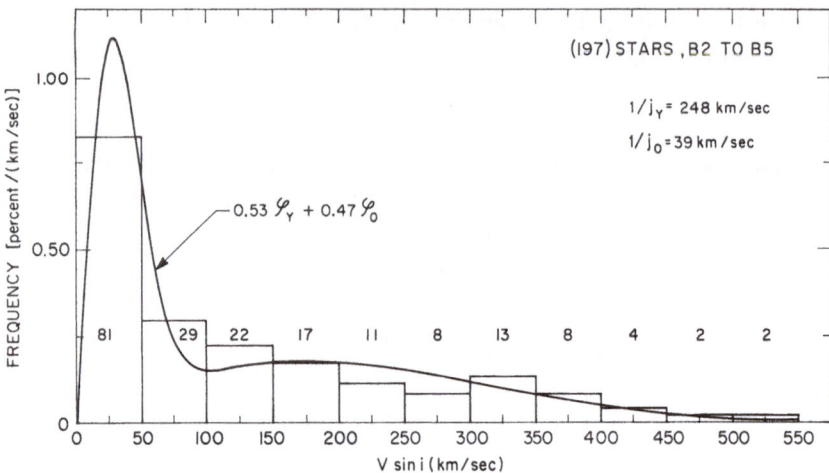

Fig. 1. Distribution of rotational velocities of early B stars in the main-sequence band, according to the measures of Walker and Hodge (1966). The smooth curve is derived from a bimodal Maxwellian distribution for the true rotational velocities V.

4. Interpretation of the Statistics

The chi-square test indicates that one cannot reject these bimodal Maxwellian distributions on statistical grounds. However, one need not accept the hypothesis that the 'extra' slow rotators comprising the O Population are really metamorphs of red giants. Indeed, a number of objections can now be lodged against this suggestion. The chief difficulty concerns the kinematics. If the stars of the O Population originally lay on the main-sequence as late A or F stars, their motions should still be distinctively larger than the Y Population stars which they now resemble. For example, in the compilation of Allen (1963), the mean velocity dispersion is 20 km/sec at A0, 29 km/sec at F0, and 42 km/sec at G0. The possibility exists that the O Population stars near A0 have somewhat brighter absolute magnitudes than the Y Population stars. Since this has not been taken into account in the calculation of their motions, the latter may have been systematically underestimated. However, after a review of the absolute-magnitude and velocity determinations for the stars near A0V (Eggen, 1963), it seems most unlikely that the O Population stars can have a velocity dispersion which is much larger than that of the Y Population stars.

Other objects that have been interpreted as possible metamorphs of red giants are the blue stragglers that are found in M 67 and several other old open clusters. The A stars in M 67 have a mean rotational velocity of nearly 100 km/sec (Deutsch, 1966) – a value which is awkwardly intermediate between the mean rotational velocities of the two populations that seem discernible among the early A stars in the field (Figure 3). Spectroscopic determinations of g and M have been reported by Sargent (1968) for three of the early A stars in M 67. His result for the mean mass is $0.7 \pm 0.2\,m_\odot$. This is low for ordinary main-sequence A stars, leading Sargent to conclude "... that these three stars do constitute the blue end of a thinly populated horizontal branch in M 67".

However, Sargent (1969) now agrees that this mass-determination is sufficiently sensitive to observational errors and to uncertainties in the broadening theory for the Balmer lines, that it cannot really rule out the larger masses corresponding to ordinary main-sequence A stars.

Rodgers (1968) has reported a similar mass-determination for the field star HD 6870. The motion of this object is such that it could be a horizontal-branch star belonging to the old moving group associated with σ Puppis (Eggen, 1964). Rodgers concludes that HD 6870 probably is a member of this moving group. However, he also concludes that its mass is nearly normal for the spectral type of a main-sequence star, and that the object is therefore probably not a horizontal-branch star.

Meanwhile, Deutsch (1969) has obtained a similar result for Fagerholm 190, an A8 star which is one of the blue stragglers in M 67. In this instance, no mass-determination has been made. However, the star appears to be a single-line spectroscopic

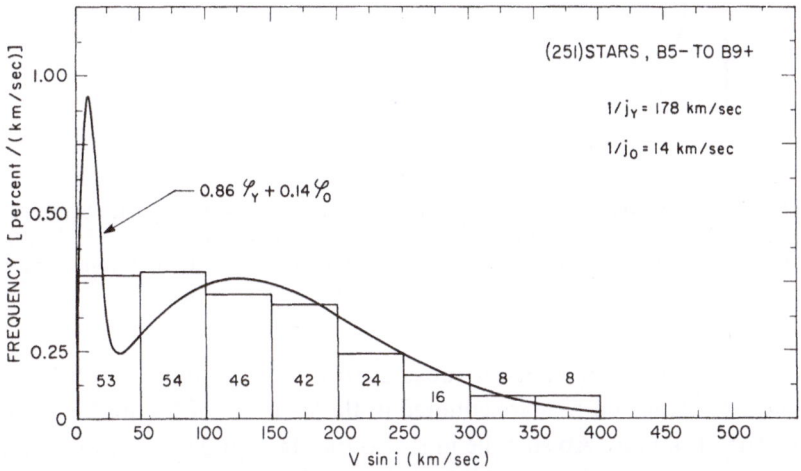

Fig. 2.   Distribution of rotational velocities of late B stars, in the main-sequence band, according to the measures of Slettebak (1954). The statistics include Preston's (1969a) measures for 43 Ap stars that have the color-classes of late B stars. The smooth curve is derived from a bimodal Maxwellian law for V.

binary, with a period of four days and a low mass-function. The object therefore cannot be a metamorph of a red giant. Most probably it has reached its present position in the color-magnitude diagram by the accretion of mass from its companion, according to the McCrea-Kippenhahn theory of evolution in close binary systems.

Van den Heuvel (1968) has shown that after the transfer of mass from the original primary of a close pair, the new primary is likely to have a rotational velocity appreciably lower than most stars of the same mass, which have not experienced accretion from an evolving companion. Moreover, he has recently (1969) concluded that "... at least 11.7% (and possibly up to 22%) of the main-sequence stars earlier than A5 are expected to be CBR" – i.e., close-binary remnants of a McCrea-Kippenhahn evolutionary process. The work of Van den Heuvel also shows that the incidence of CBR may go appreciably higher at some spectral types, depending on the incidence of close binaries and on the fraction of mass that is lost to the system when the original primary overflows its Roche lobe.

On the basis of these arguments, it therefore seems plausible to regard most of the 'extra' slow rotators of our O Population as CBR. However, we still lack a reliable

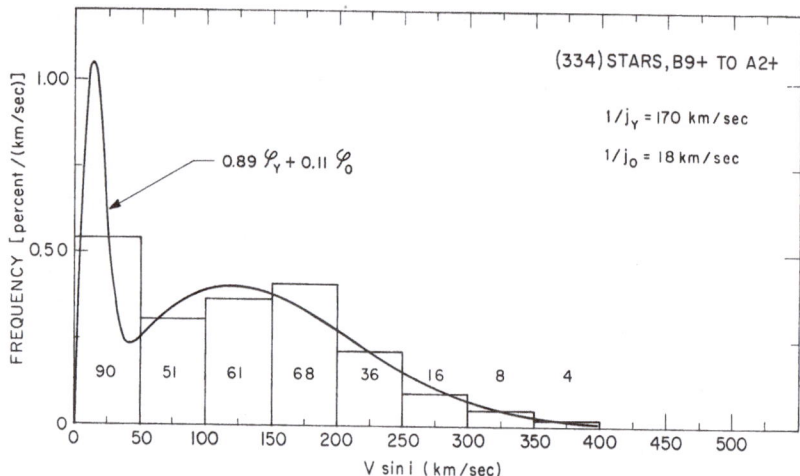

Fig. 3.   Distribution of rotational velocities of early A stars in the main-sequence band, according to the measures of Slettebak (1955). The statistics include Preston's (1969a) measures for 18 Ap stars that have the color-classes of early A stars. The smooth curve is derived from a bimodal Maxwellian law for V.

spectroscopic discriminant between those sharp-line stars that are CBR and those that are not. Thus, of the 53 stars counted in the first box of Figure 2, 29 are known to be Ap stars. Ten more Ap stars lie in the second box, and two each in the next two boxes. From the parameters for the best-fitting bimodal distribution, one ought to find 17 of the Y Population stars in the first box, and 35 of the O Population. Similarly one expects that in the first box of Figure 3 there ought to be 22 Y Population stars and 68 O Population stars. The box actually contains 14 Ap stars; three more lie in the second box and one in the third. These results hardly permit one to assign the Ap stars to one population or the other.

It remains to note that not all the 'extra' slow rotators we have found in the field need be CBR. Indeed, Preston (1969b) has shown that some of the Ap stars appear to have been subject to a different and very powerful kind of rotational deceleration, which may arise from an interaction between the stellar magnetic fields and the interstellar medium. Mestel (1969) has recently described the torques on a star that support a stellar wind and a dipole field inclined to the rotation axis. Other modes of rotational braking have been briefly reviewed by Huang and Struve (1960).

If the creation function for early-type stars is constant with time, and Maxwellian with parameter $1/j_0$; and if all stellar rotations are subject to a similar, exponential decay; then the steady-state distribution function for rotational velocities can be

shown to be $G_T(j_0 V) d(j_0 V)$,
where

$$G_T(x) = \frac{4T}{\sqrt{\pi x}} \int_x^{xe^{1/T}} z^2 e^{-z^2} dz.$$

In this expression, T is the decay time for stellar rotations, expressed in units of the lifetime in the main-sequence band. The function $G_T(x)$ is actually a limiting case of a sum of Maxwellian distributions having all values of $1/j$ between $1/j_0$ and zero (see Appendix II). The corresponding distribution of $V \sin i$ is $J_T(j_0 V \sin i) d(j_0 V \sin i)$, where

$$J_T(x) = \int_0^{\pi/2} G_T(x/\sin i) \, di.$$

The function $G_T(x)$ is a special case of the distribution functions for rotationally-braked stars, that have been discussed by Huang (1967). The corresponding function $J_T(x)$ is characterized by the two free parameters $1/j_0$ and T.

One finds that $J_T(x)$ can represent the statistics of $V \sin i$ better than can the function $\varphi(x)$ which is derived from the simple Maxwellian law, and which has one free parameter. However, $J_T(x)$ is less successful than the distribution derived from a bimodal Maxwellian law, which has three parameters. Little attention has been paid in this paper to the use of $J_T(x)$, for the work of Van den Heuvel indicates that the McCrea-Kippenhahn evolution of close binary pairs is likely to have produced most or all of the 'extra' slow rotators that lie on the early main-sequence.

Appendix I. The Distribution Function for the Magnitude of a Vector that has Random Orientation

It is required to find the distribution function of a positive scalar ω, which is the magnitude of a vector $\boldsymbol{\omega}$. We assume that the distribution of $\boldsymbol{\omega}$ is isotropic. We also assume that if $\boldsymbol{\omega}$ is decomposed into components along Cartesian axes, the distribution of any component (say, ω_x) is independent of the other components (ω_y and ω_z in this case).

Let us designate $\boldsymbol{\Omega}$ as the non-dimensional quantity $j\boldsymbol{\omega}$, where j is a parameter with the dimensions of ω^{-1}, and where

$$\boldsymbol{\Omega} = \mathbf{i}\Omega_x + \mathbf{j}\Omega_y + \mathbf{k}\Omega_z.$$

The probability that Ω_x lies in $d\Omega_x$, Ω_y in $d\Omega_y$, and Ω_z in $d\Omega_z$ is then

$$F(\Omega_x, \Omega_y, \Omega_z) \, d\Omega_x \, d\Omega_y \, d\Omega_z = h(\Omega_x^2) \, h(\Omega_y^2) \, h(\Omega_z^2) \, d\Omega_x \, d\Omega_y \, d\Omega_z.$$

The assumption of isotropy also insures that we may write

$$F(\Omega_x, \Omega_y, \Omega_z) = H(\Omega^2) = H(\Omega_x^2 + \Omega_y^2 + \Omega_z^2).$$

If

$$\Omega_x = \Omega_y = 0, \quad \text{and} \quad \Omega_z = \Omega,$$

then

$$H(\Omega^2) = h^2(0)\, h(\Omega^2);$$

and therefore by Equation (2)

$$h(\Omega_x^2)\, h(\Omega_y^2)\, h(\Omega_z^2) = h^2(0)\, h(\Omega^2).$$

Now for any $u \geqslant 0$, let

$$\xi(u) = \ln\frac{h(u)}{h(0)}.$$

Then by Equation (5)

$$\xi(\Omega^2) = \ln\frac{h(\Omega_x^2)}{h(0)} + \ln\frac{h(\Omega_z^2)}{h(0)} + \ln\frac{h(\Omega_z^2)}{h(0)}$$

$$= \xi(\Omega_x^2) + \xi(\Omega_y^2) + \xi(\Omega_z^2).$$

In particular, if $\Omega_x^2 = \Omega_y^2 = u$, and $\Omega_z = 0$,

$$\xi(2u) = 2\xi(u).$$

Again, if $\Omega_x^2 = u$, $\Omega_y^2 = 2u$, and $\Omega_z = 0$,

$$\xi(3u) = \xi(u) + \xi(2u) = 3\xi(u);$$

and, by induction, for any positive integer n,

$$\xi(nu) = n\xi(u).$$

If we let

$$u = v/n,$$

then

$$\xi[n(v/n)] = n\xi(v/n),$$

whence

$$\xi(v/n) = \frac{1}{n}\xi(v).$$

If m is another positive integer, then Equation (13) yields

$$m\xi(v/n) = \frac{m}{n}\xi(v),$$

and by Equation (10),

$$\xi\left(\frac{m}{n}v\right) = \frac{m}{n}\xi(v).$$

Since this result is valid for any positive rational number (m/n), by continuity it will also be valid for any positive irrational number x; viz.,

$$\xi(xv) = x\xi(v).$$

in particular, if $v = 1$,

$$\xi(x) = x\xi(1) = cx.$$

Therefore

$$d\xi(x)/dx = c,$$

and, by Equation (6),

$$\frac{1}{h(x)}\frac{dh}{dx} = c,$$

the solution of which is

$$h(x) = ae^{cx}.$$

Since $h(\Omega_x^2)$ is a probability function, the appropriate normalization condition is

$$1 = 2\int_0^\infty h(\Omega_x^2)\,d\Omega_x = 2a\int_0^\infty e^{c\Omega_x^2}\,d\Omega_x,$$

and the mean value of Ω_x^2 is

$$\langle\Omega_x^2\rangle = 2a\int_0^\infty \Omega_x^2\,e^{c\Omega_x^2}\,d\Omega_x.$$

If we define the parameter j in the relation

$$(1/j)^2 = 2\langle\omega_x^2\rangle,$$

we then find that Equations (21) and (22) yield the results $a = \sqrt{\pi}$, and $c = -1$.

We may now write the distribution of Ω in the form $f(\Omega)\,d\Omega$, where

$$f(\Omega) = \iint_{\Omega=\text{const}} H(\Omega^2)\,\Omega^2\,\sin\vartheta\,d\vartheta\,d\varphi$$

$$= 4\pi\Omega^2 H(\Omega^2).$$

Adducing Equation (4), we find that

$$f(\Omega) = 4\pi\Omega^2 h^2(0)\,h(\Omega^2),$$

and, from Equation (20),

$$f(\Omega) = 4\pi\Omega^2 a^2\,e^{c\Omega^2}$$

$$= \frac{4}{\sqrt{\pi}}\,\Omega^2\,e^{-\Omega^2}.$$

This is the Maxwellian distribution.

Now suppose that the distribution is bimodal, with parameters j_1 and j_2, respectively. Let a_1 be the incidence of one distribution and $a_2 = 1 - a_1$ the incidence of the other. Then the probability that ω lies in $d\omega_1$, $d\omega_2$, $d\omega_3$ may be written as

$$a_1 F(\Omega_x, \Omega_y, \Omega_z)\, d\Omega_x\, d\Omega_y\, d\Omega_z + a_2 F(\Omega'_x, \Omega'_y, \Omega'_z)\, d\Omega'_x\, d\Omega'_y\, d\Omega'_z,$$

where $\Omega_x = j_1\omega_x$, $\Omega'_x = j_2\omega_x$, etc., etc. The derivation of each distribution then proceeds as before, and we obtain for the composite distribution the bimodal Maxwellian law,

$$f(\Omega_1) + f(\Omega_2) = \frac{4}{\sqrt{\pi}}\, a_1 \Omega_1^2\, e^{-\Omega_1^2} + \frac{4}{\sqrt{\pi}}\, a_2 \Omega_2^2\, e^{-\Omega_2^2}.$$

The probability that ω lies in $d\omega$ is then

$$f(j_1\omega)\, d(j_1\omega) + f(j_2\omega)\, d(j_2\omega).$$

Appendix II. The Steady-State Distribution Function Corresponding to Decelerating Maxwellian Distributions

Suppose that in unit volume near the sun, the rate of creation of stars with (non-dimensional) angular velocities $j_0\omega_0$ in $d(j_0\omega_0)$ is $Nf(j_0\omega_0)\, d(j_0\omega_0)$, where N is a constant; f is the Maxwellian law,

$$f(j_0\omega_0) = \frac{4}{\sqrt{\pi}}\, (j_0\omega_0)^2\, e^{-(j_0\omega_0)^2}$$

and

$$1/j_0 = \left(\tfrac{2}{3}\right)^{1/2} \langle \omega_0^2 \rangle^{1/2} = \langle (\omega_0 \sin i)^2 \rangle^{1/2}.$$

Suppose, further, that the angular velocity of every star decays according to the law

$$\omega = \omega_0\, e^{-\tau/T}.$$

In this expression, τ is the age of the star, and T the decay-time for the angular velocity. We take the unit of time to be the whole main-sequence lifetime of the star, so that $0 < \tau < 1$.

For the incidence of stars with angular velocities in $d(j_0\omega)$ and ages in $d\tau$, we then have

$$g_T [j_0\omega]\, d[j_0\omega]\, d\tau = f\big[(j_0\omega)\, e^{\tau/T}\big]\, d\big[(j_0\omega)\, e^{\tau/T}\big]\, d\tau$$
$$= f(j\omega)\, d(j\omega)\, d\tau,$$

where

$$1/j = (1/j_0)\, e^{-\tau/T}.$$

Therefore, among stars created at a given epoch, the distribution of $j\omega$ remains Maxwellian, but the parameter $1/j$ decays according to Equation (5).

In a steady state, stars of all ages $0 < \tau < 1$ will contribute to the incidence in any increment $d(j_0\omega)$. We may write this distribution as $G_T(j_0\omega)\, d(j_0\omega)$, where

$$G_T(x) = \int_0^1 g(x)\, d\tau$$

$$= T \int_0^{1/T} e^{\tau/T} f(x e^{\tau/T})\, d(\tau/T),$$

which may be reduced to

$$G_T(x) = \frac{4T}{\sqrt{\pi x}} \int_x^{x e^{1/\tau}} z^2 e^{-z^2}\, dz.$$

The corresponding distribution of $V \sin i$ is

$$J_T(j_0 V \sin i)\, d(j_0 V \sin i),$$

where

$$J_T(x) = \int_0^{\pi/2} G_T(x/\sin i)\, di.$$

Acknowledgements

I gratefully acknowledge numerous helpful discussions with G. W. Preston concerning the subject of this paper. My thanks also go to D. M. Peterson who improved my proof of the theorem given in Appendix I; and to J. F. Bartlett, who obtained the bimodal Maxwellian distributions drawn in the figures.

References

Allen, C. W.: 1963, *Astrophysical Quantities*, 2nd ed., Athlone Press, London, p. 243.
Conti, P. S.: 1965, *Astrophys. J.* **142**, 1594.
Deutsch, A. J.: 1966, *Astron. J.* **71**, 383.
Deutsch, A. J.: 1967, *The Magnetic and Related Stars*, (ed. by R. C. Cameron), Mono Book Corp., Baltimore, p. 181.
Deutsch, A. J.: 1969, reported by H. W. Babcock in *C.I.W. Yrbk.* **68**.
Eggen, O. J.: 1963, *Astron. J.* **68**, 697.
Eggen, O. J.: 1964, *Astron. J.* **69**, 570.
Huang, S.-S.: 1967, *Astrophys. J.* **150**, 229.
Huang, S.-S. and Struve, O.: 1960, in *Stellar Atmospheres*, (ed. by J. L. Greenstein), Univ. of Chicago Press, p. 321.
Mestel, L.: 1969, *Monthly Notices Roy. Astron. Soc.* **140**, 177.
Preston, G. W.: 1969a, private communication.
Preston, G. W.: 1969b, this volume, p. 254.
Rodgers, A. W.: 1968, *Astrophys. J.* **152**, 109.
Sargent, W. L. W.: 1968, *Astrophys. J.* **152**, 885.
Sargent, W. L. W.: 1969, private communication.

Slettebak, A.: 1954, *Astrophys. J.* **119**, 146.
Slettebak, A.: 1955, *Astrophys. J.* **121**, 653.
Slettebak, A. and Howard, R. F.: 1955, *Astrophys. J.* **121**, 102.
Van den Heuvel, E. P. J.: 1968, *Bull. Astron. Inst. Neth.* **19**, 326.
Van den Heuvel, E. P. J.: 1969, *Lick Obs. Bull.*, No. 603.
Walker, G. A. H. and Hodge, S. M.: 1966, *Pub. Dominion Astrophys. Obs.* **13**, 2.

Discussion

Van den Heuvel: I want to make a comment on what Dr. Deutsch just said. When I said yesterday that the existence of the slow rotators of type B8 and later could perhaps be explained by the close-binary evolution process, I did not imply that the same would not work at earlier spectral types. However, my material presented yesterday did not allow such a conclusion. I already suggested some years ago (Van den Heuvel, 1967, 1968a), however, that the close-binary evolution is the explanation for the occurrence of slow rotators (Van den Heuvel, 1966, 1968b) among the B as well as the A stars.

Van den Heuvel, E. P. J.: 1966, *Proc. Koninkl. Nederl. Akademie van Wetenschappen*, Series B, **69**, 357.
Van den Heuvel, E. P. J.: 1967, *Observatory* **87**, 68.
Van den Heuvel, E. P. J.: 1968a, *Bull. Astron. Inst. Neth.* **19**, 326.
Van den Heuvel, E. P. J.: 1968b, *Bull Astron. Inst. Neth.* **19**, 309.

Abt: By grouping the Ap stars with the rapid rotators, are you implying that Ap stars are rapid rotators seen pole-on?

Deutsch: No. The hypothesis was that nearly all the slow rotators in the Y population will show the Ap characteristic, regardless of aspect. Of course, most will be seen at high inclination.

Kraft: The excess of sharp-lined B8–A2 stars is also shown in young clusters such as α Per and the Pleiades. This suggests that the two populations are not an 'old' vs. 'young' effect, but rather that the sharp-lined group results from a selective continuous decay of rotation in *some* stars as the stars age.

Deutsch: I agree.

Hardorp: The resolution in $V \sin i$ is not constant along the $V \sin i$-axis. It therefore is justified to use narrower boxes for the slow rotations. If you make the first box 25 km/sec instead of 50, you find the over-abundance of slow rotators also in Slettebak's old samples of B6–B9 stars as well as in the B2–B5 sample.

Deutsch: I agree.

ON THE ROTATION OF A-TYPE STARS

CARLOS JASCHEK

Observatorio Astronómico, La Plata, Argentina

Abstract. Based upon the statistics of $V \sin i$ among A-type stars it is shown that (1) no relationship exists between rotational velocity and UBV colours, (2) there might be an influence of rotation upon spectral classification, and (3) spectroscopic binaries are associated with low rotational velocity.

1. Introduction

Recently, Cowley *et al.* (1969) (this paper will be referred to as CCJJ) have published a study of bright *A*-type stars, which fulfil the conditions $m < 6^{m}5$ and $\delta > -20°$. The number of stars included in the paper was 1700; for all of them spectral classification in the Morgan-Keenan system was provided. The majority (to be accurate, 67%) has also colour measurements in the Johnson-Morgan UBV system.

This large material of homogeneous classifications can be used to discuss certain controversial topics of stellar rotation. For instance it is well known that Boyarchuk and Kopylov (1958) obtained in their discussion of the frequency of $V \sin i$ as a function of spectral types, a pronounced minimum at A0 and a maximum around A3. They believed that these extremes were real because their statistics are based upon a large number of stars. Recently Van den Heuvel showed (1965, 1968) that this is probably due to the omission of the Am and Ap stars from the statistics of the Soviet astronomers.

In a second paper on the bright A-stars, which is being prepared for publication, we have re-examined these and other questions. I will not deal with all of them, but will only quote one result concerning the Boyarchuk-Kopylov hypothesis, because of its connection with problems to be discussed later.

In order to decide if the extremes found by Boyarchuk and Kopylov are real, we have examined the average $V \sin i$ values as a function of the $B - V$ colour. The $V \sin i$

TABLE I

Average rotational velocities for dwarfs, Ap, Am and δ Del stars as a function of colour

$B - V$	Interval	$\langle V \sin i \rangle$ (km/sec)	N	Dwarfs	Ap	Am	δ Del
$-0^{m}09$	$-0^{m}05$	108	29	23	6		
.04	.00	144	71	70	1		
$+$ .01	$+$ .05	131	61	57	4	1	
.06	.10	135	47	39	4	4	
.11	.15	132	35	26	1	8	
.16	.20	122	33	21	–	11	1
.21	.25	111	23	14	1	8	
.26	.30	104	12	4	–	7	1

values were taken from Boyarchuk and Kopylov's (1964) catalogue and the colours
from CCJJ. The results are given in Table I.

As can be seen, the extremes have disappeared. This is a confirmation of Van den
Heuvel's result and can be interpreted as meaning that from a rotational point of view,
A-type dwarfs, Am and Ap stars belong to the same group. It is also shown in Paper II
that if we mix the dwarfs, Am and Ap stars, we obtain a very similar distribution
function of $V \sin i$ for different colours, so that it is possible to use a single distribution
function of $V \sin i$ for the whole group. Although this is probably not strictly true because of the smaller rotation of the A stars of latest type, it simplifies considerably our
considerations and we will assume it as being strictly true.

Two points will now be examined in detail, the first being the relation between
rotation and colour and the second the relation between rotation and duplicity.

2. Rotation and Colour

Theory predicts the existence of a relation between rotation and colour in the sense
that the colour of a normal star undergoes a change if its rotational velocity is changed.
The exact amount of this change is difficult to predict, however, because of the several
assumptions which go into the theory, and therefore I will examine the question from
a purely observational point of view.

Since it is impossible to separate in our observations V from $\sin i$, it is impossible to
establish directly the relation between V and the colour change. The answer to the
problem can thus only be a statistical one and consequently only the largest possible
material can be expected to yield a solution. Since for most of the stars UBV photo-

TABLE II

Average $U - B$, $B - V$ indices for groups of dwarfs of different rotation

Spectral class	Mean colour	$V \sin i$ (km/sec)		
		0–99	100–199	200–300
B9	$\langle U - B \rangle$	−0.20 (14)	−0.21 (13)	−0.21 (6)
	$\langle B - V \rangle$	−0.06	−0.05	−0.06
A2	$\langle U - B \rangle$	+0.04 (12)	+0.03 (19)	+0.04 (4)
	$\langle B - V \rangle$	+0.05	+0.03	+0.05
A5–A8	$\langle U - B \rangle$	+0.11 (5)	+0.11 (13)	+0.10 (6)
	$\langle B - V \rangle$	+0.16	+0.19	+0.15

Note: Numbers in parentheses give numbers of objects in each group.

metry is available, the easiest thing to do is to examine a sample of dwarfs, to see if
$V \sin i$ correlates with colour. The results are displayed in Table II.

It is obvious that no correlation exists between average colour and rotation. The
technique used is probably insufficient to detect small effects in the A stars, although

one should remember that among the B stars colour differences do exist between rapidly rotating Be stars and more slowly rotating B stars.

A word of caution should be said regarding the use of spectral types for rotation statistics. It cannot be assumed a priori that spectral types are uninfluenced by rotation. This has to be demonstrated, and quite to the contrary, it is easy to show that some classification systems are in fact very much influenced by rotation. A little consideration suggests that probably classifiers have a tendency to assign sharp line stars to a later type than the broad line stars. This can be seen for instance very clearly on a comparison between the Mt. Wilson spectral types taken from the list of Adams *et al.* (1935) and MK spectral types from CCJJ. In Figure 1 are plotted all the

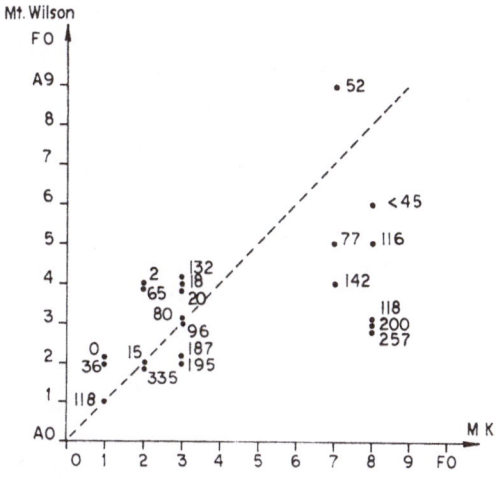

Fig. 1.   Mt. Wilson spectral types as given by Adams *et al.* (1935) vs. MK spectral types estimated by CCJJ. Rotational velocities are given in km/sec.

stars common to both catalogues and for which rotational velocities exist. It is very evident that rapid rotators tend to be classified in the Mt. Wilson system as having earlier types than the average, which is what one would expect. The explanation of this lies in the fact that the Mt. Wilson material was obtained at higher dispersion than the MK material. The visibility of the lines is thus much more influenced by rotation in the Mt. Wilson material than it is in the MK material.

One might therefore think that no influence of rotation exists in the MK system. This question can be examined if a parameter exists which substitutes for the spectral types. Assuming that narrow band photometric indices are such a substitute, we have selected samples of stars of a given spectral type for which both Hβ and K indices were measured (Cameron, 1966; Henry, 1969) and for which also $V \sin i$ has been measured. The results are exhibited in Figure 2.

It can be seen that no relation exists between the position of the objects in each plot and the rotational velocity, written at each point. This is what one would expect (Henry, 1969).

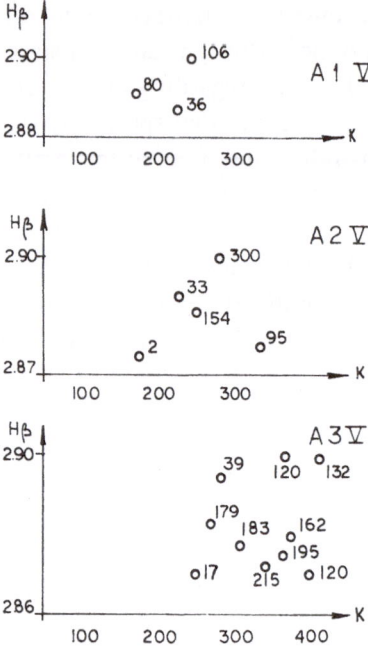

Fig. 2.   Correlation between Hβ and K indices, spectral types and rotational velocities. The V sin i
values are given in km/sec.

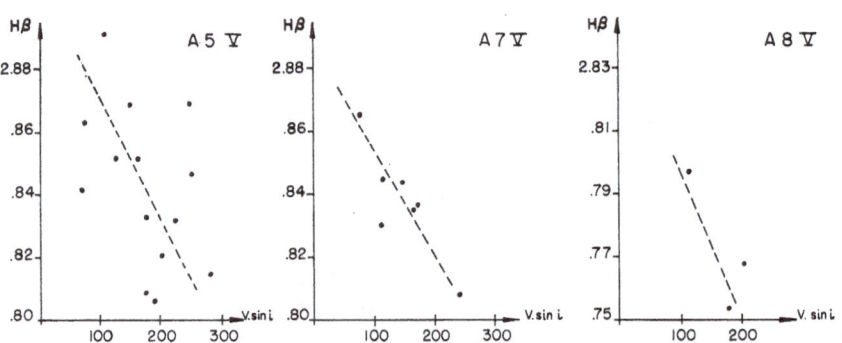

Fig. 3.   Correlation between Hβ indices, spectral types and rotational velocities. The V sin i values
are given in km/sec.

One can also examine dwarfs for which only the Hβ index has been observed (Cameron, 1966). If one assumes that Hβ indices are unaffected by rotation – which is likely but undemonstrated – one gets the situation depicted in Figure 3.

Spectral types A0–A4 do not reveal anything, probably because the Hβ index varies very little in this range. For A5–A8 dwarfs one gets the rather unexpected situation that on the average larger rotation implies lower indices, i.e. later spectral types. We have found no convincing explanation for this. Although the result is based upon few stars, and should thus be taken with caution, it must be taken as a warning that spurious effects *may* be present.

Another word of warning should be said regarding the sharp lined dwarfs. These stars are often called Am, specially when viewed at high dispersion. The southern star π Ara is a good example, and it would be easy to pick out more examples. Probably the best way out of this difficulty would be to set up suitable standard stars for both fast and slow rotators.

3. Rotation and Duplicity

Several years ago Abt and Hunter (1962), on the basis of cluster material pointed out that spectroscopic binaries are slow rotators. Since nobody examined if the same is true for field stars and if so, to what extent, the question will be examined in detail.

Table III gives the relation between $V \sin i$ and $\log P$ for spectroscopic binaries with

TABLE III

Distribution of $V \sin i$ and $\log P$ for spectroscopic binaries

$V \sin i$ (km/sec)	$\log P$ (days)			
	0–1.0	1.0–2.0	2.0–3.0	3.0–4.0
0–50	27	10		
50–100	16	4	5	1
100–150		1	1	1
150–200	1		1	2
Number	44	15	7	4

orbits. The rotational velocities were taken from Boyarchuk and Kopylov (1964), Koch *et al.* (1965) and Olson (1968) and the periods from Batten (1967).

Very clearly the majority of the systems exhibit slow rotation. The first question which might be asked is if this sample is representative of the total number of A-type spectroscopic binaries. This can be tested by examining the distribution of the systems over $\log P$.

TABLE IV

Percentage distribution over $\log P$ of A-type spectroscopic binaries

Sample		$\log P$			
	Number	−1.0–0.0	0.0–1.0	1.0–2.0	2.0–4.0
All A-type SB's	228	0.07	0.61	0.23	0.09
A-type SB from Table III	70	0.00	0.63	0.21	0.16

It follows that the sample in Table III is representative of the whole population. One can compare now the distribution of $V \sin i$ (for $P < 100^d$) with the distribution of $V \sin i$ for all A-type stars. The comparison is made in Table V.

This implies that 98% of the stars have $V \sin i < 100$ km/sec, while only 45% of the

TABLE V

Distribution of $V \sin i$ for spectroscopic binaries
and normal stars with A-type spectra

$V \sin i$ (km/sec)	SB, $P < 100^{\mathrm{d}}$	All
0–50	63%	23%
50–100	34%	22%
100–150	1%	18%
150–200	1%	20%
200–250		9%
250–300		5%
300–350		3%
Number	69	330

TABLE VI

Distribution of $V \sin i$ for spectroscopic
binaries and normal stars with B-type
spectra

$V \sin i$ (km/sec)	SB	Normal
0–100	53%	31%
100–200	39%	39%
200–300	8%	18%
300–400		8%
400–500		4%
Number	77	277

'normal stars' lie in this range. Something very similar happens among the B-type stars, for which similar statistics were made, which are given in Table VI.

In both tables we have called 'normal stars' all stars regardless of their duplicity status. This implies that in their frequency distribution one also includes spectroscopic binaries. It would evidently be preferable to compare the distribution of the rotation of spectroscopic binaries with the distribution for single stars. This is impossible, however, because many stars are considered to be single simply because we ignore if they are binaries. It is clear from the data of Table V, however, that if the spectroscopic binaries are a sizable fraction of all A-type stars (for instance one-third), it implies that essentially no single stars with low rotation exist. We are therefore led to the conclusion that in field stars also duplicity is closely related to slow rotation.

It must be remarked that this kind of relationship is quite contrary to what one would expect. In the first place one expects short periods to produce rapid rotation because of synchronism. In the second place one would expect that the discovery of spectroscopic binaries is more likely if $\sin i$ is large. If moreover rotational axes are normal to the orbital planes, one would expect spectroscopic binaries to be associated with larger than normal rotation. This is evidently not borne out by Tables III, V and VI.

We can now ask if there are possible selection effects which could explain the distribution found in Table III. The easiest one to visualize is the one implying that fuzzy lines do not permit the detection of small amplitudes. If this is true one would expect to find (1) no spectroscopic binaries with large $V \sin i$ and small amplitudes; (2) spectroscopic binaries with large $V \sin i$ associated with large amplitudes. The data assembled in Table VII show that this is not true and therefore this selection effect

TABLE VII

Distribution of semi-amplitudes as a function of $V \sin i$, for A-type spectroscopic binaries

$V \sin i$ (km/sec)	K (km/sec)			
	0–49	50–99	100–149	150–199
0–39	18	10	3	2
40–79	13	6	0	
80–119	4	4	0	1

cannot be very important. Besides this, it seems rather unlikely that if spectroscopic binaries with large $V \sin i$ do exist, none was observed until now in a sample of 70.

The conclusion is therefore that selection effects seem incapable of explaining the close association between duplicity and slow rotation. In view of the importance of this point it would be very interesting for someone to examine a sample of rapidly rotating stars for duplicity.

A final word might be added with regard to the 'breaking point' in this correlation. In other words, from which point on do we have rotational independence? We have seen already that for $P > 100^d$ higher rotations are present. One can associate this limit with the minimum distance of two protostars at the stage when rapid collapse stops. At this time the radius of the stars is given by

$$R/R_0 \sim 50(\mathfrak{M}/\mathfrak{M}_0)$$

which in our case, with $\mathfrak{M} \sim 3\mathfrak{M}_0$ gives $P = 150^d$ and with $\mathfrak{M} \sim 2\mathfrak{M}_0$ gives 65^d, thus bracketing nicely the value of $P = 100^d$.

References

Abt, H. A. and Hunter, J. H.: 1962, *Astrophys. J.* **136**, 381.
Adams, W. S., Joy, A. H., Humason, M. L., and Brayton, A. M.: 1935, *Astrophys. J.* **81**, 187.
Batten, A. H.: 1967, *Publ. Dominion Astron. Obs. Victoria* **13**, 119.
Boyarchuk, A. A. and Kopylov, I. M.: 1958, *Soviet Astron.* **35**, 804.
Boyarchuk, A. A. and Kopylov, I. M.: 1964, *Izv. Krimsk. Astron. Obs.* **31**, 44.
Cameron, R.: 1966, Georgetown Observatory Monograph No. 21.
Cowley, A., Cowley, C., Jaschek, M., and Jaschek, C.: 1969, *Astron. J.* **74**, 375.
Henry, R. C.: 1969, *Astrophys. J Suppl.* **18**, 47.
Koch, R. H., Olson, E. C., and Yoss, K. M.: 1965, *Astrophys. J.* **141**, 955.
Olson, E. C.: 1968, *Publ. Astron. Soc. Pacific* **80**, 185.
Van den Heuvel, E. P. J.: 1965, *Observatory* **85**, 241.
Van den Heuvel, E. P. J.: 1968, *Bull. Astron. Inst. Neth.* **19**, 309.

Wilson, R. E.: 1953, *General Catalogue of Stellar Radial Velocities*, Carnegie Inst. of Washington, Washington D.C.

Discussion

Abt: In your comparison of Mt. Wilson and MK types, the types for broad-lined stars are either too early on the Mt. Wilson system or too late on the MK system. The first possibility seems more likely.

Jaschek: I agree completely and I am sorry that I did not succeed in making my point clear. Certainly the figure implies that the Mt. Wilson types are influenced by rotation in the sense they should be, i.e., faster rotators are classified as being earlier. This is so because the Mt. Wilson observers classified at higher dispersion and are therefore subject to the influence of rotation.

Abt: You point to a lack of spectroscopic binaries with $V \sin i \simeq 400$ km/sec. But such binaries would have synchronous periods of $\frac{1}{4}$ day, which is impossible for A-type stars. Even periods less than one day are rare; I believe that none are known among the stars brighter than $V = 6.0$ mag.

Jaschek: I think there are some stars with $P < 1^d$ which should rotate faster than 100 km/sec. But the main point is that there should be some spectroscopic binaries with P larger than the synchronization limit, which exhibit velocities larger than 100 km/sec, and you simply do not find them.

Collins: It appears from your graph of β vs. $V \sin i$ for given spectral types that β has the expected theoretical dependence which results from changes in β not the MK type. Thus, I would like to suggest that it may not be rotational effects on MK types that you observe, but rather rotational effects on the β index.

Jaschek: This is probably true, but I would also like to emphasize that one should not take these conclusions too seriously, because they are based on too few stars. The important point is that instead of assuming that the MK types are not influenced by rotation, one must try to prove it.

Buscombe: (1) The form of your negative correlation of $H\beta$ line-strength (based on photoelectric indices) with rotational velocity for stars classified A5V, A7V, and A8V on the MK system is in the same sense as my equivalent widths for B stars.

(2) It has come to notice that some early published measures of equivalent width for A stars have not taken sufficient account of the contribution of the wings, which also are not included in the narrower Strömgren $H\beta$ filter.

Steinitz: Dr. Jaschek concludes that the velocities in binaries are independent by looking at the distribution of $v_1 \sin i_1 + v_2 \sin i_2$; however, it can be easily shown that this distribution does not give the relevant information, which can be, however, obtained from the distribution of

$$| V_1 \sin i_1 - V_2 \sin i_2. |$$

Deutsch: There certainly must be a considerable selection against the discovery of wide-line stars as spectroscopic binaries, especially when the periods are larger than a few days.

Jaschek: Yes, certainly. But in about 80 stars one would expect to observe at least one if there were many; since none is observed it is doubtful whether they really do exist.

A STATISTICAL TREATMENT OF THE STELLAR ROTATIONAL VELOCITIES WHICH CONSIDERS THE BREAK-UP LIMIT

P. L. BERNACCA

Asiago Astrophysical Observatory, University of Padua, Italy

Abstract. This paper develops a new statistical approach to the study of the axial rotational velocities of the stars, by considering the break-up limit or any upper limit for the equatorial rotational velocities. The main conclusions are as follows: (1) For random orientation of the axis of rotation, the frequency function of the inclination α to the line of sight is not $\sin\alpha$, as assumed until now. (2) For stars with the same break-up velocity v_b, the distribution of α is given by:

$$\psi(\alpha) = v_b \sin\alpha \int_0^{v_b \sin\alpha} f(y)\,(v_b^2 - y^2)^{-1/2}\,\mathrm{d}y,$$

where $f(y)$ is the distribution of the apparent velocities $y = v \sin\alpha$, v being the true rotational velocity. New integral equations governing the frequency function of α, y and v have been derived and the correct procedure to treat observations has been discussed; a simple method is also suggested to get an approximate trend of the distribution of the true velocities directly from the observed histogram. The method of analysis proposed in this paper has been applied to a re-discussion of the rotational behaviour of a group of Be stars.

1. Introduction

It is well known that the apparent (equatorial) rotational velocity of a star can be determined spectroscopically (Shajn and Struve, 1929; Slettebak, 1949; Huang, 1953). If the axis of rotation of a star is inclined at an angle α to the line of sight, the apparent velocity y is given by:

$$y = v \sin\alpha, \tag{1}$$

where v is the true (equatorial) rotational velocity.

Since the value of α for individual stars is unknown, one can attempt only to derive group properties, such as the frequency distribution of v, from observations of y in a sample of stars and this statistical approach requires some assumption as to the frequency of occurrence of a given angle of inclination among the stars in the sample under consideration. The hypothesis usually made is that the axes of rotation are randomly distributed in space so that the function

$$\psi(\alpha) = \sin\alpha \tag{2}$$

may be considered to represent the distribution of α. Let us recall here briefly the main points of the theory as developed, for instance, by Chandrasekhar and Münch (1950). The integral equations:

$$f(y) = y \int_y^\infty \frac{\varphi(v)\,\mathrm{d}v}{v(v^2 - y^2)^{1/2}} \tag{3}$$

$$\varphi(v) = -\frac{2}{\pi} v^2 \frac{\partial}{\partial v} v \int_v^\infty \frac{f(y)\,dy}{y^2(y^2 - v^2)^{1/2}} \tag{4}$$

relate the frequency function $\varphi(v)$ of the true velocities with the corresponding $f(y)$ of the apparent ones. Since the differentiation of an observed distribution (given often in the form of a histogram) can lead to results which are misleading, instead of deriving $\varphi(v)$ by (4), the function:

$$\varphi(v) = \frac{j}{\sqrt{\pi}} \left[e^{-j^2(v-v_1)^2} + e^{-j^2(v+v_1)^2} \right] \tag{5}$$

is assumed and the problem is solved by appropriate selection of the parameters j and v_1 to fit the computed distribution $f(y)$ to the observed histogram.

The central moments of $\varphi(v)$ can be derived by means of (3) without the knowledge of the function itself; for instance, the mean and the mean square deviation are given respectively by:

$$\langle v \rangle = \frac{4}{\pi} \langle y \rangle \tag{6}$$

$$\langle (v - \langle v \rangle)^2 \rangle = 1.5 \langle y^2 \rangle - \frac{16}{\pi^2} \langle y \rangle^2 \tag{7}$$

so that we can compute them directly from the observations.

In the above analysis no restrictions are imposed on the variables y and v (except of course $y \leqslant v$) and relation (2) is an 'a priori' assumption, which is considered to be valid for any observed random sample of rotating stars. Since the condition of stability requires that a star rotates with an equatorial velocity not greater than its break-up velocity v_b (determined for each star by radius and mass distribution), it follows that any statistical analysis must exclude the expectation of stars rotating with a velocity greater than some known limit; in particular, if a sample of stars with the same break-up velocity is considered, any $\varphi(v)$ must vanish for $v > v_b$ and $\langle v \rangle$ must not exceed v_b when formula (6) is used.

The question arises now whether the foregoing procedure will yield always correct results from this point of view; let us consider a sample of stars whose rotational velocities have a unique upper limit (in particular the same break-up velocity) and suppose that the observed frequency distribution be $f_0(y)$:

If one assumes a function $\varphi(v; v_1, j)$ so that:

$$\varphi(v; v_1, j) = \varphi_0(v; v_1, j) \qquad 0 \leqslant v \leqslant v_b$$
$$= 0 \qquad\qquad\qquad v_b < v$$

it may be possible to fit $f(y; v_1, j)$, as computed by (3), to $f_0(y)$; however this is not true in general. Let us consider the following two cases:

(i) When a distribution:

$$f_1(y) = \frac{y}{v_b(v_b^2 - y^2)^{1/2}} \qquad 0 \leqslant y \leqslant v_b \tag{8}$$

is observed, a Dirac δ-function $\delta(v-v_b)$ is the solution of the integral Equation (3), meaning that all stars rotate at the equatorial break-up velocity. Formula (6) gives, of course, $\langle v \rangle = v_b$ when $\langle y \rangle$ has been computed by (8). On the other hand, for a frequency function $f_2(y)$, so that $f_2(y) < f_1(y)$ in a given interval $(0, y_0)$, it is impossible to find a true distribution $\varphi(v)$ vanishing for $v > v_b$: it is obvious that (6) yields in this case $\langle v \rangle > v_b$.

It can be argued that if $f_2(y)$ is observed, the hypothesis of random orientation for the rotational axes is no longer valid but at least two conclusions can be drawn: first, one must care about the assumption 'a priori' of (2); second, the inclusion of v_b gives the possibility of detecting deviation from randomness.

(ii) Let us suppose that all apparent velocities in a sample have been observed to lie between y_0 and v_b. Since $v < v_b$, relation (1) states that $\sin^{-1}(y_0/v_b) \leqslant \alpha \leqslant \pi/2$ for each star. If the axes of rotation are randomly distributed, with α within the above limits, it may be assumed now 'a priori' (see Figure 2):

$$\psi(\alpha) = \frac{v_b \sin \alpha}{(v_b^2 - y_0^2)^{1/2}} \qquad \alpha \geqslant \sin^{-1}(y_0/v_b)$$
$$= 0 \qquad 0 \leqslant \alpha < \sin^{-1}(y_0/v_b). \tag{9}$$

The integral Equation (3) and (4) need to be modified as well as formulae (6) and (7). It follows that (2) is not always the correct frequency function for the inclination α. Further, an example analogous to case (i) can be easily constructed by means of (9) and the modified integral equations, which would show how the assumption 'a priori' of (9) may be not quite correct.

In the following sections we attempt to derive a general frequency distribution $\psi(\alpha)$ and develop thereby a different method of analysis; therefore, a conclusive judgment on the statistical meaning of (2) will be given later.

2. The Distribution Function of the Inclination of the Rotational Axes

Let us consider a sample of stars of known break-up velocity v_b and whose apparent velocities y have been determined: by (1) and the condition $v \leqslant v_b$, we have for each star:

$$y \leqslant v_b \sin \alpha. \tag{10}$$

If an angle θ is defined by the relation

$$y = v_b \sin \theta \tag{11}$$

the inclination α of each star must satisfy the condition:

$$\theta \leqslant \alpha \leqslant \pi/2. \tag{12}$$

Since y may range from 0 to one of the values v_b, θ will vary between 0 and $\pi/2$. In the (θ, α)-plane the representative points of the stars lie in the triangle defined by the straight lines $\theta = 0$, $\alpha = \pi/2$ and $\alpha = \theta$ (Figure 1). The number of stars dn with θ and α

within $(\theta, \theta+d\theta)$ and $(\alpha, \alpha+d\alpha)$ can be written:

$$dn = N\psi(\theta, \alpha)\, d\theta\, d\alpha, \tag{13}$$

where N is the total number of stars and $\psi(\theta, \alpha)$ is the distribution function in the (θ, α)-plane.

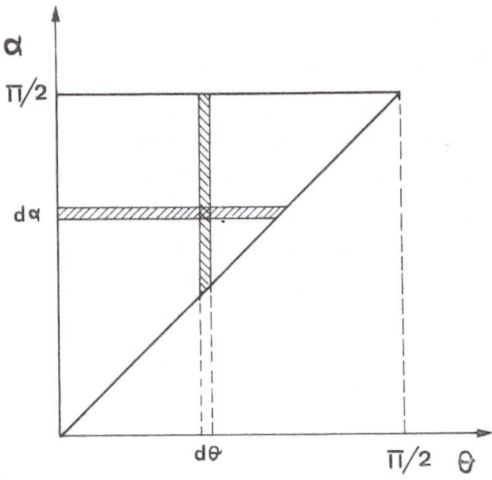

Fig. 1. The triangle defines the region occupied by the representative points (θ, α) of stars rotating with an inclination α to the line of sight and whose apparent velocity is $y = v_b \sin\theta$, v_b being the break-up limit.

The bivariate distribution $\psi(\theta, \alpha)$ can be written (see e.g. Trumpler and Weaver, 1953) as:

$$\psi(\theta, \alpha) = \psi_\theta(\theta)\,\psi(\alpha\backslash\theta), \tag{14}$$

where $\psi_\theta(\theta)$ is the marginal distribution of the sample with respect to θ, which is known from the observations, while $\psi(\alpha\backslash\theta)$ is the frequency function of α for each θ-array. If one picks from the sample all stars with a given θ and assumes that the axes of rotation of these stars are randomly distributed in space with the condition (12), it follows (see Figure 2):

$$\psi(\alpha\backslash\theta)\, d\alpha = \frac{2\pi \sin\alpha\, d\alpha}{2\pi - 2\pi(1-\cos\theta)} = \frac{\sin\alpha}{\cos\theta}\, d\alpha \qquad \alpha \geqslant \theta. \tag{15}$$

Thus the bivariate distribution (14) is known and the marginal distribution of the sample with respect to α, $\psi(\alpha)$, can be obtained simply by:

$$\psi(\alpha) = \int_0^\alpha \psi(\theta, \alpha)\, d\theta = \sin\alpha \int_0^\alpha \frac{\psi_\theta(\theta)}{\cos\theta}\, d\theta. \tag{16}$$

If only stars with the same break-up velocity are considered (or if only one upper limit to the true velocities for all stars does exist), the frequency function of θ can be written

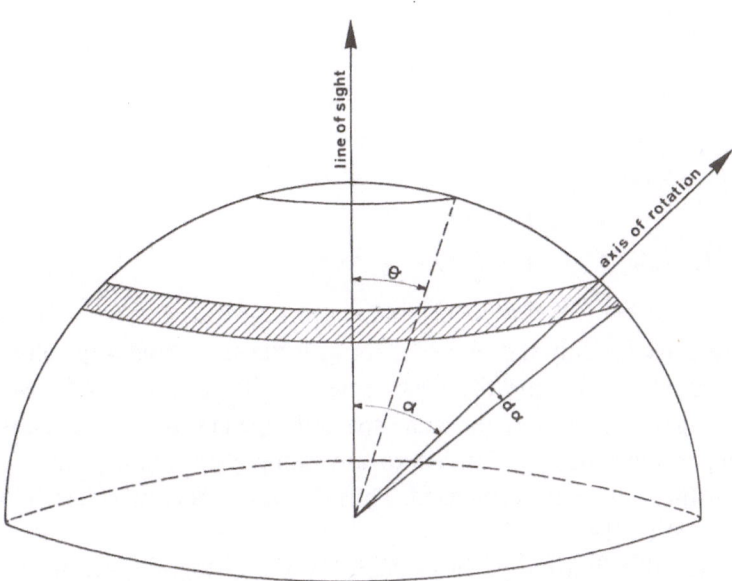

Fig. 2.   The probability of occurrence of α (ranging from θ to $\pi/2$), between α and $\alpha + d\alpha$, is given by $\sin \alpha d\alpha/\cos \theta$.

as:

$$\psi_\theta(\theta) = v_b f(v_b \sin \theta) \cos \theta \tag{17}$$

so that, by means of (16), we have:

$$\psi(\alpha) = v_b \sin \alpha \int_0^\alpha f(v_b \sin \theta) \, d\theta \tag{18}$$

or

$$\psi(\alpha) = v_b \sin \alpha \int_0^{v_b \sin \alpha} \frac{f(y) \, dy}{(v_b^2 - y^2)^{1/2}} . \tag{19}$$

Relations (16) and (18) or (19) represent the distribution of the inclination of the rotational axes; they depend on the observations through $\psi_\theta(\theta)$ or $f(y)$.

The statistical meaning of (2) and (9) now becomes clear, since they can be derived from (19) if particular observations such as $f(y) = \delta(y)$ and $f(y) = \delta(y - y_0)$ are considered respectively: the early function $\psi(\alpha) = \sin \alpha$ [and (9) too] appears to be only an array distribution and it can no longer be used in the rotational context, unless some empirical information exists showing that the fraction of stars with α between α and $\alpha + d\alpha$ is just $\sin \alpha \, d\alpha$.

3. The Moments of the Distribution of the True Rotational Velocities

The mean, the mean square deviation and the skewness of $\varphi(v)$ can be derived from the general relation:

$$\langle y^n \rangle = \langle v^n \rangle \int_0^{\pi/2} (\sin\alpha)^n \, \psi(\alpha) \, d\alpha. \tag{20}$$

Let A_n be the integral factor contained in (20); it then follows that:

$$\langle v \rangle = A_1^{-1} \langle y \rangle \tag{21}$$

$$\langle (v - \langle v \rangle)^2 \rangle = A_2^{-1} \langle y^2 \rangle - A_1^{-2} \langle y \rangle^2 \tag{22}$$

$$\langle (v - \langle v \rangle)^3 \rangle = A_3^{-1} \langle y^3 \rangle - 3(A_1 A_2)^{-1} \langle y \rangle \langle y^2 \rangle + 2A_1^{-3} \langle y \rangle^3. \tag{23}$$

Hereafter we consider only the case of stars with the same break-up velocity, so that (21) may be used to compute the mean true velocity for stars of known apparent velocity, having approximately the same spectral type and luminosity class; however the following results may be applied to any sample of rotating stars whenever a unique upper limit has been recognized. Thus the A_n are obtained from (20) using the distribution (18) or (19).

Any observed distribution, given in the form of a histogram, can be written as:

$$f(y) = I_r = \frac{N_r}{N(y_r - y_{r-1})} \qquad y_{r-1} \leqslant y \leqslant y_r \tag{24}$$

with $r = 1, 2, \ldots k$, where k is the number of intervals in which $(0, v_b)$ is divided, N is the total number of stars, N_r is the number of stars whose apparent velocity satisfies the condition $y_{r-1} < y < y_r$; further, in formula (24), $y_0 = 0$ and $y_k = v_b$.

Since the N_r stars may be considered uniformly distributed in (y_{r-1}, y_r), their distribution function $f_r(y)$ is given by:

$$\begin{aligned} f_r(y) &= 1/(y_r - y_{r-1}) \qquad y_{r-1} \leqslant y \leqslant y_r \\ &= 0 \qquad\qquad\qquad \text{elsewhere} \end{aligned} \tag{25}$$

or, by means of (24):

$$\begin{aligned} f_r(y) &= I_r N / N_r \qquad y_{r-1} \leqslant y \leqslant y_r \\ &= 0 \qquad\qquad \text{elsewhere} \end{aligned} \tag{26}$$

Each $f_r(y)$ produces a distribution function $\psi_r(\alpha)$, that, through (18) or (19), becomes:

$$\psi_r(\alpha) = \begin{bmatrix} 0 & 0 \leqslant \alpha \leqslant \theta_{r-1} \\[2mm] v_b(\alpha - \theta_{r-1}) \dfrac{I_r N}{N_r} \sin\alpha & \theta_{r-1} \leqslant \alpha \leqslant \theta_r, \\[3mm] v_b(\theta_r - \theta_{r-1}) \dfrac{I_r N}{N_r} \sin\alpha & \theta_r \leqslant \alpha \leqslant \pi/2 \end{bmatrix} \tag{27}$$

where θ_r is defined by $y_r = v_b \sin\theta_r$.

From the definition of a distribution function, it can be shown easily that the total distribution $\psi(\alpha)$ for the sample of the N stars, is given by:

$$\psi(\alpha) = N^{-1} \sum_1^k N_r \psi_r(\alpha) \tag{28}$$

so that, by (20), (27) and (28), we have for the quantities A_n:

$$A_n = v_b \sum_1^k I_r \left[\int_{\theta_{r-1}}^{\theta_r} (\alpha - \theta_{r-1})(\sin \alpha)^{n+1} \, d\alpha + (\theta_r - \theta_{r-1}) \int_{\theta_r}^{\pi/2} (\sin \alpha)^{n+1} \, d\alpha \right]. \tag{29}$$

Taking $n = 1, 2, 3$ successively, after integration and some reductions, (29) gives:

$$A_1 = \frac{v_b}{4} \sum_1^k I_r \left[\sin^2 \theta_r - \sin^2 \theta_{r-1} + \pi(\theta_r - \theta_{r-1}) - (\theta_r^2 - \theta_{r-1}^2) \right] \tag{30}$$

$$A_2 = \frac{2}{3} + \frac{v_b}{9} \sum_1^k I_r (\sin^3 \theta_r - \sin^3 \theta_{r-1}) \tag{31}$$

$$A_3 = \frac{3}{4} A_1 + \frac{v_b}{16} \sum_1^k I_r (\sin^4 \theta_r - \sin^4 \theta_{r-1}), \tag{32}$$

where for A_2 the definition of I_r and θ_r has been taken into account.

The quantities A_n ($n = 1, 2, 3$) appear to depend on the value of the histogram I_r in the interval r and on the particular subdivision of $(0, v_b)$, which defines each θ_r. Although some general rules must be followed in the drawing of a histogram, it will depend somewhat on the choice of the observer. In order to illustrate the procedure and to test the personal choice of a histogram, we consider the rotational velocities

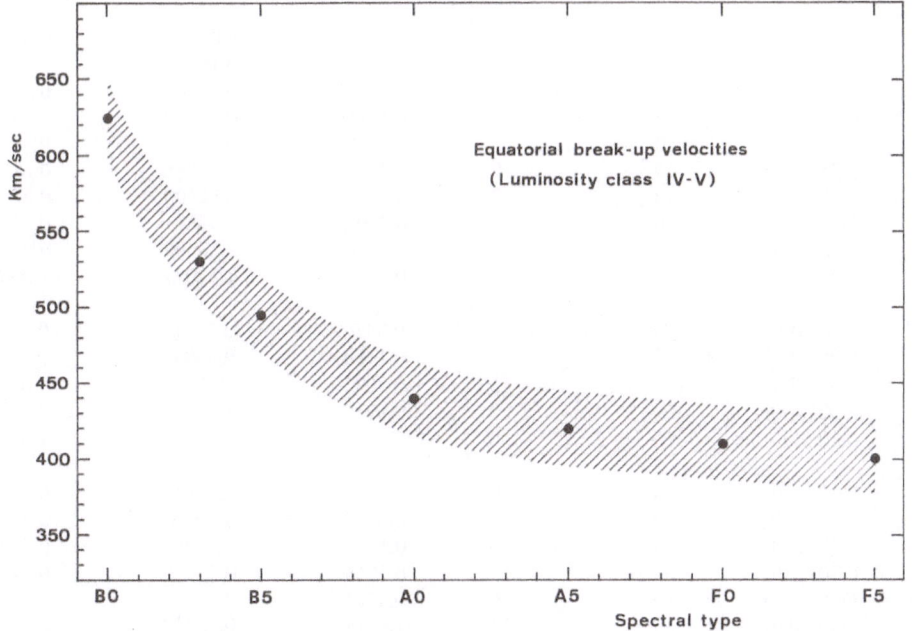

Fig. 3.   Slettebak's computed break-up velocities against spectral type (full points). The uncertainty is of the order of 25 km/sec (shaded area).

determined by Slettebak (1966a) for 42 B6-B9e stars of luminosity class IV-V.

In Figure 3 the relationship between spectral type and break-up velocity is shown according to Slettebak (1966b); the shaded region represents the uncertainty in v_b which is of the order of 25 km/sec. The break-up limit for B6-B9 stars ranges roughly from 480 km/sec to 450 km/sec, but it may be lowered for Be stars if they are considered to be evolved objects. In the following a limit of 450 km/sec is assumed for all the 42 stars, an approximation that is sufficient for our aim.

In Table I three possible frequency functions are given, computed by means of (24), together with θ_r and $\sin \theta_r$; the subdivision of $(0, v_b)$ in k intervals has been made according to the first column of Table I and histogram (b) has been constructed with the condition $y_{r-1} \leqslant y < y_r$ for all but one star, whose velocity of 400 km/sec has been arbitrarily lowered. The value of A_n ($n = 1, 2, 3$) is practically the same for histogram (a) and (c) while it is a little higher for case (b). According to formulae (21)–(23) we need first $\langle y \rangle$ and the higher moments of $f(y)$: if they are derived by means of the histogram itself, the results are slightly different in the three cases (Table II, column (i)); if the observed moments are computed directly from the values of the apparent rotational velocities, no differences occur (Table II, column (ii)).

These results encourage one to believe that different histograms will yield with

TABLE I

Three possible empirical distributions (I_r) for the same sample of Be stars (Slettebak, 1966a)

r	(y_{r-1}, y_r)	$y_r - y_{r-1}$	N_r	$10^2 I_r$	θ_r	$\sin \theta_r$
			(a)			
1	(0, 25)	25	0	0	0.05558	0.05556
2	(25, 75)	50	2	0.0952	0.16745	0.16667
3	(75, 125)	50	1	0.0476	0.28148	0.27778
4	(125, 175)	50	3	0.1428	0.39942	0.38889
5	(175, 225)	50	3	0.1428	0.52360	0.50000
6	(225, 275)	50	10	0.4761	0.65746	0.61111
7	(275, 325)	50	9	0.4285	0.80700	0.72222
8	(325, 375)	50	13	0.6190	0.98511	0.83333
9	(375, 425)	50	1	0.0476	1.23590	0.94444
10	(425, 450)	25	0	0	1.57079	1.00000
			(b)			
1	(0, 100)	100	2	0.0476	0.22410	0.22222
2	(100, 200)	100	5	0.1190	0.46055	0.44444
3	(200, 300)	100	13	0.3095	0.72973	0.66667
4	(300, 400)	100	22	0.5238	1.09491	0.88889
5	(400, 450)	50	0	0.	1.57079	1.00000
			(c)			
1	(0, 50)	50	0	0	0.11134	0.11111
2	(50, 130)	80	3	0.0892	0.29310	0.28889
3	(130, 220)	90	6	0.1587	0.51081	0.48889
4	(220, 280)	60	10	0.3968	0.67158	0.62222
5	(280, 330)	50	9	0.4285	0.82321	0.73333
6	(330, 360)	30	13	1.0317	0.92730	0.80000
7	(360, 430)	70	1	0.0340	1.27153	0.95556
8	(430, 450)	20	0	0	1.57079	1.00000

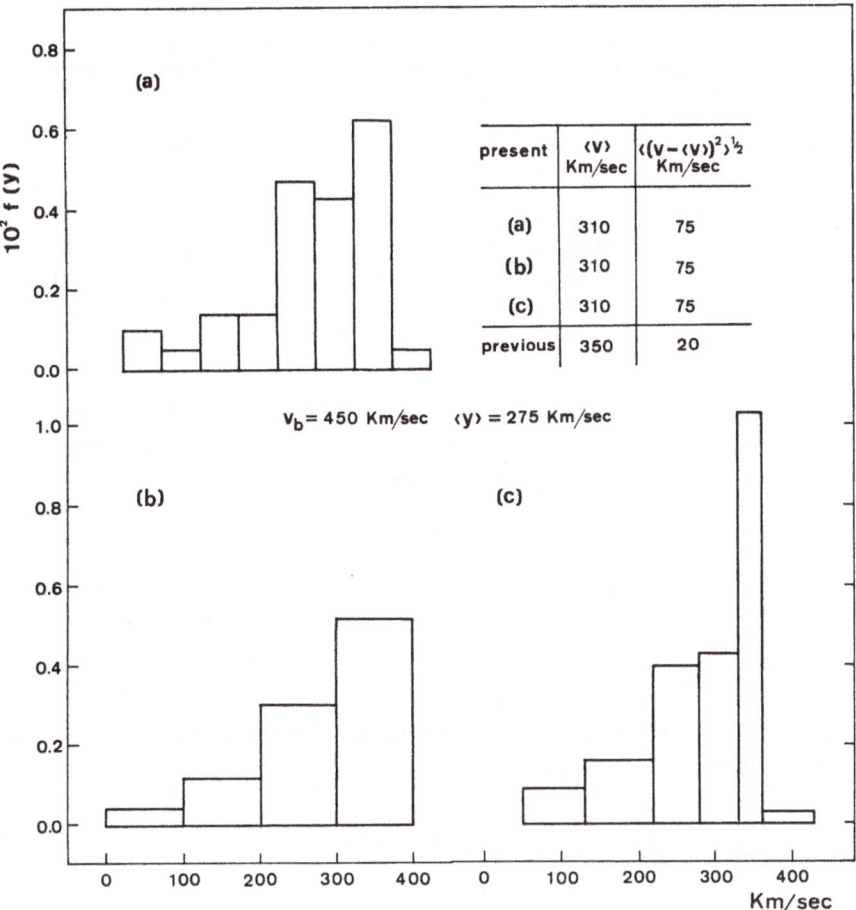

Fig. 4.   (a), (b), (c). Three possible empirical distributions for the same group of stars. Inset: comparison between the central moments of the distribution of the true rotational velocities computed by the present method and the previous one. The case is the one of the Be stars studied by Slettebak (1966a).

sufficient accuracy the same information for a given sample of stars. One must bear in mind that the variety of histograms for the same sample is not large, since the size of the intervals (y_{r-1}, y_r) can be neither too small nor too large, in order to avoid accidental errors or loss of information, respectively. In Figure 4 the three histograms of Table I have been drawn together with the present values of the first two central moments of $\varphi(v)$ and the corresponding ones determined by (6) and (7) (previous): the mean true velocity is 310 km/sec instead of 350 km/sec and the resulting root mean square deviation is much larger than the previous one: 75 km/sec instead of 20 km/sec.

4. The Distribution Function of the True Rotational Velocities for Homogeneous Samples

Hereafter we discuss the method of deriving a true distribution only for homogeneous

TABLE II

The mean and the root mean square of the true velocities v derived for the distribution of Table I: (i) by $\langle y \rangle$ and $\langle y^2 \rangle$ given by the histogram and (ii) by $\langle y \rangle$ and $\langle y^2 \rangle$ computed directly from the observations

Type	A_1	A_2	A_3	(i)			(ii)		
				$\langle y \rangle$	$\langle v \rangle$	σ_v	$\langle y \rangle$	$\langle v \rangle$	σ_v
a	0.88489	0.79983	0.73308	270	305	85	275	310	75
b	0.89112	0.81010	0.74626	280	315	90	275	310	75
c	0.88499	0.79935	0.73229	270	310	80	275	310	75

$\sigma_v = \langle (v - \langle v \rangle)^2 \rangle^{1/2}$

samples, i.e. for samples of stars which have the same equatorial break-up velocity. However, the method can be used for stars in general whenever an upper limit has been recognized, aside from the concept of break-up.

A. FORMAL SOLUTION

The true rotational velocity v, can be expressed, according to (1) and (11) as:

$$v = \frac{v_b \sin \theta}{\sin \alpha} \qquad \alpha \geqslant \theta, \tag{33}$$

where v_b is now a constant. The function $\varphi(v)$ can be obtained from the bivariate frequency function $\psi(\theta, \alpha)$ (Section 2) as follows:

$$\varphi(v) = \int\limits_0^{\sin^{-1}(v/v_b)} \psi[\alpha(v, \theta), \theta] \left| \frac{\partial \alpha(v, \theta)}{\partial v} \right| d\theta. \tag{34}$$

Using relations (14), (15), (18) and (33) in (34), results in:

$$\varphi(v) = \frac{v_b}{v^2} \int\limits_0^v \frac{y^2 f(y) \, dy}{\sqrt{(v^2 - y^2)} \sqrt{(v_b^2 - y^2)}}. \tag{35}$$

The integral Equation (35) can be easily solved with respect to $f(y)^*$ in order to

---

* Relation (35) is a Volterra's equation of the first kind which may be written

$$v\varphi(v) = \int\limits_0^v \frac{v_b y^2 f(y)}{\sqrt{(v_b^2 - y^2)} \, v \sqrt{(v^2 - y^2)}} dy. \tag{a}$$

Since the kernel $k(y, v) = 1/v \sqrt{(v^2 - y^2)}$ becomes infinite for $v = y$, solution (36) can be obtained by the following procedure:
(1) multiply both sides of the Equation (a) by $dv/\sqrt{(u^2 - v^2)}$;
(2) integrate with respect to v from 0 to u after changing the order of integration in the double integral;
(3) take the derivative with respect to u of both sides of the resulting equation;
(4) write y instead of u.

have a relation analogous to (3):

$$f(y) = \frac{2}{\pi v_b} \frac{\sqrt{(v_b^2 - y^2)}}{y} \frac{\partial}{\partial y} y \int_0^y \frac{v\varphi(v)\,dv}{\sqrt{(y^2 - v^2)}}. \tag{36}$$

If the observed histogram is of high precision, the distribution $\varphi(v)$ can be obtained by (35), through numerical integration, provided that the empirical distribution has been approximated by a continuous function $f(y)$. On the other hand, owing to the uncertainty in the measures of the apparent rotational velocities or to the small number of the stars, it would be more correct to use (36) after assuming some form for $\varphi(v)$, with two parameters, for example of the type (5). In this case, for the numerical integration, it is more convenient to write (36) as:

$$f(y) = \frac{2}{v_b} \frac{\sqrt{(v_b^2 - y^2)}}{y} \left[\frac{y_0^2}{\sqrt{(y^2 - y_0^2)}} \varphi(y_0) \right.$$

$$\left. + \int_{\sin^{-1}(y_0/y)}^{\pi/2} \sin\beta \frac{\partial}{\partial y} [y^2 \varphi(y\sin\beta)]\,d\beta \right], \tag{37}$$

where $\sin\beta = v/y$, and the function $\varphi(v)$ must be defined different from zero only in (y_0, v_b); further, y_0 ranges from 0 to v_b, depending on the interval $(0, y_0)$ where no apparent velocities have been observed.

It should be pointed out that the procedure contained in Equations (36) or (37) is analogous to that sketched in Section 1, but it has a different meaning: if we assume a function $\varphi(v) = 0$ in $(0, y_0)$ and $\varphi(v) \neq 0$ in (y_0, v_b), we can find a function $f(y) \neq 0$ in $(0, y_0)$ by means of (3), but this is impossible through (36) or (37); this fact does not imply the impossibility of observing rotational velocities smaller than y_0 for stars whose true rotational velocity is greater than y_0; it simply emphasizes that the present theory has been developed starting from the observations and gives a distribution $\varphi(v)$ different from zero in every interval where $f(y)$ does not vanish. It follows that any assumed $\varphi(v)$ has not necessarily a real physical meaning, unless we are able to find it, when the computed $f(y)$ has been fitted to the observed histogram; relations (36) and (37) are simply a mathematical tool in order to assign to the block-curve of the apparent ones, a continuous distribution of the true velocities which might be free from the accidental element contained in an empirical distribution, as far as possible.

B. SPECIAL FORMS OF DISTRIBUTIONS

With the purpose of comparing the results of the present statistical treatment with the ones derived under $\sin\alpha$, we consider in the following some special functions $f(y)$, which may conceivably represent particular observations;

(a) Suppose that the observed rotational velocities are distributed according to (8): it has already been shown (Section 1) that all stars rotate at the equatorial break-up if $\psi(\alpha) = \sin\alpha$.

Presently, the frequency function $\psi(\alpha)$ is, by (8) and (19):

$$\psi(\alpha) = - \sin \alpha \ln (\cos \alpha) \tag{38}$$

which indicates that most of the stars are viewed nearly equatorially (Figure 5, curve a); hence the observed distribution will differ little from the true one, which turns out

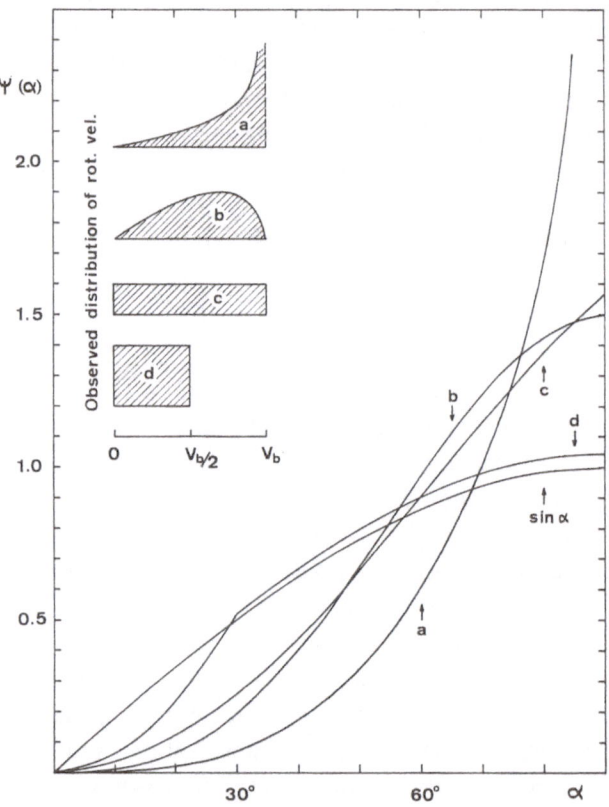

Fig. 5. The actual distribution function of the inclination α is shown for different observed distributions (a, b, c, d) of the rotational velocities. The previous function sin α is plotted for comparison.

to be:

$$\varphi(v) = \frac{v_b^2}{v^2 \sqrt{(v_b^2 - v^2)}} \sin^{-1} \frac{v}{v_b} - \frac{1}{v}. \tag{39}$$

The function (39) has been derived tediously by (35) and plotted in Figure 6a as a solid line. The difference between (39) and the δ-solution (plotted as a dashed line in the figure) is remarkable. The result can be applied to the apparent rotational behaviour of the Be stars considered in Section 3; the observed cut-off is near 350 km/sec, if one star with $v \sin \alpha = 400$: km/sec is excluded. Since formula (8) with $v_b = 350$ approximates satisfactorily the histogram ((a) and (c) of Figure 4), it has been believed

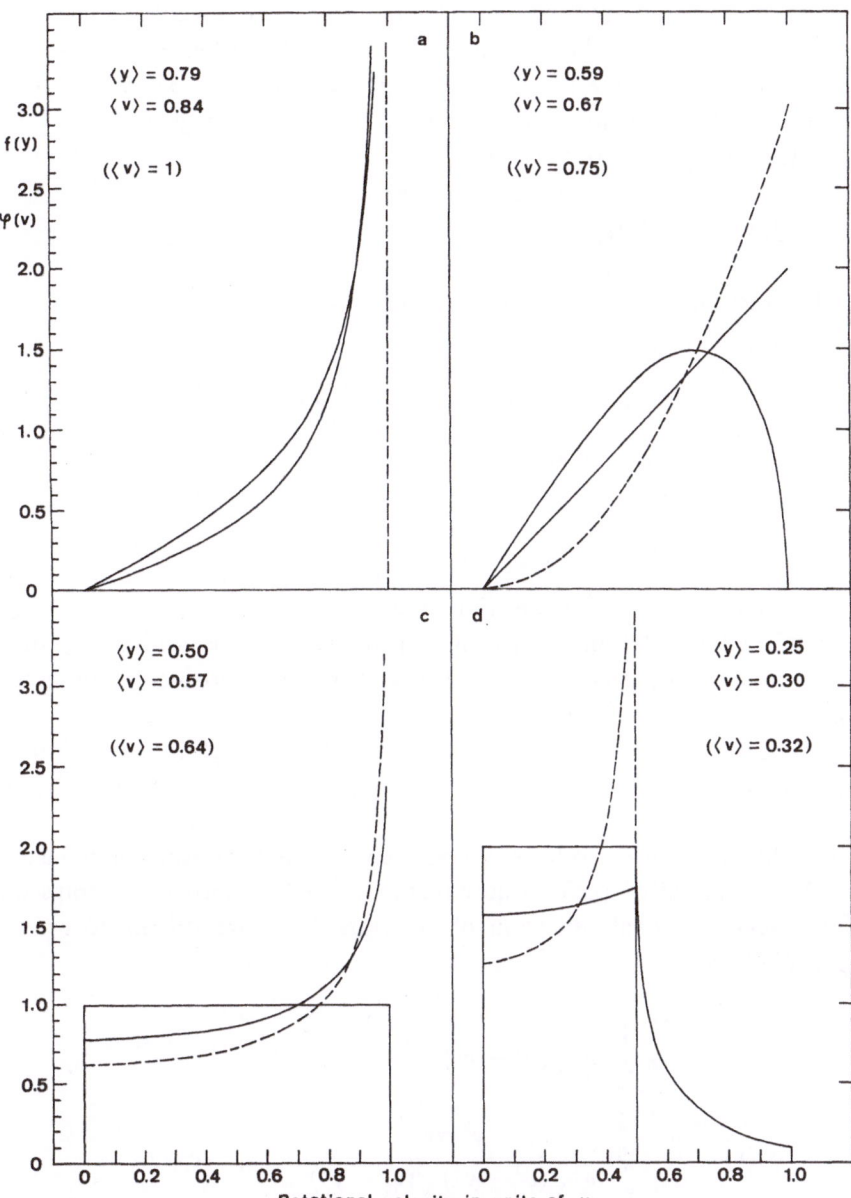

Fig. 6. The present distribution of the true rotational velocities (solid lines) is compared with that derived under sin α (broken lines). The corresponding observed distributions $f(y)$, given by the appropriate solid tracings, are the same as in Figure 5 (a, b, c, d). The mean of the apparent and true rotational velocities ($\langle y \rangle$ and $\langle v \rangle$) are also indicated in units of the break-up limit v_b. In parentheses, the value of $\langle v \rangle$ has been computed by $\langle v \rangle = (4/\pi) \langle y \rangle$. For a correct understanding of case d, see discussion in the text.

that we are concerned with stars rotating probably at the same velocity of 350 km/sec (Slettebak, 1966a; Huang, 1969).

If an effective upper limit to the true velocities near 350 km/sec is assumed, because no velocity has been observed beyond it, we have actually no longer $\varphi(v) = \delta(v - 350)$, but (39) where $v_b = 350$; the mean true velocity is now $0.84 \times 350 \approx 300$ instead of ≈ 350 km/sec and the spread around the mean becomes not negligible.

(b) Let the observed distribution be

$$f(y) = \frac{3}{v_b^3} y \sqrt{(v_b^2 - y^2)}. \tag{40}$$

The angle α turns out to be distributed according to:

$$\psi(\alpha) = \tfrac{3}{2} \sin^3 \alpha \tag{41}$$

(see Figure 5, curve b) and for the true velocities we have:

$$\varphi(v) = 2v/v_b^2 \qquad 0 \leqslant v \leqslant v_b \tag{42}$$

instead of:

$$\varphi(v) = 3v^2/v_b^3 \qquad 0 \leqslant v \leqslant v_b \tag{43}$$

which is the solution of the integral Equation (3).

The trend of both (42) and (43) is shown in Figure 6b respectively by the solid straight line and the broken one: the solid curve represents the observations according to (40).

Let us consider now an apparent distribution of the type (25); the function $\psi(\alpha)$ is given by (27) where it is now $N_r = N$ and $I_r = 1/(y_r - y_{r-1})$; hereafter write Δy_r instead of $y_r - y_{r-1}$.

Suppose that the apparent velocities have been accurately measured and the number of stars, N, is sufficiently large to believe that (25) may be a fairly good representation of the apparent rotational behaviour of our stars. If so, we attempt to use (35) to get $\varphi_r(v)$. We have:

$$\varphi_r(v) = \frac{v_b}{\Delta y_r} \int\limits_{y_{r-1}}^{v} \frac{y^2 \, dy}{v^2 \sqrt{(v^2 - y^2)} \sqrt{(v_b^2 - y^2)}} \qquad y_{r-1} \leqslant v \leqslant y_r \tag{44}$$

$$\varphi_r(v) = \frac{v_b}{\Delta y_r} \int\limits_{y_{r-1}}^{y_r} \frac{y^2 \, dy}{v^2 \sqrt{(v^2 - y^2)} \sqrt{(v_b^2 - y^2)}} \qquad y_r \leqslant v \leqslant v_b \tag{45}$$

and, of course, $\varphi_r(v) = 0$ for $0 \leqslant v \leqslant y_{r-1}$. The above integrals can be evaluated by means of elliptic integrals, as follows.

By (1) and the positions $v = xv_b$ and $y_r = s_r v_b$, (45) becomes:

$$\Delta y_r \varphi_r(xv_b) = \int\limits_{\sin^{-1}(s_{r-1}/x)}^{\sin^{-1}(s_r/x)} \frac{\sin^2 \alpha \, d\alpha}{\sqrt{(1 - x^2 \sin^2 \alpha)}} \qquad s_r \leqslant x \leqslant 1. \tag{46}$$

If the upper limit of the integration in (46) is set up equal to $\pi/2$ we obtain (44).

Consider the Legendre integral $L_r(x)$, defined by:

$$L_r(x) = \int_0^{\sin^{-1}(s_r/x)} \frac{\sin^2 \alpha \, d\alpha}{\sqrt{(1 - x^2 \sin^2 \alpha)}} \qquad (47)$$

and the following elliptic integrals of first and second kind, respectively:

$$F_r(x) = \int_0^{\sin^{-1}(s_r/x)} (1 - x^2 \sin^2 \alpha)^{-1/2} \, d\alpha \qquad (48)$$

$$E_r(x) = \int_0^{\sin^{-1}(s_r/x)} (1 - x^2 \sin^2 \alpha)^{1/2} \, d\alpha. \qquad (49)$$

Integrals (48) and (49) can be readily found in mathematical tables. In terms of (48) and (49), the Legendre integral (47) can be written:

$$L_r(x) = x^{-2} [F_r(x) - E_r(x)]. \qquad (50)$$

If the upper limit of integration in (47), (48), (49) is set up equal to $\pi/2$, formula (50) yields the complete Legendre integral $L_c(x)$, so that (44) and (45) can be evaluated by means of:

$$\Delta y_r \varphi_r(x v_b) = L_c(x) - L_{r-1}(x) \qquad s_{r-1} \leqslant x \leqslant s_r \qquad (51)$$
$$\Delta y_r \varphi_r(x v_b) = L_r(x) - L_{r-1}(x) \qquad s_r \leqslant x \leqslant 1. \qquad (52)$$

Let us consider now two particular cases of distributions (25):

(c) $y_r = v_b,$ $y_{r-1} = 0$
(d) $y_r = y_0 \neq v_b,$ $y_{r-1} = 0.$

In case (c) we soon find that the rotational axes are distributed according to $\alpha \sin \alpha$ (Figure 5, curve c) and, from the preceding analysis, the function which governs the true velocities turns out to be:

$$v_b \varphi(v) = L_c(v/v_b). \qquad (53)$$

If (2) is assumed to represent the distribution of α, $\varphi(v)$ can be obtained by (4) using in it $f(y) = 1/v_b$, resulting in:

$$\varphi(v) = \frac{2}{\pi} \frac{1}{\sqrt{(v_b^2 - v^2)}}. \qquad (54)$$

The difference between (53) and (54) is small (Figure 6c); for instance, according to (54) 32% of the stars have a rotational velocity between 0 and $0.5 v_b$, while by (53) we have 40% in the same interval.

The special distribution (d) requires particular attention if we treat it by (51) and (52).

The analytical form of $\psi(\alpha)$ is:

$$\psi(\alpha) = \frac{v_b}{y_0}\, \alpha \sin\alpha \qquad 0 \leqslant \alpha \leqslant \theta_0$$

$$= \frac{v_b}{y_0}\, \theta_0 \sin\alpha \qquad \theta_0 \leqslant \alpha \leqslant \pi/2 \tag{55}$$

and the distribution of the true velocities, through (51) and (52), where it is now $\Delta y_r = y_0$, is given by:

$$\varphi(v) = y_0^{-1} L_c(v/v_b) \qquad 0 \leqslant v \leqslant y_0$$

$$= y_0^{-1} L\left(\frac{v}{v_b};\ \sin^{-1}\frac{y_0}{v}\right) \qquad y_0 \leqslant v \leqslant v_b. \tag{56}$$

Let us fix the value of y_0 to be $v_b/2$; then (55) differs little from $\sin\alpha$ (Figure 5, curve d) so that we expect that $\varphi(v)$ derived by (3) will differ little from (56); the situation is apparently quite different since it can be easily verified that the following function:

$$\varphi(v) = \frac{2}{\pi}\, \frac{1}{\sqrt{(y_0^2 - v^2)}} \qquad 0 \leqslant v \leqslant y_0$$

$$= 0 \qquad y_0 < v \leqslant v_b \tag{57}$$

is the solution of the integral Equation (3) for the particular $f(y)$ under discussion (distributions (56) and (57) have been plotted in Figure 6d assuming $y_0 = v_b/2$).

Relation (57) excludes the existence of stars rotating beyond y_0 while one would expect to find them; on the other hand by (56) there would be $\approx 25\%$ of the stars with a velocity in (0.5, 1) but none of them has been observed near the equator; moreover, only stars with $v \leqslant 0.5$ have been observed equator-on. Such a result is clearly unreliable, if one does not pay attention to the meaning of the observed distribution. Consider the following:

If $f(y) = 1/y_0$ is the result of observations in a very large sample of stars, it is very likely that the size of the sample is sufficient to make y_0 the effective upper-limit to the true velocities; hence case d degenerates into case c and any strangeness drops.

Otherwise, when the number of stars observed is not large enough, the histogram $f(y) = 1/y_0$ must be considered necessarily as a rough empirical distribution. Then it would be more correct to fit it to a continuous function by means of (36) or (37), and the distribution (56) may indicate fairly well the form of the function $\varphi(v)$, that one must assume; an analogous procedure gives in this case a reliable true distribution under (2) (as was explained in Section 1), that would result completely different from (57) and similar to (56).

It does not seem, however, that (56) gives information about the distribution of the true rotational velocities of less accuracy than the observed histogram does about the apparent ones; using (56) we have the advantage of getting almost immediately an idea of the rotational behaviour of our stars: for instance, in the case shown in Figure 6d, we are not far from the probabilistic truth if $\approx 25\%$ of the stars are predicted

TABLE III

A pattern of theoretical distributions. The symbol L indicates a Legendre integral

Observed distribution		Distribution of rotational axes		Distribution of true velocities	Distribution of true velocities when $\psi(\alpha)=\sin\alpha$	
$f(y)$	range	$\psi(\alpha)^{a}$	range	$\varphi(v)$	$\varphi(v)$	range
(a) $\dfrac{y}{v_b \sqrt{(v_b^2 - v^2)}}$	$0 \le y \le v_b$	$-\sin\alpha \ln\cos\alpha$	$0 \le \alpha \le \pi/2$	$\dfrac{v_b^2}{v^2\sqrt{(v_b^2-v^2)}}\sin^{-1}\dfrac{v}{v_b} - \dfrac{1}{v}$	$\delta(v - v_b)$	$0 \le v \le v_b$
(b) $\dfrac{3}{v_b^3}\, y\sqrt{(v_b^2 - y^2)}$	$0 \le y \le v_b$	$\tfrac{3}{2}\sin^3\alpha$	$0 \le \alpha \le \pi/2$	$\dfrac{2v}{v_b^2}$	$\dfrac{3v^2}{v_b^3}$	$0 \le v \le v_b$
(c) $\dfrac{1}{v_b}$	$0 \le y \le v_b$	$\alpha\sin\alpha$	$0 \le \alpha \le \pi/2$	$\dfrac{1}{v_b}L_c\!\left(\dfrac{v}{v_b}\right)$	$\dfrac{2}{\pi}\,\dfrac{1}{\sqrt{(v_b^2 - v^2)}}$	$0 \le v \le v_b$
(d) $\left[\begin{array}{l}\dfrac{1}{y_0}\\[4pt] 0\end{array}\right.$	$\begin{array}{l}0 \le y \le y_0\\[4pt] y_0 < y \le v_b\end{array}$	$\left[\begin{array}{l}\dfrac{v_b}{y_0}\alpha\sin\alpha\\[6pt]\dfrac{v_b}{y_0}\theta_0\sin\alpha\end{array}\right.$	$\begin{array}{l}0 \le \alpha \le \theta_0\\[6pt]\theta_0 \le \alpha \le \pi/2\end{array}$	$\left[\begin{array}{l}\dfrac{1}{y_0}L_c\!\left(\dfrac{v}{v_b}\right)\\[8pt]\dfrac{1}{y_0}L\!\left(\dfrac{v}{v_b};\sin^{-1}\dfrac{y_0}{v}\right)\end{array}\right.$	$\left[\begin{array}{l}\dfrac{2}{\pi}\,\dfrac{1}{\sqrt{(y_0^2 - v^2)}}\\[8pt] 0\end{array}\right.$ (formally)	$\begin{array}{l}0 \le v \le y_0\\[8pt] y_0 < v \le v_b\end{array}$ b

^{a} $y_0/v_b = \sin\theta_0$
^{b} $y_0 \le v \le v_b$ for the function in the foregoing column

to rotate in (0.5, 1) and $\approx 75\%$ in (0, 0.5); it is not difficult to infer that an elegant distribution $\varphi(v)$, derived by the formally correct procedure will yield practically the same expectation.

All the frequency functions discussed so far, have been collected in Table III.

5. A Method of Getting a Rough Distribution $\varphi(v)$ from the Histogram of the Apparent Rotational Velocities

The correct procedure to derive a true distribution $\varphi(v)$ has been discussed in Section 4. However, (36) or (37) requires the tabulation of a set of functions $\varphi(v)$, each truncated at v_b, in order to fit any observed histogram. Alternatively any observed histogram must be approximated by a suitable continuous function $f(y)$ in order to integrate (35).

This procedure may be tedious and time consuming; in the following we suggest a means to get an approximate trend of $\varphi(v)$ by simple numerical calculations, starting from the case (d) considered in the foregoing section, which makes it possible to treat each block of a histogram by formulae (51) and (52).

Let us take a histogram of the type (26) and call $\varphi_r(xv_b)$ the distribution of the true velocities $v = xv_b$ for the stars which have been observed between y_{r-1} and y_r.

Analogously to what has been done in Section 3, the total distribution $\varphi(xv_b)$ is:

$$\varphi(xv_b) = \sum_r I_r \, \Delta y_r \varphi_r(xv_b). \tag{58}$$

When (51) and (52) have been used in (58), we can write for every x ranging between s_{r-1} and s_r:

$$\varphi(xv_b)_{r-1, r} = I_r L_c(x) - \sum_1^{r-1} k \, (I_{k+1} - I_k) \, L_k(x). \tag{59}$$

Although the trend of (59) may represent the distribution of the rotational velocities, it is not of very practical use: for our purpose it is sufficient to get the mean value of $\varphi(v)$ in every interval (y_{r-1}, y_r), that is:

$$\langle \varphi \rangle_{r-1, r} = (s_r - s_{r-1})^{-1} \int_{s_{r-1}}^{s_r} \varphi(xv_b)_{r-1, r} \, \mathrm{d}x. \tag{60}$$

After the calculation has been performed in (60) by simple integration of the Legendre integrals with respect to x, we find:

$$\langle \varphi \rangle_{r-1, r} = \frac{I_r}{s_r - s_{r-1}} (T_{rr} - T_{r-1r-1}) +$$

$$- (s_r - s_{r-1})^{-1} \sum_1^{r-1} k \, (I_{k+1} - I_k) \, (T_{kr} - T_{kr-1}), \tag{61}$$

where:

$$T_{kr} = s_r^{-1} E(\beta_r^k; \theta_r) - s_r^{-1} (1 - s_r^2) \, F(\beta_r^k; \theta_r). \tag{62}$$

The elliptic integral F and E, already defined in Section 4, are given now by:

$$F(\beta_r^k; \theta_r) = \int_0^{\beta_r k} (1 - \sin^2 \theta_r \sin^2 \alpha)^{-1/2} \, d\alpha \tag{63}$$

$$E(\beta_r^k; \theta_r) = \int_0^{\beta_r k} (1 - \sin^2 \theta_r \sin^2 \alpha)^{1/2} \, d\alpha, \tag{64}$$

where:

$$\theta_r = \sin^{-1} s_r = \sin^{-1}(y_r/v_b) \tag{65}$$

$$\beta_r^k = \sin^{-1}(s_k/s_r) = \sin^{-1}(y_k/y_r). \tag{66}$$

In (63) and (64) we have written $\sin^2 \theta_r$ instead of s_r^2 since in mathematical tables elliptic integrals may be tabulated for a set of values of the modular angle θ_r.

As an example of the application of the foregoing formulae (61)–(66), let us consider the Be stars partially analysed in Section 3; we may start from the histogram (b) (Table I and Figure 3), assuming again a break-up velocity of 450 km/sec for all stars, and select the first interval between 0 and 100 km/sec and the following ones of 50 km/sec in size up to 450 km/sec.

If one looks, for instance, for the mean of $\varphi(v)$ between 100 and 150 km/sec, (61) gives:

$$\langle \varphi \rangle_{12} = \frac{I_2}{s_2 - s_1}(T_{22} - T_{11}) - \frac{1}{s_2 - s_1}(I_2 - I_1)(T_{12} - T_{11}),$$

where, according to (62)–(66), it is:

$$T_{11} = s_1^{-1} E(\pi/2; \theta_1) - s_1^{-1}(1 - s_1^2) F(\pi/2; \theta_1)$$
$$T_{12} = s_2^{-1} E(\beta_2^1; \theta_2) - s_2^{-1}(1 - s_2^2) F(\beta_2^1; \theta_2)$$
$$T_{22} = s_2^{-1} E(\pi/2; \theta_2) - s_2^{-1}(1 - s_2^2) F(\pi/2; \theta_2).$$

The numerical pattern is as follows:

$$I_1 = 0.048 \times 10^{-2}$$
$$I_2 = 0.119 \times 10^{-2}$$
$$s_1 = 100/450$$
$$s_2 = 150/450$$
$$\theta_1 = \sin^{-1} s_1 \approx 13°$$
$$\theta_2 = \sin^{-1} s_2 \approx 19°\!.5$$
$$\beta_2^1 = \sin^{-1}(s_1/s_2) \approx 42°.$$

By interpolation in mathematical tables we find $F(90°; 13°)$, $F(90°; 19°.5)$, $F(42°; 19°.5)$ and the corresponding elliptic integrals of second kind E; finally we get:

$$\langle \varphi \rangle_{12} \approx 0.075 \times 10^{-2}.$$

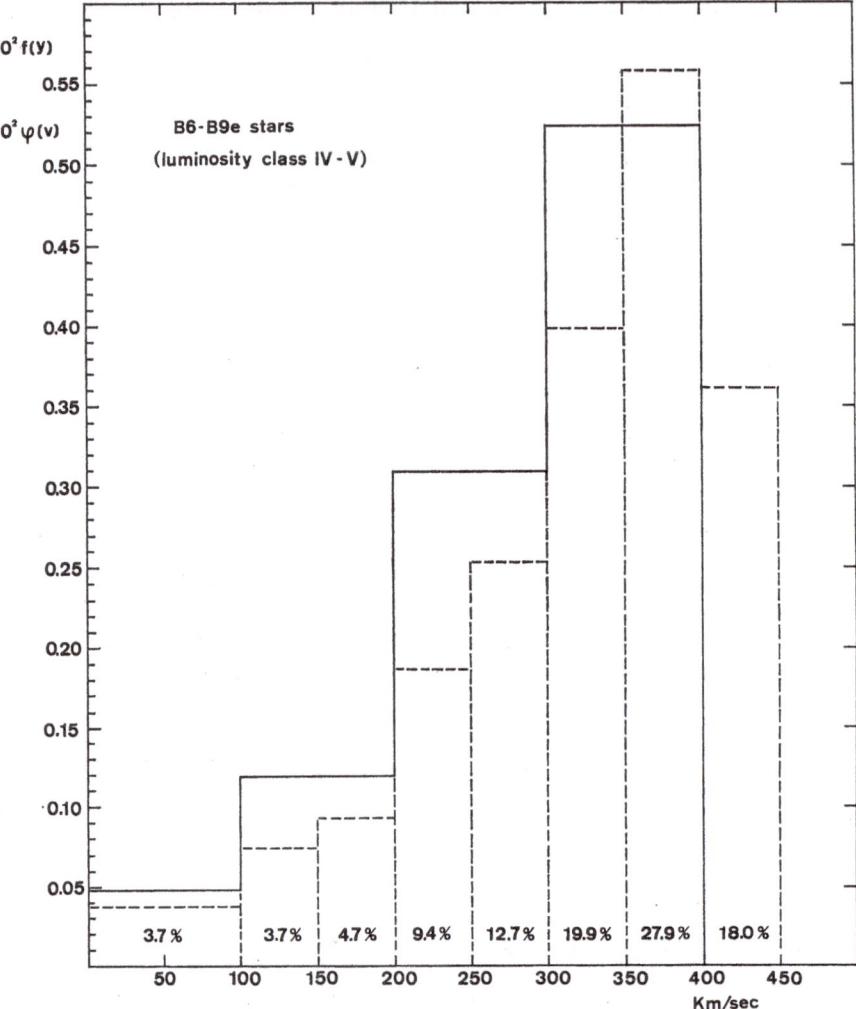

Fig. 7.   Approximate trend of the distribution of the rotational velocities v for Be stars, given in the form of a histogram (broken tracing). The solid tracing is the histogram of the apparent velocities $y = v \sin \alpha$ (case b of Figure 4). The percentage of stars in each interval of the 'true histogram' is shown. The cut-off is assumed to be at 450 km/sec.

The whole histogram of the true velocities is shown in Figure 7 by the broken tracing; the percentage of stars in each interval has also been indicated. It appears that a Be star does not need to be a rapid rotator; if the usual assumption of equatorial envelopes yielding a shell spectrum is correct, then shell stars are viewed nearly equator-on; if so, the statistical method so far suggested would explain the observation of Be stars with a shell spectrum at velocities lower than 300 km/sec (Slettebak, 1966a), a fact that a δ-solution could not explain.

It is useful to notice that the existence of intrinsically slow rotators among the Be stars has already been suggested on an observational basis by some authors (Schild, 1966; Deeming and Walker, 1967).

6. Concluding Remarks

In Sections 2–4 the statistical method of investigating the rotational behaviour of a group of stars has been established on the basis of the observations and the existence of an upper limit to the true velocities.

It may be useful to emphasize the kind of problem we are dealing with. Suppose we have a sample of bodies of known rotational velocity and spread them randomly in space. If one asks for the expectation of the apparent rotational velocity $f(y)$, the solution is just (3) where $\varphi(v)$ is known. The actual problem is quite different: apparent velocities are observed and a guess as to the distribution of the true velocities is required; the best estimate is obtained by the method developed in this paper. The usual procedure of considering valid a law when its consequences, derived under admissible general hypothesis, have been observed, cannot be applied to rotational velocities.

The present method depends on the selection of the upper limit and one must operate carefully; consider, for instance, the following: it is well known that Ap stars are observed to be slow rotators; suppose that we are dealing with a sample for which the maximum apparent velocity is 100 km/sec: if an upper limit of, say, 400 km/sec is assumed for A-stars (Figure 3), there is no difference in practice between using the earlier treatment (with $\sin\alpha$) or the present one; if, however, Ap stars are considered to be intrinsically slow rotators, it may be quite correct to take the observed maximum as the effective upper limit so that the use of the present method becomes important.

Finally, let us consider briefly again the approximate method of Section 5: since T_{kr} depends only on the selection of the intervals (s_{r-1}, s_r), one can construct once for all a square matrix:

$$T = \begin{pmatrix} T_{11} & T_{12} & T_{13} & \cdot & \cdot & T_{1n} \\ 0 & T_{22} & T_{23} & \cdot & \cdot & T_{2n} \\ 0 & 0 & T_{33} & \cdot & \cdot & T_{3n} \\ \cdot & \cdot & \cdot & \cdot & \cdot & \cdot \\ \cdot & \cdot & \cdot & \cdot & \cdot & T_{nn} \end{pmatrix}$$

where n is the number of intervals. Taking $n=20$, for instance, we have the possibility by T and (61) of transforming immediately a large variety of observed histograms yielding results of sufficient accuracy, within the errors by which apparent velocities have been determined.

Acknowledgements

I am grateful to Drs. R. Barbon, M. Capaccioli and T. Minelli for helpful discussions. I wish to thank also Mr. M. Franceschi, who did most of the numerical computation and drawings.

References

Chandrasekhar, S. and Münch, G.: 1950, *Astrophys. J.* **111**, 142.
Deeming, T. J. and Walker, G. A. H.: 1967, *Nature*, February 4, 479.

Huang, S. S.: 1953, *Astrophys. J.* **118**, 285.

Huang, S. S.: 1969, *Vistas in Astronomy*, vol. 11 (ed. by A. Beer), Pergamon Press, p. 217.

Schild, R. E.: 1966, *Astrophys. J.* **146**, 142.

Shajn, G. and Struve, O.: 1929, *Monthly Notices Roy. Astron. Soc.* **89**, 222.

Slettebak, A.: 1949, *Astrophys. J.* **110**, 498.

Slettebak, A.: 1966a, *Astrophys. J.* **145**, 121.

Slettebak, A.: 1966b, *Astrophys. J.* **145**, 126.

Trumpler, R. J. and Weaver, H. F.: 1953, *Statistical Astronomy*, University of California Press, Berkeley and Los Angeles.

Discussion

Steinitz: Would an observer in some other location in space obtain the same $\psi(\alpha)$?

Bernacca: In general, he will obtain a different $\psi(\alpha)$. Assume for a moment that the axes of Be stars are nearly perpendicular to the galactic plane. For a sample of such stars, all rotating near the break-up velocity, an observer located on the galactic plane and far from the sample would derive $\psi(\alpha) \simeq \delta(\alpha - \pi/2)$. On the other hand, an observer located far from the galaxy in the direction of the galactic pole would always measure small $v \sin \alpha$, say $\leqslant 100$ km/sec. If he thinks that Be stars are normally very rapid rotators and recognizes pole-on characteristics in all stars, he might conclude that $\psi(\alpha) \simeq \delta(\alpha)$. Otherwise he may assume that the Be phenomenon is independent of rotation and take 100 km/sec as the effective upper limit because no larger velocities have been observed: $\psi(\alpha)$ would turn out to be different again, depending on $f(y)$.

Steinitz: Would you agree that some of our classification schemes are aspect dependent?

Bernacca: I agree.

Collins: How sensitive are your results to the choice of V_b (i.e., the critical velocity)?

Bernacca: I can give an answer, at the moment, only for a sample of stars with the same critical velocity. The choice of V_b affects mostly the mean square deviations of a true distribution; however, if you have a large sample of stars, the upper limit V_b cannot be too far from the maximum observed apparent velocity, unless you are aware of selection effects. If the number of stars is small, any statistical method will yield unreliable results, of course. Let me point out that V_b is not necessarily the critical velocity, but any other upper limit which can be recognized or assumed reasonably.

ON THE UV CET-TYPE STAR ROTATION

R. E. GERSHBERG

Crimean Astrophysical Observatory, U.S.S.R.

(Paper read by R. P. Kraft)

Abstract. The upper limits of flare star rotational velocities are evaluated from the photometric features of the stars and on the basis of the widths of the Hα emission line in the quiet chromospheres. The rotational velocity of UV Cet does not exceed 22 km/sec, YZ CMi: 37 km/sec, the value $v \sin i$ of AD Leo does not exceed 52 km/sec, and that of EV Lac: 25 km/sec.

1. Introduction

The observational data and some theoretical considerations suggest that the UV Cet-type flare stars should have an appreciable rotation. Krzeminski [1] has found periodic brightness variations of dMe-stars while the dM-stars investigated did not reveal such features. Krzeminski concluded that the most probable interpretation of such events is a rotation effect of a spotted star. This hypothesis was discussed earlier by Chugainov [2] in connection with similar photometric features of the emission dwarf HDE 234677. Then, the flare quasi-periodicity of UV Cet-type stars discovered by Andrews [3] and by Chugainov [4] can also be explained by the rotation effect of a star with an active region (or regions) where flares occur. Finally, spectroscopic and radio investigations of UV Cet-type star flares give more and more evidence of the similarity between those stellar processes and solar chromospheric flares [5] i.e. they give new arguments in favor of the electromagnetic nature of the UV Cet-type star flare activity which is probably connected with stellar rotation in the end [6].

Rotational velocity determinations for flare stars have not been carried out to date. In this paper probable rotational velocities of UV Cet, YZ CMi, AD Leo and EV Lac – the stars being under detailed photometric investigation – are discussed.

2. UV Cet

UV Cet spectrograms obtained to date do not permit evaluation of the rotational velocity of the star by the classical method. Therefore nothing else is left but the method of indirect evaluations. The UV Cet rotational velocity can be evaluated on the basis of photometric features of this star.

In the wide spectral regions of the photometric UBV system, mean deviations in the dM-star energy distribution from the Planck function are small; therefore the spectrophotometric temperature of radiation which can be determined from the $B-V$ color may be supposed to be close enough to the photovisual brightness temperature of the star. Then, the color $B-V = +1\overset{m}{.}9$ gives $T(\text{UV Cet}) = 2700\,\text{K}$. This value is smaller than the effective temperature for a dM 5.5 star in Johnson's scale [7]: $T_{\text{eff}}(\text{dM } 5.5) \approx 3000\,\text{K}$, but according to Tsuji [8], this scale is overstated by 200 K for M-stars.

A. Slettebak (ed.), Stellar Rotation, 249–253. All Rights Reserved

Spectral classes of the L-726-8 system components are very close, therefore the UV Cet part of the total luminosity of the system can be found as follows:

$$L_B(\text{UV Cet}) = \frac{10^{-0.4\Delta m}}{1 + 10^{-0.4\Delta m}} \cdot L_B(\text{L-726-8 AB}) \tag{1}$$

and $\Delta m = 0^m.5$ for this system. Finally, the photometric radius of UV Cet may be calculated from the expression

$$R(\text{UV Cet}) = \sqrt{\frac{L_B(\text{L-726-8 AB}) f(\Delta m)}{4\pi \int B_\lambda(2700\,\text{K})\, \varphi_B(\lambda)\, d\lambda}} = 5.6 \times 10^9 \text{cm}\,(0.08 R_\odot) \tag{2}$$

where

$$f(\Delta m) = \frac{10^{-0.4\Delta m}}{1 + 10^{-0.4\Delta m}},$$

$B_\lambda(T)$ is the Planck function and $\varphi_B(\lambda)$ is the response curve of the B-band.

According to Chugainov [5], the characteristic time interval between the flares of UV Cet is $4^h.4$. If one identifies this time interval with a stellar rotational period, the equatorial velocity of rotation will be equal to 22 km/sec; if the real brightness temperature is within the range $2700 \pm 200\,\text{K}$, the rotational velocity will be from 15 to 35 km/sec. But the main uncertainty of that evaluation is due to identifying the characteristic time interval between flares with the rotational period. On the one hand, the $4^h.4$ period may be a multiple of the real one, and Australian astronomers were inclined to draw such a conclusion when they found two strong flares of UV Cet separated by a time interval of less than 2 hours [9]; in that case the real rotational velocity may be a few times higher. On the other hand there may be several active regions on the stellar surface and then the real rotational velocity may be several times lower than the one evaluated above. The second case seems to be more likely; therefore the value 22 km/sec must be regarded as an upper limit of the real rotational velocity.

3. YZ CMi

The photometric radius of YZ CMi is equal to 1.4×10^{10} cm ($0.21\ R_\odot$). The characteristic time interval between flares is $6^h 50^m$ [4]. Therefore a 'photometric' rotational velocity is 37 km/sec. But YZ CMi is one of several flare stars whose duplicity is not known. The magnitude $f(\Delta m)$ is then taken to be unity. If YZ CMi is an inseparable binary the photometric radius and 'photometric' rotational velocity must be multiplied by the function $[f(\Delta m)]^{1/2}$ which is equal to 0.71–0.24 for $\Delta m = 0^m - 3^m$.

In order to obtain an independent upper limit of $v \sin i$ for YZ CMi, 4 spectrograms in the red region of the spectrum were taken with the Shajn 2.6 m reflector; the short focus camera (24 Å/mm) and high sensitive A-700 emulsion were used. The Hα stellar emission line and 2–4 neon lines were measured photometrically on each film; neon lines were chosen close to Hα in terms of wavelengths and intensities. In principle, an observed width of Hα at half-intensity level is determined by an instrumental profile, by Doppler and Stark effects and by self-absorption. Because hydrogen and neon line profiles have no appreciable wings on our spectrograms, the real half widths were calculated as a square root of an observed Hα half width squared minus a neon half width squared. Half widths of Hα in YZ CMi spectrograms reduced in this way are equal to 1.0–1.4 Å. Neither thermal motion nor Stark effect can supply such wide lines under chromospheric conditions. We have no information on the character of macroscopic motion in the chromosphere of YZ CMi nor on its optical thickness at the Hα line; therefore one can take the entire reduced half widths to be due to rotation effects and one is able to evaluate $(v \sin i)_{max}$ as less than 46 km/sec.

4. AD Leo

The photometric radius of AD Leo is equal to 2.8×10^{10}cm. But AD Leo is a known binary [10, 11]. According to [10], the satellite mass is $0.03 M_\odot$ and it is natural to identify this satellite as a flare star of the AD Leo system. Supposing $\Delta m = 2^m-3^m$, one can find $R = 0.15-0.10 R_\odot$. Thus the flare star of the AD Leo system is probably very similar to UV Cet. But the available flare statistics do not give a characteristic time interval between AD Leo flares.

An upper limit of $v \sin i$ for AD Leo was evaluated on the basis of spectral data: $(v \sin i)_{max} < 52$ km/sec. According to Herbig's communication, widths of absorption spectrum lines give $(v \sin i)_{max} < 15$ km/sec for AD Leo, but the absorption spectrum of this system is determined by the main component and the magnitude mentioned may bear no relation to the flare star.

5. EV Lac

The photometric radius of EV Lac is equal to $0.22 R_\odot$, but EV Lac is a probable binary [12] and that figure may be an upper limit to a flare star radius. The search for a characteristic time interval between EV Lac flares has not been undertaken to date, but the latest observations of this star give a new argument in favor of the spotted star hypothesis [13]: slow and considerable brightness variations were found during monitoring of the star in the UV region and stellar flares were occurring only in brightness minima; one may think again that there is an active region (or regions) with a lower temperature – stellar spots – where flares occur. A similar photometric feature was found during the latest UV Cet brightness monitoring in the ultraviolet [14].

In 1960 Wilson [15] obtained an EV Lac spectogram with dispersion 9 Å/mm with the Palomar reflector. He has estimated $v \sin i < 25$ km/sec from the hydrogen emission lines. The Crimean spectrogram taken with dispersion 24 Å/mm gives a higher limit: $v \sin i < 54$ km/sec. However this result must not lead to the suspicion that all our spectrographic evaluations of $v \sin i$ are overestimated erroneously. The fact is that the flare activity of EV Lac increased considerably in the last years in comparison with 1960 [16]; therefore it is natural now to expect the star to have a more powerful chromosphere radiating optically thicker and therefore wider lines. Furthermore, Wilson has measured Hγ and higher members of Balmer series where self-absorption is less than at Hα.

6. Conclusion

The available data allow us to affirm that the rotational velocities of UV Cet-type flare stars are considerably smaller than those of young OB-stars; probably they do not exceed 25–30 km/sec. Note that according to Krzeminski [1], the 'photometric' rotational velocity of the emission dwarf BD + 34° 106 (dM0e) is 15 km/sec and that of HDE 234677 (dK7e) is 10 km/sec. If one extrapolates the dependence of mean rotational velocity of main-sequence stars on spectral classes from G through K up to M-stars with a smooth curve according to statistics [17] then one would expect a red dwarf rotational velocity to be not higher than a few km/sec.

Acknowledgements

I am very grateful to Dr. P. F. Chugainov, Dr. A. A. Bojarchuk, Dr. G. H. Herbig, Dr. R. P. Kraft, Dr. W. Krzeminski, Prof. A. B. Severny and Dr. A. Slettebak for stimulating discussions.

References

[1] W. Krzeminski: 1969, *Low Luminosity Stars* (ed. by S. S. Kumar).
[2] P. F. Chugainov: 1966, *IBVS* N122.
[3] A. D. Andrews: 1966, *Publ. Astron. Soc. Pacific* **78**, 324, 542.
[4] P. F. Chugainov: 1968, *Proceedings of the IVth Colloquium on Variable Stars*, Budapest.
[5] R. E. Gershberg: 1968, *Proceedings of the IVth Colloquium on Variable Stars*, Budapest.
[6] E. Schatzman: 1965, *Kleine Veröffentlichungen der Remeis-Sternwarte Bamberg*, Bd IV, N40, 17.
[7] H. L. Johnson: 1965, *Astrophys. J.* **141**, 170.
[8] T. Tsuji: 1969, *Low Luminosity Stars* (ed. by S. S. Kumar).
[9] C. S. Higgins, L. H. Solomon, and F. M. Bateson: 1968, *Austr. J. Phys.* **21**, 725.
[10] D. Reuyl: 1943, *Astrophys. J.* **97**, 186.
[11] P. van de Kamp and S. L. Lippincott: 1949, *Astron. J.* **55**, 16.
[12] P. van de Kamp: 1969, *Publ. Astron. Soc. Pacific* **81**, 5.
[13] S. Cristaldi, G. Godoli, M. Narbone, and M. Rodonó: 1968, *Proceedings of the IVth Colloquium on Variable Stars*, Budapest.
[14] P. F. Chugainov *et al.*: 1969, *IBVS* N343.
[15] O. C. Wilson: 1961, *Publ. Astron. Soc. Pacific* **73**, 15.
[16] P. F. Chugainov: 1969, *Izv. Crimean Astrophys. Obs.* **40**.
[17] A. A. Bojarchuk and I. M. Kopylov: 1958, *Astron. J. U.S.S.R.* **35**, 804.

Discussion

Jordahl: Kunkel (unpublished dissertation, 1967) has done an extensive study of the statistics of flares on three stars: AD Leo, YZ CMi, and Wolf 359, with 25, 25 and 29 flares recorded in total observing times of 90, 59, and 30 hours, respectively. He finds that a Poisson model fits the observations of AD Leo and YZ CMi quite well, where the fit is to the probability that *n* flares brighter than magnitude *m* occur per unit time. For Wolf 359 (the faintest of the three objects) the flares tend to be much shorter in time scale and hence difficult to identify unambiguously with the sequential filter-sampling technique used. In any case the flare frequency increases rapidly as the detection threshold improves. The concept of a 'mean time between flares' is meaningless unless accompanied by a statement of the detection limit. These statistics are from his study in the V filter of the UBV system, which is much more sensitive to flares than the B or V filters.

These statistics would seem to argue against any interpretation of a 'mean time between flares' in terms of stellar rotation.

Gershberg: Chugainov's statistics which suggest a quasiperiodicity of flares involve 28 flares of UV Cet and 24 flares of YZ CMi (see *Non-Periodic Phenomena in Variable Stars* (ed. by L. Detre), Budapest, 1969, p. 112 and 127). Therefore the volumes of data in Chugainov's and Kunkel's statistics are similar, but Chugainov has considered stronger flares on the average. He has obtained a periodicity not in the strict sense but the existence of flare groups which are prolonged up to 1–2 hours and separated by a characteristic time interval. If on a rotating star you have one active region where flares occur stochastically you shall have such a time distribution of flares. But if there are several active regions on the star the modulation effect of stellar rotation on the total time distribution of flares can be found with the thin analysis only. This modulation effect can also be smoothed out by the flare detection threshold improvement which allows the registration of flares in weaker active regions and acts as a real increase of active region number on the star. It is known that the flare detection threshold can be improved by using the U-band and by observing absolutely fainter stars. Up to this time Kunkel's extensive observations were not analyzed by the autocorrelation method and therefore they cannot be regarded now as an argument against the interpretation of time distribution of flares in terms of stellar rotation. It should be noted that according to Chugainov the UV Cet flare statistics (28 events) can be represented by a Poisson model and quasiperiodicity manner with the same success.

THE ROTATION OF THE Ap STARS FROM THE POINT OF VIEW
OF THE RIGID ROTATOR MODEL

GEORGE W. PRESTON

Hale Observatories, Carnegie Institution of Washington,
California Institute of Technology, Calif., U.S.A.

Abstract. Deutsch's period vs. line-width relation for the periodic Ap stars is re-examined from the point of view of the rigid rotator with the aid of recently determined values of $v \sin i$ for these objects. The agreement between computed rotational velocities and observed values of $v \sin i$ is satisfactory if the radii of the Ap stars are twice those of zero-age main sequence stars of the same color. The computed rotational velocities also agree with the mean rotational velocities of *all* Ap stars if suitable allowances are made for observational limitations. However, a difficulty arises in connection with the number of 'long-period' stars that have been discovered. Their small computed rotational velocities and their frequency are such that they cannot be regarded as part of any rotational distribution function that describes normal stars or even most Ap stars. It is concluded that if the periods of these objects are rotational periods, then a powerful rotational deceleration mechanism must be operative in at least some Ap stars, primarily those of the SrCrEu group.

1. Introduction

During the past few years the number of known Ap stars with well-determined periods of light, spectrum, or magnetic variations has been increased sufficiently to make feasible a re-examination of Deutsch's (1956) period vs. line-width relation. Such a study is of interest from two points of view: (1) to determine the radii of Ap stars that are required to satisfy the observations on the assumption that these objects are rigid rotators, (Stibbs, 1950; Deutsch, 1954) and (2) to compare the rotational velocities calculated by means of these radii with the rotations of *all* Ap stars as inferred from their mean projected rotational velocities. The latter comparison provides an important consistency check on the rotator model.

Arguments for or against the rigid rotator have been advanced in several summary discussions (Deutsch, 1958; Babcock 1960; Ledoux and Renson, 1966) by authors of dissimilar points of view. However, it is generally agreed that the rotator model explains the period vs. line-width relation in a natural way, predicts the reversals of magnetic polarity that are commonly observed, and explains the order of magnitude, sign convention, and phases of the crossover effect (Babcock, 1951, 1956). Objections to the hypothesis that these stars possess 'apparent abundance patches' have recently been countered by strong evidence that the magnetic field variations of different elements are not alike in at least two stars, HD 188041 (Preston, 1967a; Wolff, 1969) and β CrB (Preston and Sturch, 1967). This result requires a spatial separation of the elements involved. There remain two long-standing objections to the rigid rotator model. The first is that it does not predict irregular magnetic activity or irregular spectrum variability. The first counter to this objection appeared when Steinitz (1964) discovered the period of β CrB, the prototype of irregular magnetic variables with reversing polarity. Since then magnetic periods for HD 10783 (Steinitz, 1964),

A. Slettebak (ed.), Stellar Rotation, 254–263. All Rights Reserved

HD 133029 (Renson reported by Herbig, 1967), and HD 215441 (Preston, 1969a) have been reported. Even more recently, the periodic spectrum variability of 21 Per (heretofore regarded as an irregular spectrum variable) and the periodic magnetic and spectrum variability of 17 Com A and 78 Vir (the classic example of an apparently irregular magnetic star) have been established (Preston 1969b, c; Preston et al., 1969). Therefore, at this juncture it is my opinion that the burden of proof lies with those who believe that irregular magnetic or spectrum variations exist.

The final objection to the rigid rotator concerns the fact that the photometric and magnetic periods of any particular Ap star are always identical. Up to the present time there has been no explanation of why surface brightness should depend on magnetic polarity. However, a reason may be forthcoming in view of recent observations of the mean surface fields of β CrB (Wolff and Wolff, 1970) and HD 126515 (Preston, 1970f). The mean surface field is the absolute value of the total surface field averaged over the visible hemisphere of a star; it can be derived accurately for a number of stars that possess fields sufficiently large to produce splitting of certain favorable Zeeman doublet patterns (Preston 1969d, e). For both β CrB and HD 126515 this quantity also varies with a single wave in the period of the effective field, and visual inspection of spectrograms indicates that this is the case for 53 Cam as well. In the case of HD 126515 the mean surface field varies between 10 and 17 kilogauss while the effective field varies between $+2$ and -2 kilogauss. This behavior can be explained qualitatively by a decentered field distribution, i.e. the two magnetic hemispheres are dissimilar with respect to their total magnetic field strengths. The disparity amounts to a factor of $\simeq 2$ in H and hence to a factor of $\simeq 4$ in the mean magnetic energy density, $H^2/8\pi$ of opposing hemispheres. Thus it is conceivable, though not yet demonstrated, that this asymmetry in the field distribution is related to the aspect-dependant luminosity that is required by the rigid-rotator model.

In summary then, it does not appear that the rigid rotator can be discounted by virtue of any of the objections discussed above. We, therefore, proceed to the present investigation.

2. The Observations

During the course of a survey of rotational velocities of the Ap stars, values of $v \sin i$ were obtained for all of the well-studied periodic Ap stars known to the writer. These are listed in Table I. With one or two exceptions, for which a *prima facie* case for variability could be made from one set of observational data, the list is restricted to stars whose variations have been confirmed by independent sets of observations or for which a single period has been derived for two different kinds of variations (light, spectrum, magnetic field). The values of $v \sin i$ were derived from a comparison of measured line-widths on coudé spectrograms with those of a set of rotational standard stars. The projected rotational velocities of these standard stars were derived by a procedure to be described in detail elsewhere. The linear dispersions employed in the survey ranged from 1.3 to 10 Å mm^{-1} and for the periodic stars were generally matched to the stars so that the instrumental profiles did not seriously distort the

TABLE I

Date for well-studied periodic Ap stars

HD	Name	Type	B − V	U − B	P	v sin i (km/sec)	R sin i	v = 162/P	References[b]
10 783	–	SrCr	−0.05	−0.19	4.16	24	1.97	39	
15 089	ι Cas	SrCr	+0.12	+0.03	1.74	47	1.62	93	
18 296	21 Per	SiEuSrCr	−0.01	−0.23	2.88	22	1.25	56	1, 2
19 832	56 Ari	Si4200	−0.09	−0.39	0.73	–	–	220	
32 633	–	SiCr	−0.06	−0.40	6.43	23	2.92	25	
34 452	HR 1732	Si	−0.20	−0.61	2.47	62	3.03	66	
65 339	53 Cam	SrCrEu	+0.14	+0.05	8.02	14:	2.22:	20	
71 866	–	SrCrEu	+0.08	+0.02	6.80	≤17	<2.28	24	
98 088	HR 4369	SrCr	+0.20	+0.12	5.90	25	2.92	27	
108 662	17 ComA	SrCrEu	−0.06	−0.12	5.09	20	2.01	32	3
112 185	ε UMa	Cr	−0.02	+0.03	5.09	35	3.52	32	
112 413	α² CVn	SiCrEu	−0.11	−0.32	5.47	24	2.59	30	
118 022	78 Vir	SrCrEu	+0.04	−0.02	3.72	10	0.74	44	4
124 224	HR 5313	Si	−0.11	−0.38	0.52	160:	1.65:	312	
125 248	HR 5355	CrEu	−0.01	−0.04	9.30	≤17	<3.12	17	
126 515	–	SrCr	+0.01	−0.02	130.	≤5	–	1.3	5
137 909	β CrB	SrCrEu	+0.27	+0.11	18.5	≤3	≤1.10	9	6
140 160	χ Ser	SrCr	+0.02	+0.05	1.60	66	2.09	101	
153 882	HR 6326	Cr	+0.06	0.00	6.01	26	3.09	27	
173 650	HR 7058	SiSrCr	+0.03	−0.11	10.1	16	3.19	16	
187 474	HR 7552	CrEu	−0.05	−0.23	2500.	≤8	–	0.065	
188 041	HR 7575	SrCrEu	+0.20	+0.07	224.	≤3	–	0.72	
196 502	73 Dra	SrCrEu	+0.07	+0.10	20.3	8	3.21	8	
215 038[a]	–	Si4200	(−0.14)	−0.45	2.04	31	1.25	80	1
215 441[a]	–	Si4200	(−0.18)	−0.49	9.49	≤5	0.94	17	7
221 568[a]	–	SrCrEu	+0.23:	−0.10:	160.	≤5	–	1	8
224 801	HR 9080	SiCrEu	−0.06	−0.38	3.74	38	2.81	43	

a Appreciable interstellar reddening. Unreddened B − V values, given in parentheses, are taken from Stępień (1968)

b References to recent discoveries or confirmations of periods not listed by Ledoux and Renson (1966) are as follows: (1) Stępień (1968); (2) Preston (1969b); (3) Preston et al. (1969); (4) Preston (1969c); (5) Preston (1970); (6) Preston and Sturch (1967); (7) Preston (1969a); (8) Osawa (1967).

stellar profiles except in the cases of HD 126515, β CrB, HD 188041, HD 215441, and HD 221568; for these objects I could only establish approximate upper limits to $v \sin i$, as indicated in column 7 of Table I. The values of $v \sin i$ for β CrB and HD 188041 are derived from the width of the null line Fe I $\lambda 4065.40$ (see Preston, 1967b); for HD 126515 and HD 215441, the widths of resolved Zeeman line components were used. The Zeeman effect is known to produce measurable line broadening in some magnetic stars (Preston 1967a, b; 1969d) and it seriously affects the line widths of 53 Cam, HD 71866, HD 125248, and HD 187474. The upper limits on $v \sin i$ for these objects were obtained from the widths of polarized line components as observed with a Babcock-type differential circular analyzer.

The stars in Table II were omitted from the discussion because their reported periods lack confirmation (if no notes are appended) or are doubtful or in dispute as indicated by the Notes in column 8. Hopefully, many of these reported periods will be confirmed by observers in the near future.

3. Radius Estimates

For a rigid rotator the equatorial rotational velocity v, the period P, and the stellar radius R are related by $2\pi R = Pv$, or

$$R = Pv/50.6 \tag{1}$$

if R, v, and P are expressed in solar radii, km sec^{-1}, and days, respectively. The stars are viewed at unknown inclination i so that Equation (1) takes the form

$$R \sin i = (P/50.6)(v \sin i) \tag{2}$$

in a discussion of projected rotational velocities. If the i values of the periodic Ap stars are randomly distributed, then for a set of stars with constant R we have $R = (4/\pi) \langle R \sin i \rangle$ or

$$R = 0.025 \langle Pv \sin i \rangle. \tag{3}$$

If we apply Equation (3) to the data in Table I we obtain $R = 2.9 \, R_\odot$. However, inspection of Figure 1 indicates that this value is not satisfactory because too many stars with well-determined periods and values of $v \sin i$ exceed it. This result could be anticipated from the observed frequency distribution of $R \sin i$ which is not proportional to $\tan i$, as it should be. This result can arise if $R \not\equiv$ constant or if there is an observational bias in favor of short periods at a given $v \sin i$ and vice versa. I suspect that both biases exist because photometric and spectroscopic observing programs frequently have been designed to study short – rather than long – period variations, and there is an obvious bias in favor of detection of spectrum variations in sharp-lined stars. Therefore, I have used an upper-envelope method to estimate the value $R = 3.2$ $R_\odot$; a zero-age main sequence curve scaled by a factor of 2, as indicated in Figure 1 would serve as well or better. In either case the rigid rotator requires that a major fraction and perhaps all of the Ap stars have radii that exceed those of zero-age main sequence stars by a factor of $\simeq 2$. This places them $\simeq 1.5$ magnitudes above the zero-

TABLE II

Ap stars with published periods in need of confirmation

HD	Name	Type	$v \sin i$ (km/sec)	P (days)	P^a based on	Source of P	Notes
8441	—	SrCrEu	≤ 5	2.96	m	Steinitz (1965)	b
				106.		Renson (1966)	
9996	HR 465	CrEu	≤ 5	8400.	$l:s:m:$	Preston and Wolff (1970)	
11 503	γ Ari (S)	Cr	69	2.61	s	Deutsch (1947)	
25 354	—	SrCrEu	17	3.90	l	Rakos (1962)	
25 823	41 Tau	Si	21	11.94	l	Rakos (1962)	c
68 351	15 Cnc	SiCr	38	4.12	l	Stępień (1968)	
74 521	49 Cnc	SiCr	19	5.43	l	Stępień (1968)	b, c
78 316	κ Cnc	HgMn	≤ 5	5.00	l, m	Preston, Stępień and Wolff (1970)	b
108 945	21 Com	SrCr	66	1.03	s	Deutsch (1955)	
				1.10		Bahner and Mawridis (1957)	
				2.20		Bahner and Mawridis (1957)	
129 174	π BooA	HgMn	16	2.24	s	Deutsch (1947)	d
133 029	HR 5597	SiCr	20	1.05	m	Renson (Herbig, 1967)	
140 728	HR 5857	Cr	75	1.31	l	Wehlau (1962)	
176 232	10 Aql	SrCr	≤ 5	9.78	l	Stępień (1968)	c, e
192 678	—	Cr	<10	18.	l	Stępień (1968)	c
219 749	—	Si	90	2.60	l	Rakos (1962)	b, f
				0.72	l	Renson (1965)	
220 825	κ Psc	CrSr	34	0.58	l	Rakos (1962)	

a l = light variation; s = spectrum variation; m = magnetic variation; all based on single sets of observational data except for 21 Com.
b Two periods have been derived from a single set of observations
c Range of variation is comparable with the scatter in the observations
d The variation and its period are doubtful
e The magnetic field showed no variation during 10 days in 1967
f The large line-width reported by Babcock (1958) is probably due to observation of the wrong star

age main sequence, a result that is in reasonable agreement with the meager data available (see, for example Kraft, 1967, and Wolff, 1967).

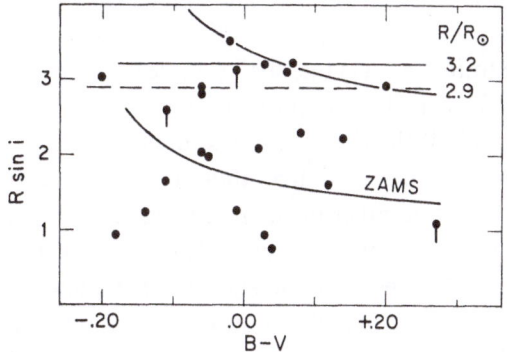

Fig. 1. A plot of $R \sin i$ in units of solar radii vs. $B - V$ for the periodic Ap stars listed in Table I. The radii of zero-age main sequence stars calculated from the data of Morton and Adams (1968) is indicated by ZAMS. The average and upper envelope estimates of $R/R_{\odot} = 2.9$ and 3.2 are indicated by dashed and solid horizontal lines. A ZAMS curve, scaled by a factor of 2 is also shown. Vertical bars below three points denote upper limits on $R \sin i$.

4. The Mean Rotational Velocities of the Periodic Ap Stars

In Table III the rotational velocities of the periodic Ap stars are compared with those obtained from the survey of all Ap stars mentioned above. The survey results are preliminary ones based on about 90% of the total sample. Because the mean rotational velocities of various Ap subgroups are not alike, I have presented the results separately for the two main groups, the Si stars and the SrCrEu stars. Any star with enhanced Si II lines is treated as a Si star regardless of its other properties. Thus, in this paper α^2 CVn and 21 Per are regarded as Si stars in spite of the fact that their spectra show other more striking peculiarities. The types are taken primarily from Osawa (1965). Some of them will be revised in the final discussion of the survey data but it is not likely that such revisions or the addition of $\simeq 30$ more stars will affect the results qualitatively. The quantities $\langle v \sin i \rangle$ and $\langle v \rangle = (4/\pi)\langle v \sin i \rangle$ for the survey stars are

TABLE III

Rotational velocity data for Ap stars

		Si	SrCrEu [a]
$\langle v \sin i \rangle$	All Ap	53	< 29
	Periodic Ap	42	< 19
$\langle v \rangle$	All Ap	67	< 37
$\langle v(P) \rangle$	Periodic Ap	86	28
No. of	All Ap	100	88
stars	Periodic Ap	10	17

[a] The inequalities result from the fact that the line-widths in many members of the SrCrEu group are smaller than the instrumental width. Upper limits on $v \sin i$ for such stars were included in the averages.

based on substantial samples as indicated at the bottom of Table III. The mean rotational velocities of the periodic stars were obtained from the data in Table I by means of Equation (1) with $R = 3.2\,R_\odot$. The agreement between the results for all Ap stars and for the periodic Ap stars is satisfactory. In this assessment allowance must be made for the following: (1) the small sample of periodic stars, (2) the inclusion of 56 Ari in the $\langle v(P) \rangle$ calculation and its exclusion in the $\langle v \sin i \rangle$ calculation for the periodic Si stars, and (3) the fact that the periodic SrCrEu stars have been systematically observed at higher dispersion than those in the survey. This results in smaller upper limits on $v \sin i$ for a number of the periodic stars (no account has been taken of inequality signs in the computation of average values). In summary, then, it is possible to choose radii for the periodic stars so that their rotational velocities, as calculated for the rigid-rotator model, are in no way anomalous with respect to the rotations of all Ap stars.

5. The Long-Period Ap Stars

There remains one disquieting feature of the observational data from the point of view of the rigid rotator model – the number of long-period Ap stars and the rotational velocities inferred from their periods. There are now six known or suspected Ap stars with periods greater than 100 days as indicated in Table IV. The

TABLE IV

HD	mag.	P (days)	$v(P)$
8441	6.6	106 (?)	1.5 (?) km/sec
126 515	7.0	130	1.2
221 568	8.0	160	1.0
188 041	5.6	224	0.72
187 474	5.4	2500: = 6.7 years	0.065
9996	6.3	long: 23 years (?)	0.019 (?)

suggested magnetic period for HD 8441 is also the spectroscopic binary period of that star according to Renson (1966); the period of HD 126515 is of recent origin (Preston, 1970) and definite; and those for HD 221 568, HD 188 041, and HD 187 474 all seem secure. The extraordinary period of HD 9996 given above must be regarded as tentative because of an inadequate number and distribution of observations, but the variation is certainly slow as indicated not only by a rare earth spectrum variation, but by fragmentary magnetic and photometric data as well; this star is also a spectroscopic binary. None of these stars are extraordinary among the Ap stars from the spectroscopic point of view. Their lines are very sharp, but this is a common feature of the SrCrEu group to which they all belong. Five of the six star are spectrum variables and for at least three of them the spectrum and magnetic variations proceed in phase as is generally the case for magnetic stars.

Astronomers will generally view reports of rotational velocities less than 0.1. km

sec^{-1} with some reluctance and the following illustrative calculations indicate why. The Maxwellian distribution gives a fair representation of rotational velocities for normal upper main sequence stars (see, for example Babcock, 1960, and Deutsch, 1967). This distribution may be written

$$f(x) = \frac{4}{\sqrt{\pi}} x^2 e^{-x^2}, \tag{4}$$

where $x = v/v_m$ is the velocity in units of the modal velocity of the distribution. The fraction of a Maxwellian population that has x in $0 \leqslant x \leqslant X$ is

$$F(X) = \int_0^X f(x)\,dx \approx 0.75X^3 \quad \text{if} \quad X \ll 1. \tag{5}$$

There are $\simeq 10^3$ B5–A3 stars brighter than $m_v = 6$ and thus there are certainly $< 10^4$ such stars brighter than $m_v = 7$. For these objects $v_m \simeq 150$ km sec^{-1}. There are $\simeq 4 \pm 1$ known long-period Ap stars with $v \leqslant 1.5$ km sec^{-1} and $m_v \leqslant 7$. The expected number for $X = 1.5/150 = 0.01$ is $\simeq 10^4 F(0.01) \simeq 10^{-2}$; there is a discrepancy of a factor of 400. Suppose that the long-period stars are the slowest rotators of an Ap distribution with $v_m = 30$ km sec^{-1} If the Ap stars comprise 10% of the normal stars, then the expected number with $v < 1.5$ km sec^{-1} is $\simeq 10^3 F(0.05) \simeq 10^{-1}$; there is a discrepancy of a factor of 40. The actual discrepancies must be much larger than those calculated above because the list of long-period stars can hardly be expected to be complete. Thus it does not seem possible to regard these objects as members of any known velocity distribution. If their periods are rotational periods, then a powerful deceleration mechanism must be operative in a substantial fraction of the Ap stars, primarily those of the SrCrEu group. If their periods are not rotational periods, then, by virtue of the similarity of their variations to those of other Ap stars of shorter period, I believe it will be necessary to abandon the rigid-rotator model.

Acknowledgement

The writer wishes to acknowledge helpful discussions with Dr. A. J. Deutsch during the preparation of this report.

References

Babcock, H. W.: 1951, *Astrophys. J.* **114**, 1.
Babcock, H. W.: 1956, *Astrophys. J.* **124**, 489.
Babcock, H. W.: 1958, *Astrophys. J. Suppl.* **3**, No. 30.
Babcock, H. W.: 1960, in *Stars and Stellar Systems* 6, p. 282.
Bahner, K. and Mawridis, L.: 1957, *Z. Astrophys.* **41**, 254.
Deutsch, A. J.: 1947, *Astrophys. J.* **105**, 283.
Deutsch, A. J.: 1954, *Trans. IAU* **8**, 801.
Deutsch, A. J.: 1955, *Publ. Astron. Soc. Pacific* **67**, 342.
Deutsch, A. J.: 1956, *Publ. Astron. Soc. Pacific* **68**, 92.

Deutsch, A. J.: 1958, *Handbuch der Physik* **51**, 689.

Deutsch, A. J : 1967, in *The Magnetic and Related Stars* (ed. by R. Cameron), Mono Book Corp., Baltimore, p. 181.

Herbig, G. H.: 1967, *Trans. IAU* **13A**, 523.

Kraft, R. P.: 1967, in *The Magnetic and Related Stars* (ed. by R. Cameron), Mono Book Corp., Baltimore, p. 303.

Ledoux, P. and Renson, P.: 1966, *Ann. Rev. Astron. Astrophys.* **4**, 293.

Morton, D. C. and Adams, T. F.: 1968, *Astrophys. J.* **151**, 611.

Osawa, K.: 1965, *Ann. Tokyo Astron. Obs.*, Ser. 2, **9**, 123.

Osawa, K.: 1967, in *The Magnetic and Related Stars* (ed. by R. Cameron), Mono Book Corp., Baltimore, p. 363.

Preston, G. W.: 1967a, in *The Magnetic and Related Stars* (ed. by R. Cameron), Mono Book Corp., Baltimore, p. 3.

Preston, G. W.: 1967b, *Astrophys. J.* **147**, 804.

Preston, G. W.: 1969a, *Astrophys. J.* **156**, 967.

Preston, G. W.: 1969b, *Astrophys. J.* **158**, 243.

Preston, G. W.: 1969c, *Astrophys. J.* **158**, 251.

Preston, G. W.: 1969d, *Astrophys. J.*, **157**, 247.

Preston, G. W.: 1969e, *Astrophys. J.* **158**, 1081.

Preston, G. W.: 1970, *Astrophys. J.* **160**, in press.

Preston, G. W. and Sturch, C.: 1967, in *The Magnetic and Related Stars* (ed. by R. Cameron), Mono Book Corp., Baltimore, p. 363.

Preston, G. W. and Wolff, S. C.: 1970, *Astrophys. J.* **160**, in press.

Preston, G. W., Stępień, K., and Wolff, S. C.: 1969, *Astrophys. J.* **156**, 653.

Rakos, K.: 1962, *Lowell Obs. Bull.* **5**, 227.

Renson, P.: 1965, *Bull. Soc. Roy. Sci. Liège*, **34**, No. 5–6, 302.

Renson, P.: 1966, *Bull. Soc. Roy. Sci. Liège*, **35**, No. 3–4, 244.

Steinitz, R.: 1964, *Bull. Astron Inst. Netherl.* **17**, 504.

Steinitz, R.: 1965, *Bull. Astron. Inst. Netherl.* **18**, 125.

Stępień, K.: 1968, *Astrophys. J.* **154**, 945.

Stibbs, D. W. N.: 1950, *Monthly Notices Roy. Astron. Soc.* **110**, 395.

Wehlau, W.: 1962, *Pub. Astron. Soc. Pacific*, **74**, 286.

Wolff, S. C.: 1967, *Astrophys. J. Suppl. Ser.* **15**, 21.

Wolff, S. C.: 1969, *Astrophys. J.* **157**, 253.

Wolff, S. C. and Wolff, R.: 1970, *Astrophys. J.*, in press.

Discussion

Van den Heuvel: Is there any evidence that the last six stars you listed are considerably older than the other Ap stars? If a braking mechanism is going on, they must be the oldest Ap stars present.

Preston: These stars are all members of the SrCrEu group and hence are systematically cooler and presumably older than, say, the Si stars. However, I have no way of determining the relative ages of stars within the SrCrEu group. In any event, I think it would be premature to regard the long-period stars as the oldest ones until we understand the nature of the braking mechanism.

Van den Heuvel: What braking mechanism for the rotation of Ap stars would you like to propose?

Preston: For some time I have been looking for observational evidence of an interaction with the interstellar medium. During these meetings Dr. Ostriker has suggested to me one such interaction which involves the dissipation of rotational energy by Alfvén waves generated at the boundary of a stellar magnetosphere.

Abt: Could you please explain more carefully how you obtain two different measures of the magnetic field?

Preston: The effective field is the mean value of the *longitudinal component* of the magnetic field averaged over the stellar disk. It is derived from the displacement between the centroids of unresolved Zeeman patterns that have been analyzed for left and right circular polarization. The positions of the centroids are determined by the relative intensities of the Zeeman components as viewed through a differential circular analyzer. If the stellar magnetic field is sufficiently strong, then certain favorable

Zeeman patterns are resolved and the separation of the resolved σ components provides a direct measure of the *total* field averaged over the stellar disk. A detailed discussion is given in the references cited in the text.

Bernacca: You derived mean true values such as $\langle v \rangle$ or $\langle R \sin i \rangle / \langle \sin i \rangle$ by means of $\langle \sin i \rangle = \pi/4$. Now, according to my paper this averaging factor is not the most correct one that you can have for an observed sample of stars. If, for your Ap stars it can be assumed that the upper limit to the true velocities is the maximum observed apparent velocity, you get a different value for $\langle \sin i \rangle$. The question is the following: How sensitive are your conclusions to the above predicted mean values?

Preston: The two procedures gave radii that differed by about 10%. However, as I indicated, observational selection effects and/or a dispersion in radii may be responsible for this difference.

Jaschek: Do magnetic fields repeat exactly cycle after cycle?

Preston: I do not know of any well-documented case of cyclic or random fluctuations in the shapes or ranges of periodic magnetic variations.

Jaschek: Do you have any mechanism for forming patches of Cr or Eu and holding them fixed on the surface?

Preston: I know of no mechanism for the formation of abundance patches. On the other hand a static magnetic field will act to preserve the identity of such patches once they have been formed.

ON THE THEORY OF ROTATING MAGNETIC STARS

G. A. E. WRIGHT

The University of Manchester, Manchester, England

All observations of magnetic stars necessarily yield information only about their surface features. We are ignorant of the nature of the fields in the interiors of such stars, and equally we cannot be sure of the non-existence of interior fields in stars which are superficially non-magnetic. In fact, if we assume the truth of the 'fossil' theory – that the magnetic flux of an Ap star is a relic of the flux initially present in the gas cloud from which the star condensed – then it is surprising that magnetic stars are not observed to be much more common, since magnetic fields appear to be ubiquitous in interstellar gas clouds. For those stars with strong surface convection zones, we might expect that a fossil field of low energy would be expelled by the turbulence and would possibly be trapped in the interior. However, the majority of early-type stars with radiative envelopes also do not exhibit any observable magnetic field.

The work described in this paper was motivated by the possibility that the appearance or non-appearance of magnetic fields might depend on whether there are mass-motions within the star, which would tend to drag the field-lines beneath the surface.

Such motions are likely to be much more effective in rapidly rotating early-type stars (Mestel, 1965, 1967). If this idea is correct, then the non-appearance of magnetic fields in most rapidly-rotating stars is not just an aspect effect, due for example to Doppler broadening. There would then be a two-way interaction between a primeval magnetic field and stellar rotation. A strongly magnetic star which spends long enough in a pre-main sequence convective phase could suffer sufficient braking by a magnetically-controlled stellar wind for its rotation to be abnormally low, even after subsequent contraction to the main sequence. On the other hand a weakly magnetic star, one which has only a short-lived stellar wind phase, could reach the main sequence with a rotation rapid enough to cause the surface field to disappear well within the star's lifetime.

In the radiative envelope of a uniformly rotating star, the divergence of the radiative flux **F** is not zero, but is a prescribed function of position (Von Zeipel, 1924). The consequent buoyancy forces drive a circulation that is slow over the bulk of the star, but can become considerably faster in low-density surface regions (Eddington, 1929; Sweet, 1950; Opik, 1951; Baker and Kippenhahn, 1959; Mestel 1966). This circulation tends to convect both angular momentum and magnetic flux; however, it is possible to construct approximate self-consistent models in which the magnetic field is quite strong enough to keep the rotation uniform in spite of the circulation, but is too weak to affect sensibly the thermal-gravitational field over the bulk of the star (Roxburgh, 1963; Mestel, 1965). In particular, we expect such an inexorable rotationally-driven circulation steadily to distort a weak primeval field and to reduce the net flux emanating from the surface.

A. Slettebak (ed.), Stellar Rotation 264–268. All Rights Reserved
Copyright © 1970 by D. Reidel Publishing Company, Dordrecht-Holland

In this paper we invert the problem and look for magnetic fields that not only keep the star rotating uniformly, but are strong enough to suppress the Eddington-Sweet circulation. In a linear perturbation theory centrifugal and magnetic forces cause independent disturbances to the pressure-density-temperature field, and so also to the flux of energy. We demand that (in an obvious notation)

$$(\nabla \cdot \mathbf{F})_\Omega + (\nabla \cdot \mathbf{F})_H = 0$$

over the whole star. This imposes a constraint on the magnetic field structure through the star. In these models there are no motions, so that field-lines emanating from the surface do not suffer a progressive distortion.

The models described in this paper are all assumed to possess axisymmetric magnetic fields which limits their observational applications, since most magnetic stars appear to have large angles of obliquity. The assumption of uniform rotation is more than is strictly required by Ferraro's Law of Isorotation, which could permit relative shearing of individual field-lines. Any departure from the axisymmetric assumption would however tend to maintain uniform rotation, so we select this case in order to best approximate the non-axisymmetric situation. It was further assumed that the magnetic field permeated the convective core of the star, i.e. that the turbulence was not strong enough to expel it. The calculations were in fact repeated with the condition that the field should not enter the core and, although the field-structures were different, the integrated results, which form the main conclusions of this paper, were not affected.

We write the radial and transverse components of the field in terms of a stream function ψ:

$$H_r = -\frac{1}{r^2}\frac{\partial \psi}{\partial \mu} \quad \text{and} \quad H_\theta = -\frac{1}{r \sin \theta}\frac{\partial \psi}{\partial r},$$

where $\mu = \cos\theta$. It is found that if ψ is taken to have the basically dipolar form

$$\psi = (1 - \mu^2)\, B(r)$$

then the angular dependence in the basic equations of stellar structure can be eliminated, simplifying the problem considerably. This simplification is made solely on mathematical grounds and we have to assume that it will yield results qualitatively similar to those which would be given by a full analysis of the evolution and decay of the field. The method of formulating the problem by means of separating the radial and $P_2(\mu)$ components and comparing the coefficients of P_2 has been described by Roxburgh (in Lüst, 1965) and Monaghan (1966) in the case of zero rotation and only a slight extension is required to include the effects of uniform rotation (Wright, 1969). There results a set of 5 non-linear ordinary differential equations which have now been solved both by ourselves and independently by Davies (1968).

In order to describe these results we define two integrals over the equatorial plane:

$$F_t = \left| \int_0^{r_0} 2\pi r H_\theta \, dr \right|$$

(where r_0 is the radius of the neutral point) is the total flux of the field, and

$$F_s = \left| \int_0^{R} 2\pi r H_\theta \, dr \right|$$

is the flux which emerges from the surface. In all models it was found that the field-lines were concentrated towards the centre of the star and that if F_t was regarded as

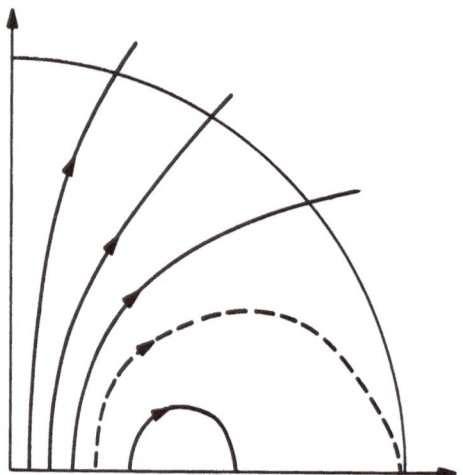

Fig. 1.   Field-lines for zero angular velocity.

fixed and the angular velocity, Ω, gradually increased, then F_s tended to zero for a *finite* value of Ω.

The *structure* of the models is fixed by the value of $\Omega/\bar{H}$, where $\bar{H}$ is the surface polar field. For $\Omega = 0$, the structure is as in Figure 1; and Figure 2 represents the structure when Ω and $\bar{H}$ have values corresponding to a period of $4\frac{1}{2}$ days with a surface field of 10^3 gauss. In the first case the central field strength is about 40 times the surface field, and in the second case about 1600 times. The flux inside the dotted field-lines never emerges above the surface of the star. Figure 3 shows F_t plotted against Ω for differing constant values of $\bar{H}$. For a given value of Ω there corresponds a unique minimum total flux, which has $\bar{H} = 0$.

These models have all been computed under the idealisation of infinite conductivity. In fact, the value of F_t for any particular star must monotonically decrease as field-lines slowly contract into the neutral point. If radiative equilibrium is to be maintained,

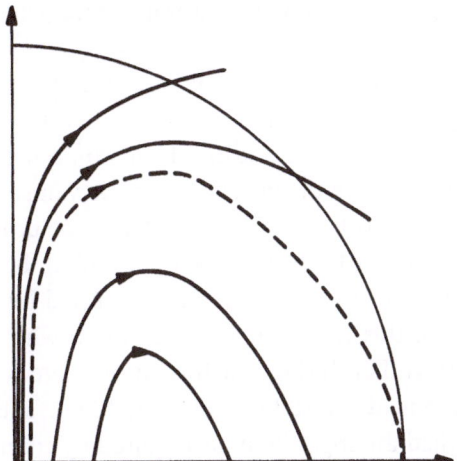

Fig. 2.   Field-lines for $\bar{H} = 10^3$ gauss and a period of $4\frac{1}{2}$ days.

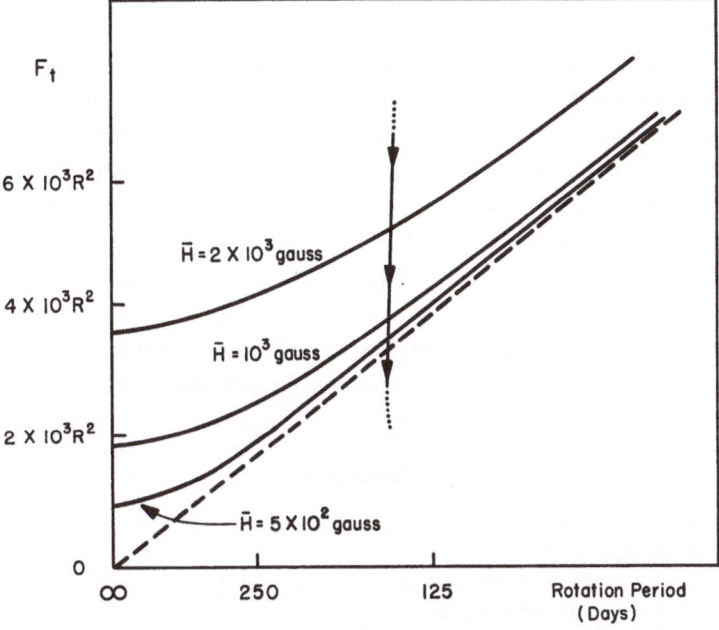

Fig. 3.   Total field flux plotted against rotation period for differing constant values of
the surface polar field.

with a basically P_1-field and with no change in Ω, then we must assume that the small
circulation currents which arise as F_t decays will restore the field-lines to the new
equilibrium configuration. As this evolution proceeds, the value of the surface field is
gradually reduced to zero, an effect represented by the downward vertical line in
Figure 3. Once the flux is reduced below this critical value, then either radiative
equilibrium is maintained by fields of a more complex structure than the simple
P_1-form assumed so far, or slow circulation currents are started which will steadily dis-

tort the field. In either case it is unlikely that strong surface fields will re-appear above the surface.

When compared with observation, assuming that these computed structures are to some degree a good approximation to the real state of affairs inside magnetic stars, most stars appear to have values of $\bar{H}$ and Ω which would indicate that their values of F_t are no more that 10% above the critical value for vanishing surface fields, and that their ratios of magnetic to rotational energies are of the order of 0.07. A decay of about 10% would lead ultimately to the disappearance of all the surface flux. Further, these values of F_t are much below the maximum permitted by the virial theorem. If the stellar flux is primeval, this raises the question as to why such a small fraction remained. One possibility is that during star formation there was substantial motion across as well as along the field (Mestel and Spitzer, 1956). The difficulty is to find a reason why this should yield an upper limit, as required by comparison of our models with observation. Another possibility is that a star with too much flux would be thermally unstable, leading to motions of flux-tubes and accelerated Ohmic decay. Fricke (1969) has shown that uniformly rotating stars with poloidal magnetic fields are secularly stable to axisymmetric perturbations, but this work has yet to be extended to non-axisymmetric perturbations and the inclusion of finite conductivity. If these hypothetical instabilities slow up as flux is lost, it may be possible to explain the upper limit (however it may turn out that no magnetic star is sufficiently stable unless there is a negative gradient of mean molecular weight in at least part of the star).

This concludes the analysis and discussion of the results so far computed. Work is now proceeding on the investigation of the secular stability of these models, and on the production of plausible 'quasi-steady' solutions representing stars with radiative equilibrium in the surface regions but with a slow circulation within. In addition we are analysing the models of this paper to see how the perturbations to the luminosity and surface temperature, caused by the mixture of rotation and magnetism, affect their position on the H-R diagram.

References

Baker, N. and Kippenhahn, R.: 1959, *Z. Astrophys.* **48**, 140.

Davies, G. F.: 1968, *Aust. J. Phys.* **21**, 294.

Eddington, A. S.: 1929, *Monthly Notices Roy. Astron. Soc.* **90**, 54.

Fricke, K.: 1969, *Astron. Astrophys.* **2**, 309.

Lüst, R. (ed.): 1965, Stellar and Solar Magnetic Fields, North-Holland Publ. Co., Amsterdam.

Mestel, L.: 1965, 'Meridional Circulation in Stars' in *Stars and Stellar Systems* 8 (ed. by L. Aller and D. McLaughlin), Chicago University Press, Chicago, Ill.

Mestel, L.: 1966, *Z. Astrophys.* **63**, 196.

Mestel, L.: 1967, 'Stellar Magnetism' in *Rendiconti della Scuola Internazionale di Fisica, Enrico Fermi* (ed. by P. A. Sturrock), Academic Press, New York.

Mestel, L. and Spitzer, Jr., L.: 1956, *Monthly Notices Roy. Astron. Soc.* **116**, 503.

Monaghan, J. J.: 1966, *Monthly Notices Roy. Astron. Soc.* **132**, 1.

Opik, E. J.: 1951, *Monthly Notices Roy. Astron. Soc.* **111**, 278.

Roxburgh, I. W.: 1963, *Monthly Notices Roy. Astron. Soc.* **126**, 67.

Sweet, P. A.: 1950, *Monthly Notices Roy. Astron. Soc.* **110**, 548.

Von Zeipel, H.: 1924, *Festschrift für H. von Seeliger*, p. 144.

Wright, G. A. E.: 1969, *Monthly Notices Roy. Astron. Soc.* **146**, 197.

MAGNETIC BRAKING AND THE OBLIQUE ROTATOR MODEL

L. MESTEL and C. S. SELLEY

The University of Manchester, Manchester, England

This work investigates the dynamical evolution of a rotating magnetic star which drives a stellar wind. The basic magnetic field of the star is supposed symmetric about an axis, which is inclined at an angle χ to the rotation axis $\mathbf{k}$ (Figure 1). We adopt the familiar equations of an inviscid perfectly conducting gas. In a steady state, the velocity as seen in a frame rotating with the star is taken as

$$\mathbf{v} = \kappa \mathbf{H} \tag{1}$$

where κ is a scalar. In the inertial frame, the velocity $\mathbf{v}$ is no longer parallel to $\mathbf{H}$, but has the velocity of corotation superposed.

Close to the star, the strong magnetic field controls the flow of the wind, in particular maintaining approximate corotation with the star. The required torque on the gas

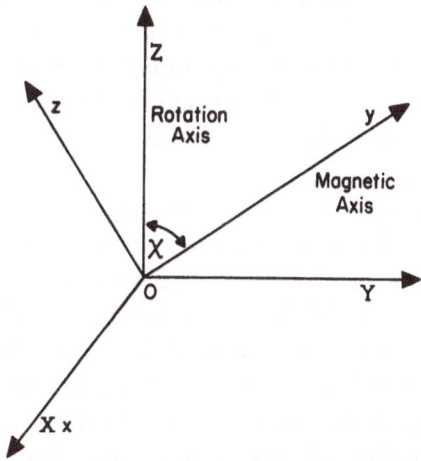

Fig. 1.   Coordinate axes describing the stellar rotation and magnetic field.

arises from a perturbation $\mathbf{H}'$ to the basic magnetic field $\bar{\mathbf{H}}$. When $\chi = 0$, $\mathbf{H}'$ is a twist about the rotation axis. By (1), the gas must have an azimuthal component of velocity in the rotating frame, but close to the star this will be small compared with the rotational velocity of the frame. Further out the lag becomes more and more pronounced. At great distances from the star, corotation ceases entirely: the field exerts effectively no torques on the gas, which therefore flows with approximate conservation of angular momentum, drawing out the field into a spiral configuration (Parker, 1958). The change-over occurs near the critical surface S_A at which the wind-speed in the inertial frame equals the speed of Alfvén waves along the poloidal field.

A. Slettebak (ed.), Stellar Rotation 269–273. All Rights Reserved
Copyright © 1970 by D. Reidel Publishing Company, Dordrecht-Holland

In the absence of a magnetic field, the star experiences a slight braking torque, parallel to the rotation axis, due to transport by the stellar wind of the angular momentum it had at the stellar surface. When $\chi=0$ or $\pi/2$ the principal effect of the magnetically enforced corotation is to increase greatly the angular momentum flow per unit mass loss. For general χ, the reduction in symmetry yields an additional component of torque in the direction OY perpendicular to the rotation axis OZ (Figure 1). With the field frozen into the star, the principal effect of this Y-torque is to cause the instantaneous axis of rotation to precess about OX, changing the angle χ; simultaneously the magnetic axis rotates steadily in space. If the Y-torque is negative (like the Z-torque), then $\dot{\chi}>0$; if positive, $\dot{\chi}<0$ (Mestel, 1968b). The problem is at once seen to be relevant to the oblique rotator model.

The axisymmetric case has been extensively studied, and the results extrapolated to apply to the sun. The magnetic field is conveniently resolved into a poloidal part, present if the star is non-rotating, and the toroidal perturbation $\mathbf{H}'$ due to the rotation. The toroidal equation of motion has a simple integral, from which $\mathbf{H}'$ can be determined. This integral is easily interpreted: there is no interchange of the $\mathbf{k}$-component of angular momentum between adjacent field-streamlines, but there is a steady flow of $\mathbf{k}$-angular momentum along the field-streamlines, carried jointly by the gas and the magnetic tensions (Lüst and Schlüter, 1955). It turns out that the total torque on the star is given by assuming that the gas is kept in a state of *strict* corotation out to the Alfvén surface S_A, even though the actual rotation of the gas at S_A is well below the stellar rotation (Mestel, 1966, 1967, 1968a; Weber and Davis, 1967; Modisette, 1967; Ferraro and Bhatia, 1967).

With general χ there is no simple integral relating the perturbed and unperturbed fields, and $\mathbf{k}$-angular momentum will be interchanged between adjacent field-lines. Analogues of the axisymmetric poloidal and toroidal fields – called $\bar{\mathbf{H}}$ and $\mathbf{H}'$ – can be conveniently defined in terms of their properties under reflection in the plane OYZ in Figure 1. The actual method of solution is by a perturbation procedure in the stellar rotation α, the term of order α being the first approximation to $\mathbf{H}'$. Terms of the second order in the equation include the centrifugal force, and the magnetic force term $(\nabla \times \mathbf{H}') \times \mathbf{H}'/4\pi$, and they will modify the poloidal field (seriously if the rotation is rapid); however, we have not as yet gone beyond first-order terms. Once $\mathbf{H}'$ is known, the Y- and Z-components of torque can be computed directly by integrating the moment of the Maxwell stresses over the stellar surface. Before performing this expansion, however, we derive an exact transform for the torque $\mathbf{L}$ on the star. It can be written as the sum of three integrals:

$$-\mathbf{L} = (\alpha\mathbf{k} \times \mathbf{h}) + \int_{S_A} \left(P + \frac{\mathbf{H}^2}{8\pi}\right)(\mathbf{r} \times \mathbf{n})\,dS$$

$$+ \int_{S_A} \varrho(\mathbf{v}\cdot\mathbf{n})(\mathbf{r} \times (\alpha\mathbf{k} \times \mathbf{r}))\,dS, \tag{2}$$

where

$$\mathbf{h} = \int_{\tau} \mathbf{r} \times \varrho \, (\mathbf{v} + \alpha \mathbf{k} \times \mathbf{r}) \, d\tau, \tag{3}$$

and τ is the volume between the star and the Alfvénic surface S_A. The vector $\mathbf{h}$ is the angular momentum of the gas flowing through τ, and the first term on the right of (2) gives the torque that must be exerted to rotate this 'flywheel' at the rate $\alpha \mathbf{k}$. The integrated moment of the total pressure $(P + H^2/8\pi)$ would vanish if the Alfvénic surface were spherical. The last term is seen to be the generalization of the axisymmetric result quoted above; and in fact if the magnetic field is axisymmetric, and $\chi = 0$, the first two terms in (2) vanish by symmetry.

The simplest possible poloidal field is the split monopole, with the field-lines radial and the field-strength independent of angle, and with a sheet of gas across the cut keeping the oppositely-directed field-lines from diffusing into each other. It is easily shown that the first-order perturbation in this field due to rotation is again just a twist about the rotation axis, with an associated distortion in the cut. The field still has enough symmetry for the net torque to be parallel to the rotation axis. The Alfvénic surface is a sphere, and the first-order perturbation $(P' + \bar{\mathbf{H}} \cdot \mathbf{H}'/4\pi)$ in the total pressure is zero.

We now suppose the flux distribution over the stellar surface departs from that of a split monopole by a small part of order ε, which is expanded in surface harmonics. In addition to the Z-torque – of order 2 – there is now a Y-component with a dominant part of order $\alpha\varepsilon$. Because of the properties of the split monopole solution, only the flow surface integral in (2) contributes to this order, with the velocity field $\mathbf{V}$ and the distortion of the Alfvénic surface S_A computed to zero order in α but to the first order in ε. Thus we have a result analogous to the axisymmetric result: the transform (2) enables us to compute the torque to order $\alpha\varepsilon$ without a prior computation of the magnetic field perturbation of this order.

It is convenient to write the flux distribution over the stellar surface R in the form

$$\frac{\bar{H}_{1r}(R)}{\bar{H}_0(R)} = \varepsilon \sum Y_n(\theta_1 \lambda) \tag{4}$$

where $\bar{H}_0$ is the basic (split monopole) field, and

$$Y_n(\theta_1 \lambda) = a_0(n) P_n(\cos\theta)$$

$$+ \sum_{k=1}^{n} \{a_k(n) \cos k\lambda + b_k(n) \sin k\lambda\} P_n^k(\cos\theta) \tag{5}$$

in standard notation. We then find that the Y and Z components of the torque $\mathbf{L}$ are related by

$$\frac{L_Y}{L_Z} = -\varepsilon A \left(\frac{r_c}{r_A}\right) \{[\tfrac{1}{2}a_0(2) - a_2(2)] \sin 2\chi - a_1(2) \cos 2\chi\} \tag{6}$$

where the coefficient $A(r_c/r_A)$ depends on the ratio of the zero-order Alfvénic and Sonic spheres. For the parameter range in which A is not so small that the effect is negligible, A is positive and about 0.04. The maximum change in χ is then found to be of order

$$|\Delta\chi| \simeq .04\varepsilon \log\{(C\alpha)_i/(c\alpha)_f\} \tag{7}$$

where C is the moment of inertia of the star, and the suffixes i and f refer respectively to an initial state (e.g. on the Hayashi track), and to a final (main-sequence) state. Thus for the field structure studied $|\Delta\chi|$ is not large. However, the perturbed split monopole field was adopted purely for mathematical convenience. A more realistic basic field structure will allow the field to be curl-free near the star, with the wind channelled parallel to the strong field rather than with the field lines pulled out by the wind; for such structures $|L_Y/L_Z|$ and so also $\Delta\chi$ will be much larger.

The sign of χ depends on χ and on the flux distribution over the stellar surface, as measured by the coefficients $a_k(2)$. If $a_1(2)=0$ i.e. the Y_2 part of $\bar{H}_{1r}$ is antisymmetric in the equator – then χ will steadily approach 0 or $\pi/2$, depending on whether $[\frac{1}{2}a_0(2)-a_2(2)]\gtrless 0$ respectively. If $a_1(2)\neq 0$ the same criterion decides whether χ approaches a small or large value. We expect that with the more realistic approximately curl-free structure, the sign of χ will still be largely determined by the photospheric flux distribution.

The present work suggests the possibility that the same magnetic coupling with a stellar wind that may explain the slow rotations of the magnetic stars may also explain the large inclinations of the magnetic and rotation axes required by the oblique rotator model. It is hoped to extend the work in several ways. We need to compute the torques for more plausible field structures. One can ask whether those surface flux distributions that do yield large asymptotic values are also consistent with observed Zeeman shifts and cross-over effects. There is the formidable problem in internal stellar hydromagnetics of explaining the required flux distribution. Finally, we note that any dissipation of kinetic energy of stars (at fixed angular momentum) will tend to reduce the angle χ (Spitzer, 1958); the time-scale of this process should be estimated and compared with that for the present process.

References

Deutsch, A. J.: 1958, in *Electromagnetic Processes in Cosmical Physics* (ed. by B. Lehnert), Cambridge University Press, Cambridge, England.
Ferraro, V. C. A. and Bhatia, V. B.: 1967, *Astrophys. J.* **147**, 220.
Lüst, R. and Schlüter, A.: 1955, *Z. Astrophys.* **38**, 190.
Mestel, L.: 1966, Liège Symposium.
Mestel, L.: 1967, in *Stellar Magnetism in Plasma Astrophysics*, 39th Enrico Fermi School, Varenna, 1966.
Mestel, L.: 1968a, *Monthly Notices Roy. Astron. Soc.* **138**, 359.
Mestel, L.: 1968b, *Monthly Notices Roy. Astron. Soc.* **140**, 177.
Modisette, J. L.: 1967, *J. Geophys. R.* **72**, 1521.
Parker, E. N.: 1958, *Astrophys. J.* **128**, 664.

Spitzer, Jr., L.: 1958, in *Electromagnetic Processes in Cosmical Physics* (ed. by B. Lehnert), Cambridge University Press, Cambridge, England.
Weber, E. J. and Davis Jr., L.: 1967, *Astrophys. J.* **148**, 271.

Discussion

Deutsch: The A stars are not known to support stellar winds, of course. One can argue that the magnetic A stars are the least likely of any stars to support winds.

Selley: Yes, on the main sequence the early-type stars lack the violent surface convection zones necessary for the generation of an expanding corona. We are postulating however, that magnetic braking is operative during the Hayashi phase. We suppose that the magnetic field is primeval and while it is sufficiently strong to resist continuous deformation by the Hayashi turbulence it does not suppress convection. In the Hayashi phase while we expect all A-type stars to possess a dynamo-built surface field, a magnetic A star will experience far greater magnetic braking if it has, in addition, a primeval magnetic field.

The Ap stars are abnormally slow rotators; at the same time many of them show periodic field reversal which indicates oblique rotators with large values of the angle χ.

Magnetic braking on the Hayashi track due to a primeval magnetic field explains both the slow rotation and the postulated high angles of obliquity of these Ap stars. It remains to be seen whether the same result is given by more realistic models of the magnetic fields.

Nariai: It may be possible that the rotation of a star with a strong magnetic field is decelerated during a part of the Hayashi contraction phase when the star has a convection zone and corona, it then becoming a slowly rotating peculiar A star in the main-sequence phase.

SIGNIFICANCE OF TIME VARIATIONS
IN THE Be PHENOMENON

D. NELSON LIMBER

University of Virginia, Charlottesville, Va., U.S.A.

1. Preliminary Remarks

When I prepared the discussion in the subsequent sections I did not realize just how subject to question were certain basic ideas relating to the interpretation of the Be phenomenon. In view of the points already raised in the course of this Colloquium, I should like to preface my prepared remarks with certain additional ones.

There appears to be little question that the Be phenomenon has important things to tell us of stellar rotation. It appears almost equally clear that too superficial an approach could well lead to invalid conclusions at any one of a number of points. Some caution therefore, appears advisable. Nevertheless, there appear to be a number of reasons for believing that a relatively simple rotational interpretation of the phenomenon rests on much more solid ground than does any other yet put forward.

To give us something concrete on which to fix our attention, let us consider a mechanical model which, though much oversimplified, contains what I believe is the essence of the dynamics of the Be phenomenon. This model consists of an array of rigid spokes fixed in, and radiating outward from, a hub, the whole of which can be rotated about an axis through the hub and perpendicular to the plane of the spokes. Along each spoke a frictionless bead is constrained to move subject to three forces in addition to the force of constraint: (1) a $1/r^2$-force of attraction directed toward the center of the hub, (2) an outward-directed force falling off faster than $1/r^2$ due to a bumper spring coaxial with the spoke and attached to the hub, and (3) the centrifugal force consequent to the rotation of the whole (cf. Figure 1). The bead represents a mass element in the vicinity of a star's equator, the above three forces representing the gravitational force, the force due to the atmospheric pressure gradient, and the centrifugal force, respectively, on such a mass element. The force of constraint imposed on the bead by the spoke represents some kind of viscous force, that provided by magnetic field lines coupling the matter in the atmosphere and circumstellar envelope with the body of the star appearing to be the most likely candidate.

Consider now the stability of such a bead with respect to radial perturbations. In the limit of zero rotation, an outward-directed radial perturbation of the bead from its equilibrium position leads to a situation in which the inward-directed gravitational force exceeds the outward directed force due to the bumper spring; and the bead moves back toward its equilibrium position, the equilibrium being stable with respect to such perturbations.

In the case of intermediate angular velocities, a small outward perturbation from the equilibrium position will lead to a situation in which the net force after the perturbation will again be directed inward in spite of the increase in the centrifugal force on the mass element subsequent to the perturbation, and the equilibrium here, too, will be stable.

The situation is completely different for the case in which the system is rotating with that critical angular velocity at which the centrifugal force balances the gravitational force on the bead, the force due to the bumper spring being vanishingly small in equilibrium in this case. Here, an infinitesimal outward perturbation leads to a situation in which the centrifugal force becomes larger. Since the gravitational force becomes smaller, the net result is a continuing motion of the bead outward along the spoke and, hence, an instability with respect to such perturbations.

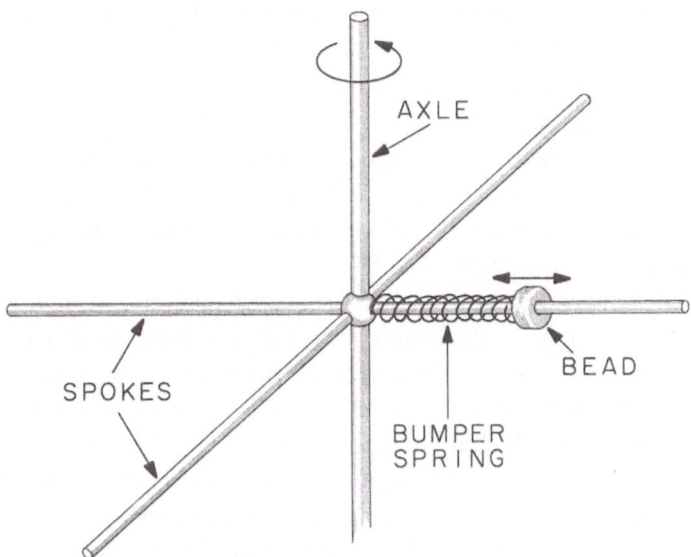

Fig. 1. Schematic mechanical model to illustrate the nature of the Be phenomenon as it is interpreted in this paper.

It is a phenomenon of this latter kind that I believe plays a major role in the Be phenomenon. To be sure, the viscosity is infinite in the model considered, but the nature of the situation with respect to stability is not altered in its essentials so long as there is a finite rigidity to the spokes, i.e. so long as there is a finite viscosity.

It is of interest here to point out that a star need not rotate such that its photosphere is at the rotational limit (that limit at which the centrifugal force balances gravity on mass elements at the equator) in order that the Be phenomenon take place. All that is required is that at points in the atmosphere where the rotation is at the rotational limit there be a density large enough to give rise to a significant mass outflow as a result of the above described instability. Rough considerations suggest that because of the relatively small density gradients in the atmospheres near the rotational limit, the rotation parameter, $\Omega r^3 / GM$, characterizing the photospheres in the equatorial planes may well be as small as 0.9 when this condition is satisfied. This effect would act to help bridge the gap between the critical rotational velocities for the Be stars and the observed maximum values for their equatorial velocities, the latter characterizing their photospheres.

In concluding this introductory section, let me note briefly a number of arguments that appear to support – or at least are quite consistent with – the overall model of the Be phenomenon presented above.

(1) The observational evidence strongly supports the view that the Be stars are stars close to the rotational limit.

(2) Efforts to understand the dynamics of the circumstellar envelopes suggest that the transfer of angular momentum into the envelopes from the bodies of the stars appears much more probable than any alternative mechanism thus far proposed.

(3) The difficulties associated with alternative mechanisms is made even more acute by the Pleione-type of envelope behavior in which the circumstellar envelope is observed to come into being and to dissipate over a period of from 5 to 30 years, all without the rotational state of the star as a whole undergoing any significant change.

(4) The evidence for nearly uniform rotation in stars of moderate rotation in early evolutionary states, from work such as that described by Mrs. Faber (this volume, p. 39) lends support to the belief that there are significant viscosities operating at least within the interiors of such stars.

(5) The model calculations for the main-sequence evolution of uniformly rotating B stars by Miss Sackmann (1968) appear to provide quite good semi-quantitative agreement with the observations concerning the frequency of occurrence and position of occurrence in the main-sequence band of the Be stars on the assumption that a model for the Be stars of the kind described above is correct.

(6) The observed symmetry and geometry of Be envelopes supports the view that rotation over and above that consequent to the conservation of angular momentum within the envelopes is present and, hence, that there is a significant angular-momentum coupling of mass elements within such an envelope with the body of the star.

2. General Background

The Be stars are stars of spectral type B whose spectra show hydrogen lines in emission. It has long been known that this emission originates in tenuous circumstellar envelopes that extend outward from the stars' surfaces some 10 or more stellar radii. The stars themselves appear to be normal main-sequence B stars except that they possess rotational velocities at their surfaces that are considerably greater than those of the typical non-emission B stars on the main sequence.

It was realized more than a third of a century ago that the observed rotational velocities of the Be stars approached the limiting velocity at which such a star's gravitational force on a mass element at its equator is just balanced by the centrifugal force on that element. This suggested that the Be phenomenon was to be understood in terms of a model in which a star – as the result of evolutionary changes – finds itself rotating at (or near) the rotational limit. With the help of centrifugal effects, matter is then able to move outward from the star's equator, giving rise to a circumstellar envelope of the kind observed. Studies of the line profiles have strongly supported this picture. They indicate that the velocity fields in the envelopes are for the most part those corresponding to differential rotational motion, the circular velocity decreasing outward from the equator. The radial component of velocity is, in general, less than 10 km/sec – often only a few kilometers per sec or smaller.

The general picture outlined above is suggestive but is far from complete. That the rotation of a star at, or near, the rotational limit will make the problem of forming a circumstellar envelope simpler, other things being equal, is clear enough; but mass elements must still gain in moving outward from the equator to infinity as much energy again as they possess at the equator in the form of rotational energy. A recent study (Limber and Marlborough, 1968) of the role of the different possible forces that might act in the envelope to accomplish this energy transfer has strongly supported the view that the relevant force is a centrifugal one consequent to the transfer outward through the envelope of angular momentum from the body of the star either through magnetic fields coupling the envelope with the body of the star or through turbulent viscosity, the former appearing to be the more likely.

In the foregoing picture rotation not only brings the star to the rotational limit but also plays the decisive role in the envelope dynamics, as well. If this picture is correct, it then appears that information concerning the observed time variations in the envelopes of Be stars may well shed light on problems concerning the present rotational state and rotational evolution of the bodies of these stars and help us, as well, to better determine the properties of the magnetic fields at their surfaces and within their circumstellar envelopes.

3. The Observed Time Variations in the Be Phenomenon

At the present time it appears convenient to divide the observed kinds of time variations in the circumstellar envelopes into 3 classes on the basis of the time scales for the variations involved. These time scales are: (1) a main-sequence hydrogen-burning time-scale (a time scale ranging from 10^6 to 10^8 years, depending upon the spectral type of the Be star), (2) a time scale of the order of years, and (3) a time scale of the order of days – or less.

A. TIME VARIATIONS ON THE MAIN-SEQUENCE TIME-SCALE

Though we have no direct evidence pertaining to the first – and longest – of these time scales for individual stars, the indirect evidence appears conclusive at several important points. First, for the early type B stars, Be stars make up somewhat more than 10% of all main-sequence B stars. From this alone it follows that the early B stars spend *on the average* more than a tenth of their main-sequence lifetimes in the Be phase. If by reason of their rotational states at the times of their reaching the main sequence only a fraction f of all early B stars can become Be stars, then such stars would spend somewhat more than $0.1/f$ of their main-sequence lifetimes in the Be phase. At the present time there seems to be no sure way of estimating the fraction f. What can be said is that it must exceed 0.1 and, of course, be less than or equal to unity. The two limiting possibilities here are either that all early B stars pass through the Be phase and spend somewhat more than a tenth of their main-sequence lives in this phase or that only somewhat more than a tenth of all early B stars pass through the Be phase, those that do spending almost all their main-sequence lifetimes in this phase.

B. TIME VARIATIONS ON THE TIME-SCALE OF YEARS

There is a large body of direct observational evidence available concerning changes on the intermediate time scale of the order of a few years. Among a group of 40 of the brightest Be stars observed at Ann Arbor over many years, 26 have shown conspicuous changes in the spectral properties of their envelopes, 6 more have shown relatively small changes, while only 8 have failed to show convincing evidence of changes (McLaughlin, 1961). The changes observed include changes in the equivalent widths of the emission lines, changes in the forms of the line profiles, and changes in the radial velocities for the lines. Strictly periodic variations are not observed on this or any other time scale – except insofar as they simply reflect the orbital motion of Be stars that are components of binary systems. One rather characteristic type of variation on the intermediate time scale involves the alternation of 'active' intervals of 10 to 30 years, during which quasi-periodic changes with periods of 2 to 5 years are the rule, with more nearly time-independent intervals of roughly the same length. Representative stars showing this type of behavior are π Aquarii (McLaughlin, 1962) and HD 20336 (McLaughlin, 1961). A second type of variation, while not so common, appears important because of its overall simplicity; it may, in fact, be but a variation of the type just described. In this second type of variation – and here we shall simply describe the behavior of Pleione (Limber, 1969) – the star was first observed to have a circumstellar envelope in 1888. This envelope was observed through 1903 but was gone in 1905. From 1905 until 1938 there was no evidence for the existence of a circumstellar envelope, the star's spectrum being that of a rapidly rotating but non-emission B star. Beginning in 1938 a circumstellar envelope was observed to come into being, reaching its greatest strength in 1945 or 1946. The envelope then began to fade away and was essentially completely gone by 1954. Here the evidence strongly supports the view that we have on two occasions seen this particular rapidly rotating B star produce a circumstellar envelope and have it dissipate over an interval of some 15 years, the interval between the two 'shell' or envelope episodes being about 35 years. It appears quite possible that many or all of the Be stars that have not yet been observed to show significant time variations will show them over intervals of from 50 to 100 years.

C. TIME VARIATIONS ON THE TIME SCALE OF DAYS

The body of observational data pertaining to time variations on the time scale of days is not at all large. Indeed, at this stage of our understanding it requires a rather arbitrary act to neatly separate the time variations of the intermediate and short time scales. I shall confine myself to a very brief description of 3 different studies that have revealed significant time variations over intervals ranging from one hour to one week. The variations include both significant variations in the equivalent widths of Hα as well as relatively small changes in the details of the line profiles for Hβ and Hγ. A study of HD 174237 by Lacoarret (1965) indicated that there were significant variations in the equivalent widths of Hα with a characteristic quasi-period of about 7 days. A study by Miss Doazon (1965) of the Be star HD 50238 gave evidence for significant changes in the radial velocities for the Balmer lines over a few days, for the appearance and disappearance of 'satellite' absorption features to the envelope lines in a day, and for 10% changes in the equivalent width of Hβ in a day and for 50% changes in a week. Finally, Hutchings (1967) has detected significant changes is the Hγ profile of γ Cassiopeiae over intervals of an hour through a photoelectric scanning technique.

4. Implications

A. TIME VARIATIONS ON THE MAIN-SEQUENCE TIME-SCALE

Let us now turn to a brief consideration of some of the implications – or possible implications – of these observed time variations. Consider first those of the time variations on the main-sequence time scale. Here the evidence strongly supports the view that a fraction f of all early B stars are brought reasonably close to the rotational limit and kept there during at least a fraction $0.1/f$ of their main-sequence lifetimes, where $0.1 < f \leq 1$. Several additional points should now be noted. First, as a result of hydrogen burning in the core, it appears highly probable that the equatorial radius will increase in time subsequent to such a star's contraction to the main sequence. If there were no angular momentum coupling between the different spherical or cylindrical shells making up the body of the star, this evolutionary change would act to move the star *away* from the rotational limit. Second, there appears to be no evidence that the Be stars are stars only in their earliest main-sequence stages. On the contrary, there appears to be some evidence that the Be stars represent stars either in the later main-sequence stages only or, in the extreme, stars spread over all main-sequence stages. This suggests that there is at least some continuing transfer of angular momentum from the inner parts of these stars to their equatorial regions that either moves the latter regions close to the rotational limit or keeps it there in spite of the reverse tendency consequent to its surface expansion. With further study it may become possible to determine this average rate of angular momentum transfer with considerable accuracy.

Two non-mutually exclusive possibilities have been suggested as sources of the angular momentum flows to the equatorial regions. The first was suggested by Crampin and Hoyle (1960). Model calculations for spherically symmetric, non-rotating stars during main-sequence and early post-main-sequence phases showed that although their surfaces expanded, their inner parts contracted more than enough to compensate, with the net result that – at least in the limit of infinitesimal rotation – the main-sequence hydrogen burning phase and, particularly, the stage of rapid core contraction subsequent to hydrogen exhaustion are phases in which for uniform rotation the changes are such as to move the equatorial regions in the direction of the rotational limit. More recent model calculations, which include the effects of uniform rotation, appear to support this conclusion, giving evidence that the effects even during the hydrogen burning phase prior to the state of rapid core contraction are large, indeed (Sackmann, 1968).

An alternative possibility arises from the circumstance that at least many stars, upon reaching the zero-age main sequence, may be in a state of differential rotation in which their inner parts rotate with angular velocities significantly larger than those in their equatorial regions as a consequence of their pre-main-sequence histories. If this is the case, then the operation of any 'viscous' agents (including magnetic fields and turbulent viscosity) will act to transfer angular momentum to the equatorial regions at the expense of that in the inner regions and may, conceivably, proceed at a rate more than sufficient to overcome the effects of the evolutionary expansion of the equatorial regions and move such stars toward the rotational limit.

In either case, the observations suggest that here are viscous agents operating in the interiors of main-sequence B stars that can significantly alter the rotational states of their equators over time scales of the order of the main-sequence one.

B. TIME VARIATIONS ON THE TIME-SCALE OF YEARS

The time variations observed over a time scale of the order of years – especially those involving variations in which envelopes are observed to come into being and to disappear – provide the strongest kind of evidence that the rate at which matter leaves these stars' equators often varies markedly over the course of 5 or 10 years and that the matter that forms an envelope dissipates in the course of a few years when left to itself. The important thing to be noted here is that individual envelopes characteristically come and go on a time scale of from 5 to 50 years and not on anything like the main-sequence time scale. Since the mass lost during such an envelope episode is of the order of $10^{-6} M_\odot$ or less and since the time scale is so much shorter than the hydrogen-burning one, the rotational state of the star as a whole can scarcely change in any significant way during such an envelope episode. This has been taken by some as evidence that, although rotation plays a significant role in the Be phenomenon, there must be something else that actually triggers it. If the stars were constrained to rotate in detail as solid bodies, this argument would be compelling indeed. However, in the absence of any direct evidence for a nonrotational triggering mechanism, it appears reasonable to seek to understand the situation in terms of rotation's doing the triggering – but for non-uniformly rotating stars in which the flow of angular momentum outward from the inner portions of the stars to their equators is not smoothly continuous – at least in the vicinity of the stars' equators – but is, instead, rather 'lumpy'. Rough calculations have been carried out for a simple model in which an equatorial band of arbitrary thickness rotates uniformly out to that distance from the rotation axis at which centrifugal force balances gravity and on the assumption that the rate of mass loss is proportional to the density at this critical surface. The calculations show that if at time zero the equatorial band possesses angular momentum enough to lead to an appreciable rate of mass loss and if the angular momentum supplied to this band is abruptly cut off at this time, then the subsequent mass loss will have effectively come to an end some ten or twenty years later for the case in which the equatorial band includes only that part of the star lying more than eight-tenths of the equatorial radius from the star's rotation axis; the decay time for the mass loss is of the order of a week or so if the equatorial band includes only that part of the star's mass exterior to 0.96 of the star's equatorial radius. In both cases the decay times for the rate of mass loss from the equators set lower limits to the times required for the envelopes to disappear, the latter depending, as well, upon the velocity field within the envelopes. (For purposes of comparison, the decay time-scale for the case in which the star as a whole rotates uniformly is a thousand years or more.) In none of the above cases (except that for uniform rotation throughout) would the stars' observed rotational velocities be perceptibly smaller at the termination of the envelope formation than at its peak.

What is suggested to me is that the variations on the time scale of years are to be understood in terms of the transfer of angular momentum into the equatorial regions in a manner exhibiting variations over intervals of this same order – or smaller. Work is in progress that may help to put certain aspects of this problem on a more quantitative footing.

C. TIME VARIATIONS ON THE TIME SCALE OF DAYS

Time variations on a time scale of a day are in all probability the manifestations of rather localized phenomena insofar as the envelope as a whole is concerned; this conclusion appears all the more certain for the variations that have been observed by Hutchings to take place within an hour. At the sound speed within these envelopes the time required to cover one stellar radius is about a week; a time-scale of an hour would imply an active volume whose characteristic dimension was one one-hundredth of a stellar radius or smaller. (Use of the Alfvén velocity rather than the sound speed doesn't appear likely to alter this conclusion drastically.) Consequently, it is tempting to think in terms of a rather localized instability's being responsible for these short-time variations. The time scales involved suggest flare-like activity. Indeed, a simple extension of a mechanism suggested for explaining solar flares, involving the electromagnetic energy release consequent to the reconnection of magnetic field lines in the differentially rotating envelopes appears promising. If this – or a related mechanism – is found to explain these short-term variations, we shall have further support for the view that magnetic fields play an important role in the dynamics of the envelopes of Be stars.

References

Crampin, J. and Hoyle, F.: 1960, *Monthly Notices Roy. Astron. Soc.* **120**, 33.
Doazan, V.: 1965, *Ann. Astrophys.* **28**, 1.
Hutchings, J. B.: 1967, *Observatory* **87**, 289.
Lacoarret, M.: 1965, *Ann. Astrophys.* **28**, 321.

Limber, D. N.: 1969, *Astrophys. J.* **157**, 785.
Limber, D. N. and Marlborough, J. M.: 1968, *Astrophys. J.* **152**, 181.
McLaughlin, D. B.: 1961, *J. Roy. Astron. Soc. Can.* **55**, 73.
McLaughlin, D. B.: 1962, *Astrophys. J. Suppl.* **7**, 65.
Sackmann, I. J.: 1968, *The Structure and Evolution of Rapidly Rotating B-Type Stars*, University of Toronto Thesis.

Discussion

Roxburgh: It is clear that the material in the shell has to be ejected with considerable energy. It is probably necessary that some instability sets in for high angular velocity, such that the equatoria regions can acquire this energy.

Another possibility is that a slow pressure-driven 'wind' is possible for high angular velocity.

Buscombe: Two aspects of the variation of flux from Be stars in an interval of weeks or months need more intensive observational surveillance – the ultraviolet brightening and variations in intrinsic polarization. Some Be stars in high galactic latitudes are especially favorably located for the separation from interstellar effects. Is there a sporadic burst of electron plasma, or what?

Limber: The mechanism that has suggested itself to me for accounting for the very short-term time variations of the order of hours is one involving the reconnection of magnetic field lines with a consequent conversion of magnetic energy into the kinetic energy of charged particles. This latter energy would presumably manifest itself insofar as the observations are concerned as a sudden heating of relatively small elements within the circumstellar envelopes.

THE ROTATION OF THE Be STAR γ CAS

J. B. HUTCHINGS

Dominion Astrophysical Observatory, Victoria, British Columbia, Canada

(Paper read by T. R. Stoeckley)

Abstract. A period of 0.7 days has been found in the peak separation and V/R ratio of the double emission profiles of Hγ and Hβ in this star. The phenomenon is believed to arise from short lived condensations near the base of an equatorially extended atmosphere, and the period to be that of rotation of the star. The deduced stellar diameter is $7\,R_\odot$. Further considerations on the emission line profiles lead to a model of an equatorial ring extending 1–2 stellar radii with angular velocity decreasing outwards, and viewed at an axial inclination of some 60°.

1. Introduction

γ Cas is a well-known bright Be star, whose spectrum was investigated quite extensively some 30 years ago, when it showed prominent shell lines (Beer, 1956). Recent photoelectric line profile studies made by Hutchings (1967, 1968a) have shown that the double emission peaks in the Balmer lines vary rapidly and irregularly from minute to minute, as well as showing larger scale changes in intensity over periods of the order of weeks. An intensive study was made of the Hγ and Hβ profiles during the winter of 1968–9 and from some 140 scans there is now also evidence for a period in the separation and V/R ratio of the peaks, which is about 0.7 days.

2. The Observations

The observations were made with the coude scanner of the Victoria 48″ telescope (Hutchings, 1968a) and have a probable error of $\pm 1\%$ of the continuum and a time

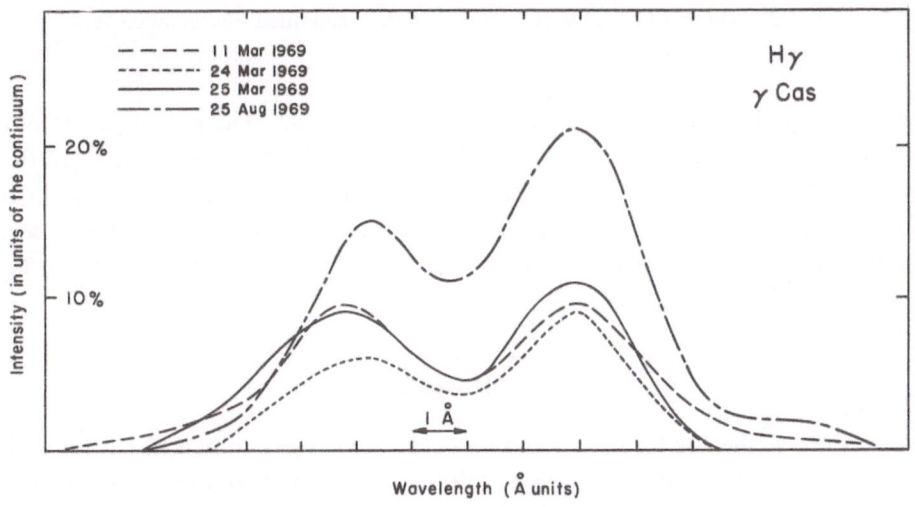

Fig. 1. Observed Hγ profiles in γ Cas with 20 min time resolution.

A. Slettebak (ed.), Stellar Rotation, 283–286. All Rights Reserved

resolution of some 5 min. Figure 1 shows some typical Hγ profiles which represent mean profiles over periods of about 20 min. The minute-to-minute variations are therefore eliminated and the diagram illustrates the day-to-day and longer term changes observed. The lower three profiles show the type of V/R and peak separation changes which have been found to be periodic and have typical amplitudes in these changes. The upper profile shows how the whole feature strengthened between March and August 1969.

The periodic variation, partly hidden in the rapid irregular fluctuations, was found in all series of observations taken within some 10 days of each other. Over longer times discontinuous phase shifts were found. Figure 2 shows the observations and

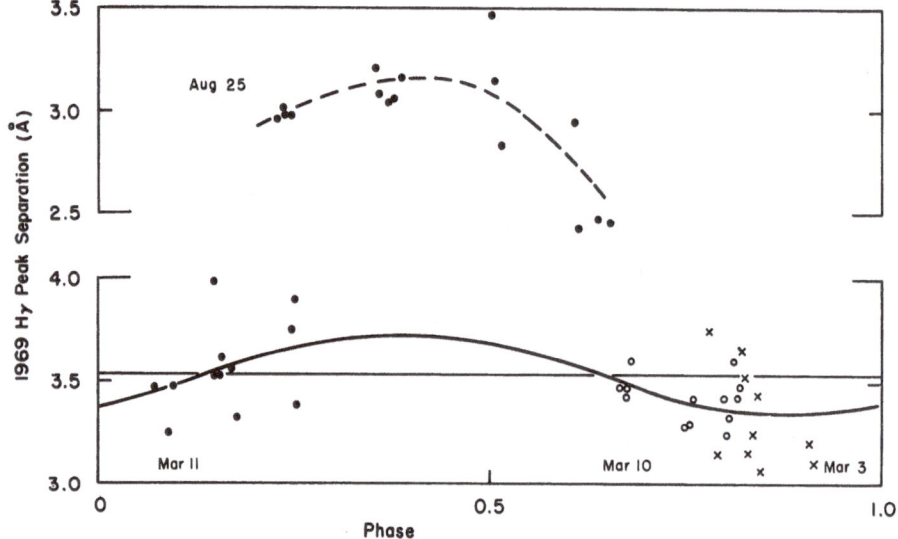

Fig. 2.   Observed Hγ peak separation variation and fitted sinusoidal curves.

fitted sinusiodal curve for two typical sets of observations. The period was found by means of a computer program provided by Hill (1969). The mean separation of the peaks is 3.2 Å for Hβ and 3.4Å for Hγ, with a standard deviation of 0.05 Å in each case. The mean amplitude of the periodic variation is 0.5 Å for all coherent sets of observations. The value of V/R varies more widely from month to month, but has a typical periodic amplitude of about 10%.

3. Discussion

The star is classified as B0 IVe by most workers and is evidently a fast rotator. Following Roxburgh and Strittmatter (1965) and Stoeckley (1968) it seems likely that the star is a class IV or V star rotating at equatorial breakup velocity, with an equatorially extended envelope in which the emission lines are formed. This approach has been developed by Hutchings (1968b), in which it is suggested that the star is viewed from within 30° of the equator-on position.

The breakup velocity of rotation of the star is close to 500 km/sec at the equator, so that if we assume the period of 0.7 days to be the rotation period of the photosphere we derive a stellar radius of 7 $R_{\odot}$. This is in close agreement with accepted values for the radius of such a star.

We now attempt to justify this assumption by checking that it leads to a self con-sistent model of the star. If the star is losing matter through an equatorial region of zero gravity we may expect that this part of the photosphere fluctuates irregularly in brightness and density and may release occasional condensations of gas whose density is higher than the mean. These condensations would rise slowly and disperse in the outer layers, possibly losing angular momentum. Such condensations would appear irregularly and may last for several days during which they are detectable as a strength-ening of the main emission peak alternately on the V and R sides as the mass rotates. The mass-loss mechanism in such a star is not yet understood (Limber and Marl-borough, 1968) but it is here suggested that radiation pressure may be sufficient in the zero gravity equatorial region. The P Cygni profile of C IV at 1550 Å observed in γ Cas by Morton (1970) supports this. Loss of angular momentum in the disc at some dis-tance (1–2 R*) from the photosphere therefore need not result in the collapse of the disc, and is in any case demanded by the observed Hγ and Hβ peak separations, which correspond to a $V \sin i$ of some 200 km/sec. The postulated lower-lying condensations therefore give rise to the observed periodic variation in peak separation by having a higher angular velocity.

There are two more points in support of this model. The observed profiles never show the R component weaker than the V. This is incompatible with a static rotating envelope, but as shown by Hutchings (1968b) is easily explainable if the entire equatorial envelope has a small outward velocity. Some preliminary line profile calcu-lations have been made, which cannot be described in full here. They are similar to those previously published, but use an excitation-height relation similar to that deduced for the Orion supergiants (Hutchings, 1970). It was possible to match the observed Hγ

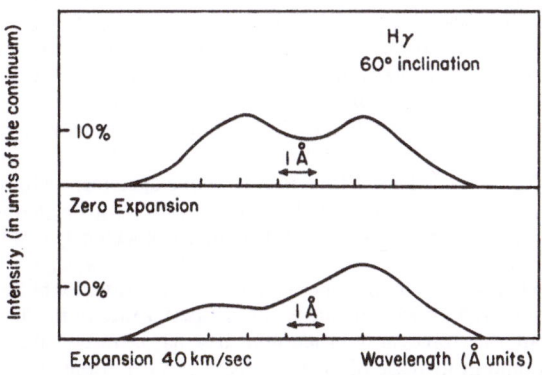

Fig. 3.   Computed Hγ profiles for 60° inclined star expanding at 0–40 km/sec.

and Hβ profiles using an axial inclination of 60° to the line of sight and a surface expansion velocity of $\simeq 20$ km/sec (Figure 3).

The final point of agreement is that of the mean separation of the peaks. These correspond to velocities of 240 km/sec for Hγ and 200 km/sec for Hβ. This sort of discrepancy is found in several Be stars of this type (e.g. κ Dra) and can be explained by two effects. The stronger Balmer lines are formed through more extended regions of the disc than the weaker ones, so that if the star is equator-on the effect must be caused by a lower angular velocity in the outer disc; if the axis is inclined the central part of the emission feature may be filled in by emission seen over the pole of the star. This filling in will be stronger for the stronger Balmer lines. It is possible to distinguish between these effects by comparing computed and observed profiles, and optimum fit in the case of γ Cas is indicated at about a 60° inclination and angular velocity decreasing by a factor of 2 in 1 R^*.

At present it is difficult to see how any other model can explain all the observations. Further computations are being made to refine the above results and to derive a mass-loss estimate. These will be published in full in due course.

References

Beer, A.: 1956, *Vistas in Astronomy*, Vol. 2, Pergamon, London, p. 1407.
Hill, G.: 1969, Private communication.
Hutchings, J. B.: 1967, *Observatory* **87**, 289.
Hutchings, J. B.: 1968a, in *I.A.U. Colloquium on Variable Stars*, Budapest.
Hutchings, J. B.: 1968b, *Monthly Notices Roy. Astron. Soc.* **141**, 329.
Hutchings, J. B.: 1970, *Monthly Notices Roy. Astron. Soc.* (in press).
Limber, D. N. and Marlborough, J. M.: 1968, *Astrophys. J.* **152**, 181.
Morton, D. C.: 1970, in *I.A.U. Symposium* **36**, Reidel, Dordrecht, The Netherlands.
Roxburgh, I. W. and Strittmatter, P. A.: 1965, *Z. Astrophys.* **63**, 15.
Stoeckley, T. R.: 1968, *Monthly Notices Roy. Astron. Soc.* **140**, 141.

Discussion

Collins: What sort of error estimates would you put on your results ($i = 60°$, etc.)?

Hutchings: The figures quoted in the paper are preliminary ones, but work in the few weeks since writing them down indicates that i has a p.e. range of 50°–75°, $\omega \sim (R/R^*)^x$ where $0.8 < x < 1.2$, and a mean expansion velocity of 20 km/sec $\pm$ 10 km/sec (which is indistinguishable from a velocity field rising exponentially to 50 km/sec at $\simeq 3R^*$). With so many free parameters it seems possible to reproduce the profiles observed with combinations of figures within the above ranges.

Jordahl: An observed $V \sin i$ of 300 km/sec for a breakup velocity of 500 km/sec would give $\sin i = 0.6$ whereas $\sin 60° = 0.833$. How good is the observed $V \sin i$?

Hutchings: $V \sin i$ is quoted as 300 km/sec by Boyarchuk and Kopylov but I cannot comment on the accuracy of this. Dr. Stoeckley has shown that this should be corrected for gravity darkening to yield a value of some 400 km/sec, which is then in reasonable agreement with my results. I would in any case regard an analysis of emission line profiles, such as mine, as giving a more reliable value for i.

Stoeckley: Do you mean it when you suggest that angular *momentum* is lost by the shell? Why? Do you mean angular *velocity*?

Hutchings: What I meant was angular velocity. There seems little doubt about this, but whether momentum is lost is a more tricky question, and we need a better idea of the density-height relation to answer it. If viscous or radiative forces are considerable they may be responsible for converting angular to outward momentum, and in the case of higher density condensations, they may transfer angular momentum to the remainder of the envelope.

PART IV

THE ROTATION OF THE SUN

THE ROTATION OF THE SUN*

(Review Paper)

R. H. DICKE

Palmer Physical Laboratory, Princeton University, Princeton, N.J., U.S.A.

Abstract. The author's 1964 article in *Nature* on the sun's rotation is rediscussed in the light of new data. This article suggested that the sun might be oblate because of a gravitational quadrupole moment induced by a core rotating with a period < 2 days. Angular momentum lost from the core by molecular diffusion was assumed to be transferred to the solar wind which kept the sun's surface rotating slowly. The estimated solar wind torque was found to be in good agreement with the torque calculated from a solution of the diffusion equation.

Subsequent to the 1964 paper the oblateness was observed. Also the solar wind torque was 'observed' to be in good agreement with the early estimate. New observations discussed here seem to be important if the sun can be safely assumed to be a typical star and not an exception. It has been found by Kraft (1967) that very young solar type stars (Pleiades) are rotating with roughly the same angular velocity postulated for the solar core. As determined from observations on the Hyades in comparison with the Pleiades, the rotation of young solar-type stars is slowing at a rate consistent with a stellar wind torque equal to that of the solar wind acting on the outer 20–30% of the star by radius. The slowing of the rotation in young stars is accompanied by a depletion of lithium, but not berylium. This implies that only the outer 45% of the star by radius, or 5% by mass, is slowed by the solar wind. The rapid rotation of the inner 95% of the mass is sufficient to generate the observed oblateness. The rate of depletion of lithium, determined from observations on solar-type stars of various ages, is consistent with the rate of angular momentum loss assuming a reasonable model for the transport of angular momentum to the convective zone.

1. Introduction

In a brief note published in 1964 I suggested that the sun might have a rapidly rotating core. (Dicke, 1964.) This possibility was also discussed by Roxburgh (1964), Plaskett (1965) and Deutsch (1967). In the *Nature* article it was noted that such a rapidly spinning core would induce a gravitational quadrupole moment in the sun and that the resulting perturbation of Mercury's orbit could account for 5–10% of the classical 43″ arc/century excess motion of Mercury's perihelion. The resulting solar oblateness, as large as 6×10^{-5}, was considered measurable. In collaboration with H. Hill and H. M. Goldenberg, the design and construction of a special instrument had been launched a year earlier, in the spring of 1963. The first version of this telescope was put in operation during the summer of 1963. Two years were required to study the systematic errors of the instrument and to correct and improve its design. (See Figures 1, 2, and 3.) The first useful measurements, made during the summer of 1966 (Dicke and Goldenberg, 1967), gave an oblateness of $(r_{eq} - r_{pole})/r = (4.8 \pm .9) \times 10^{-5}$. (See Figure 4.) The measurements made during the summer of 1967 are not yet published, but they yield the same value for the oblateness with comparable precision.

* This research was supported in part by the National Science Foundation and the Office of Naval Research of the U.S. Navy.

Fig. 1.   The solar oblateness telescope located on the grounds of the Princeton Observatory. This instrument was designed by H. Hill, H. M. Goldenberg and the author. As the picture clearly shows, the ratio of number of pieces of electronic equipment to telescope aperture in inches is probably greater for this instrument than any other in existence (reprinted courtesy American Philosophical Society).

I do not propose to discuss the observations here. This will be the subject of a full treatment by Goldenberg and me. Rather I shall return to the old publication in *Nature*. It may seem strange to devote space to a paper that is over 5 years old, but much of this story has not been told and some of it now requires updating.

The note in *Nature* was extremely condensed and I had intended immediately to follow it with several detailed papers, but this was precluded by the press of the observational program. One of these papers was finally written and is soon to be published (Dicke, 1970a).

This paper discusses the effects of surface stresses on the sun's shape and brightness distribution. This contribution to oblateness and equatorial brightening is explicitly calculated. The constraint imposed on the theory by the lack of equatorial brightening is examined. The effects on the oblateness of observed magnetic and velocity fields are

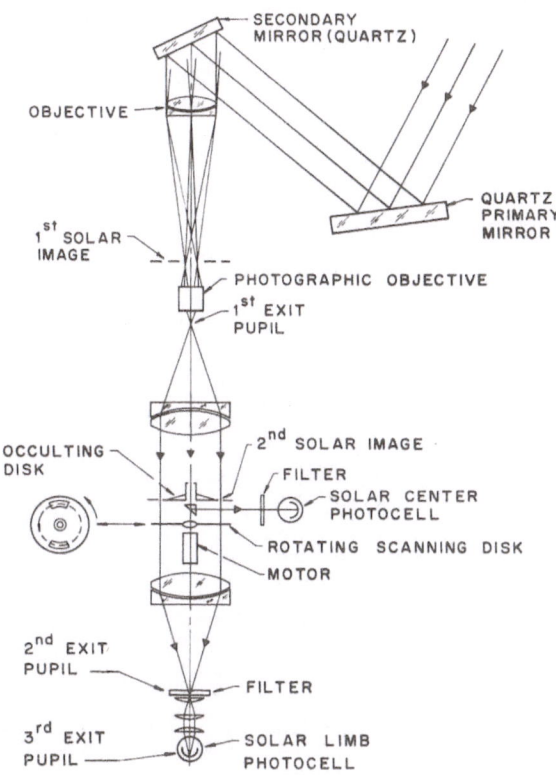

Fig. 2. The optical system of the oblateness telescope. The first mirror tracks the sun by means of a motor drive of a gimbal-mounting of the mirror. A fast-acting solenoid-actuated servo-system provides precision pointing of the mirror. To avoid stick-friction bearing noise, bent-hinge bearings are provided for the servo-system. An image of the sun is projected on an occulting disk which passes the outer 6″–20″ arc of the sun's disk to a rapidly spinning scanning wheel. Two apertures of slightly different size in the scanning wheel pass light to the main photocell. The error signal to the servo-system is derived from this photo-cell as the fundamental rotation frequency of the scanning wheel. The oblateness signal is derived from the 2nd harmonic of this frequency (reprinted courtesy Physical Review).

considered with the conclusion that, as yet, there is no explanation for the excess oblateness, other than the effect of a quadrupole moment.

Another paper, intended to be joint with P. J. E. Peebles, is no longer needed. The note in *Nature* contained the main result derived from our theory of the solar wind torque. This theory, based on Schatzman's (1962) original idea, has been independently developed by three other groups (Modiesette, 1967; Weber and Davis, 1967; and Alfonso-Faus, 1967). Unfortunately, through an error in placing a reference number, Modiesette's description of our calculation is inaccurate, having been meant for another reference.

I propose to organize the present article along lines similar to the *Nature* article, to discuss a number of important points in more detail, and to introduce a substantial amount of information new since 1964.

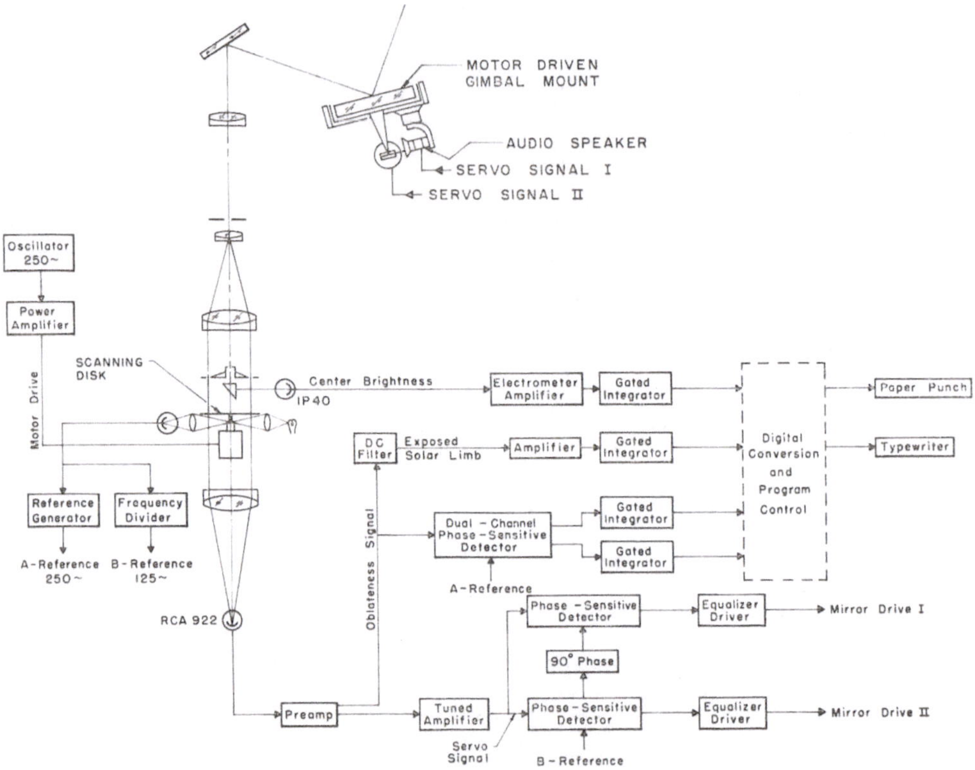

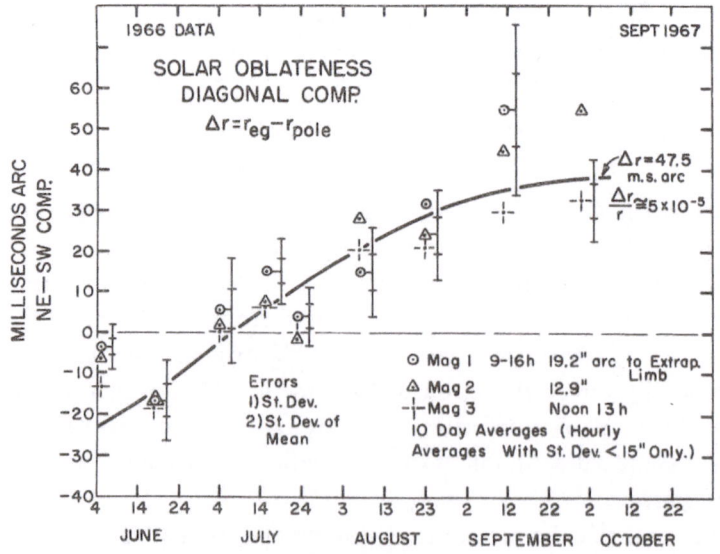

Fig. 3.   A block diagram of the system showing the chief parts of the instrument (reprinted courtesy
American Philosophical Society).

Fig. 4.   The fit of the curve for the 'diagonal' component of the oblateness to the observed '10-day
averages' of 1966. The exposed limb for magnification #3 (Mag 3) was 6.6″ arc. The 'diagonal'
component is the part of the oblateness associated with a shortening of the sun's disk along the
NE-SW line (reprinted courtesy American Philosophical Society).

In pre-relativity days the observed excess motion of Mercury's perihelion ($\sim 43''$ arc/century) was enigmatic and lead to several unsuccessful attempts to find a source of a gravitational field which could advance the perihelion. Among the sources considered were Vulcan, a hypothetical planet whose orbit was thought to lie between the sun and Mercury. It has not been discovered. To be large enough to be significant in perturbing Mercury's orbit Vulcan should have been visible against the sun's disk. Interplanetary debris is likewise an unpromising source of such a gravitational field (Chazy, 1928).

Of these old proposed sources, only a solar quadrupole moment (Newcomb, 1897) remains today as an interesting possibility, and it is unsuitable as a source of the full excess motion of Mercury's perihelion. Such a quadrupole moment would also induce a $43''$ arc/century regression of Mercury's node on the equatorial plane of the sun. Such a large error in orbital wobble can be excluded by the observations.

With the advent of Einstein's General Relativity, a relativistic explanation was available for the excess motion of Mercury's perihelion and the search for conventional sources ceased. But the scalar-tensor theory, an alternative general relativistic theory of gravitation, (Jordan, 1948, 1959; Thirry, 1948; Bergmann 1948; Brans and Dicke, 1961; Dicke, 1962) predicts a slightly smaller effect, $(1 - \frac{4}{3}s)$ times Einstein's value. Here s is the fraction of a body's weight due to the scalar field under the Dicke (1962) version of the theory. Expressed in terms of ω, the coupling constant of the Brans-Dicke (1961) theory, $s = 1/(2\omega + 4)$. On various grounds ω had been estimated to fall in the range 4–7 (Brans and Dicke, 1961; Dicke and Peebles, 1965; Dicke, 1966).

The scalar-tensor theory with $\omega = 5$ yields a relativistic rotation of the perihelion of $38.7''$ arc/century. This is compatible with the observations, providing the sun has an oblateness of $\sim 5 \times 10^{-5}$ and the connection of this oblateness with a quadrupole moment has been properly interpreted. It had been noted (Dicke, 1964) that the resulting motion of the node of Mercury's orbit on the equator of the sun, $4.3''$ arc/century, becomes mainly a $0.21''$ arc/century decrease in the inclination, when referred to the ecliptic. This is a tricky point, that the coordinate transformation has such a large effect. It has been missed on several occasions and as often rediscussed by others (Shapiro, 1965; Gilvarry and Sturrock, 1967; O'Connell, 1968). The expected residual in the rate of increase of the inclination ($-0.21''$ arc/century with $\omega = 5$) is to be compared with an observed residual of $-0.12'' \pm 0.16''$/century. It is evident that a quadrupole moment large enough to rotate the perihelion by $3''$–$4''$ arc/century can be tolerated but not one 3–4 times larger.

At the time that C. Brans and I published our paper on the scalar-tensor theory we considered the uncertainty in the indirectly determined mass of Venus large enough to permit a few percent correction to the 'observed' excess motion of Mercury's perihelion, but after the mass of Venus had been directly determined from the Mariner II orbit, this possibility was excluded.

The mass of Venus is now known with even more precision and the classical perturbations (except for that of an oblate sun) must be assumed to be well known. The masses of the earth and Jupiter have long been accurately known. If the observations of Mercury's motion are as accurate as believed by the experts, the observed classical excess motion of Mercury's perihelion, 43" arc/century, has an accuracy of 1%. Prudence would require that we permit an observational error as large as 5%, but this is not enough to admit the relativistic perihelion rotation of 39" arc/century expected under the scalar-tensor theory. If the sun has an appropriate quadrupole moment, the scalar-tensor theory is favored. If not, it is excluded with $\omega < 10$.

It must be emphasized that a rapidly rotating solar core is not the only possible source of a substantial quadrupole moment in the sun. A strong, deeply buried field along the rotation axis could also generate such a distortion (Sturrock and Gilvarry, 1967). My reasons for rejecting this in 1963 still seem compelling. A magnetic field of the order of 10^6 gauss must stay buried or it will appear at the surface as a very strong permanent dipole. Such a strong dipolar field oriented along the rotation axis would be expected to diffuse to the surface. But a permanent 'general magnetic' field does not seem to exist at the solar surface.

2. Physical Requirements for a Rapidly Rotating Core

A rapidly rotating solar core is impossible unless two conditions are satisfied:

(a) The core must be able to spin with very little friction.

(b) A frictional drag on the solar surface must keep the surface rotating slowly.

The theory of the solar wind torque requires a knowledge of the magnetic field strength in the solar wind for a calculation of the torque and this information was not directly available in 1964. In lieu of an observed field strength, I estimate an equivalent field strength at the sun's surface and used it to calculate the torque. The basic conception of the solar wind torque is due to Schatzmann (1962).

The torque density at the equator on the solar surface is (Dicke, 1964; Modiesette, 1967; Weber and Davis, 1967; Alfonso-Faus, 1967)

$$K = Jr^2\omega_0 \tag{1}$$

where J is the mass flux density at the solar surface, expressed in gm/sec cm^2 lost to the solar wind. ω_0 is the angular velocity of the solar surface and r is the critical radius for which B^2 satisfies the equation

$$\varrho v^2 = \frac{1}{4\pi} B^2 \tag{2}$$

namely, the radius at which v, the radial component of the solar wind velocity equals the Alfvén velocity calculated from B, the radial component of the magnetic field.

The magnetic field is trapped in the solar wind and B falls off inversely as the square of the radial distance. Expressed in terms of the equivalent radial component of magnetic field strength at the sun's surface, B_0, the critical radius is such that the density at this radius satisfies

$$\varrho = 4\pi J^2 / B_0^2. \tag{3}$$

In Equation (2) the square of magnetic field strength should not be interpreted as a mean squared value of the field over the whole surface but rather as an average over such weak unipolar regions as are pulled out by the solar wind. The r.m.s. of this field was estimated to be $\frac{3}{4}$ gauss. Subsequently, when the interplanetary field was measured at 1 AU using the Mariner II space probe (Coleman, 1966), the field was found to have a value consistent with this estimate. As expected, the field was found to be twisted into a spiral pattern. The radial component of B fluctuated some and the r.m.s. value of this component was 3.5×10^{-5} gauss. This is in good agreement with the value of B obtained from $B_0 = \frac{3}{4}$ gauss, namely $\frac{3}{4} \times \frac{1}{200}^2 = 2 \times 10^{-5}$ gauss assuming that the field falls off inversely as the square of the distance, a somewhat simplified assumption as it ignores centrifugal concentration to the equatorial plane.

From the same space probe, the rate of mass loss, or mass flux density at the solar surface, was determined to be $J = 1.7 \times 10^{-11}$ gm/cm^2 sec assuming that the flux density varies inversely with distance squared (Neugebauer and Snyder, 1966).

Substituting these results in (3) gives 3.9×10^3 for the proton number density at the critical radius. Making use of a standard model of the solar corona (e.g. Allen, 1963) gives $r = 20\, r_0$ for the critical radius and, from Equation (1), torque density of $K = 9.3 \times 10^7$ dyne/cm.

The close agreement between this value for the torque density and the value published in 1964 (1×10^8 dyne/cm) is fortuitous for the value of the effective solar field strength, $\frac{3}{4}$ gauss, was only an estimate. Also, because of concentration to the ecliptic, this 'observed' value may be much too large.

The basis for the estimate is worth some discussion, for, if correct, it can be used to extrapolate the solar wind torque into the past when the sun may have been much more active magnetically.

It is well known that a magnetic flux tube carrying a fluid moving along the field lines possesses a magnetic pressure $B^2/8\pi$, in excess of its gas pressure, and a tension along the field lines of $+(B^2/4\pi - \rho v^2)$. For a steady state of a cylindrical flux tube surrounded by field free gas of pressure P, the magnetic field must satisfy the condition

$$|B| < \sqrt{8\pi P}. \tag{4}$$

Equation (4) states that a negative gas pressure in the tube is impossible. The connection between the magnetic field and the gas properties inside the tube is actually stronger than that indicated above. For an arbitrary flux tube (not necessarily cylindrical) the distribution of both density and pressure inside the flux tube are uniquely determined (relative to their values outside) by the magnetic and velocity field distributions in the flux tube (Dicke, 1970b).

The above inequality must be imposed sufficiently low in the corona that the tension associated with the gas flow is small compared with the magnetic tension. From Allen's model of the corona at the equator the upper limit for B_0 given by Equation (4) takes on the values 0.8, 0.7, 0.5 and 0.35 gauss at $r/r_0 = 1.01$, 1.1, 1.4, and 2.0 respectively, and $r/r_0 = 2$ is probably too high in the corona. Apparently the strength of the coronal magnetic field is near its upper limit. Thus, assuming that the sun was more active magnetically in earlier times, the field B_0 was probably little greater than at present, perhaps by less than a factor of 2.

It can also be argued that the particle flux density in the solar wind has changed little during the past 4.5×10^9 years. It is now well recognized that this flux is determined by the rate of heating of the corona, in turn determined by the turbulence near the solar surface. But this turbulence is fixed by the luminosity which presumably has not changed drastically. Admittedly these arguments are crude, but they suggest that over the life-time of the sun the ratio of solar wind torque to angular velocity ω_0 can be assumed to have decreased little, by perhaps a factor of $\frac{1}{4}$. Furthermore for stars similar to the sun, the ratio of stellar wind torque to angular velocity would be expected to be approximately equal to that of the sun, providing the star is sufficiently active magnetically.

If the above equatorial torque density is correct, the total solar wind torque now is $(8\pi/3)r_0^2 \, K = 3.8 \times 10^{30}$ dyne cm. If the whole sun is slowed by such a torque, proportional to ω_0, the e-folding time is 14×10^9 years. If only an outer shell is slowed in its rotation, the decay time is much shorter. Table I gives these times for a variety of assumptions about the inner radius of the outer, slowly rotating shell.

The second necessary condition requiring some discussion concerns the stability of, and frictional torque acting on, a rapidly rotating core. The outer convective zone cannot support a large angular velocity gradient as it is convectively unstable and turbulent.

While not certain, it seems likely that the observed latitude dependence of the surface angular velocity represents 'rotation on cylinders' in the convective zone, i.e.

TABLE I

Decay time in years and fractional moment of inertia as functions of the radius of an outer solar shell assuming that only the shell is slowed by a solar wind torque of 3.8×10^{30} dyne cm. The decay time for the sun rotating rigidly is 15×10^9 years

r/r_0	I_s/I	T (year)
0.86	0.0122	0.184×10^9
0.78	0.034	$0.51 \ \times 10^9$
0.70	0.074	$1.12 \ \times 10^9$
0.62	0.140	$2.11 \ \times 10^9$
0.54	0.241	$3.63 \ \times 10^9$

angular velocity a function of distance from the axis of rotation. If so, the observed variation with latitude implies that the angular velocity increases approximately linearly with distance from the rotational axis and that

$$\omega = \omega_0 + \omega_1 (r/r_0) \sin \theta \tag{5}$$

with

$$\omega_0 = 2.68 \times 10^{-3} (g/r_0)^{1/2}$$
$$\omega_1 = 1.94 \times 10^{-3} (g/r_0)^{1/2}.$$

Assuming that the magnetic and turbulent-viscous forces are small compared with the centrifugal force, we have for rotational motion approximately

$$0 = \nabla P + \varrho \nabla \varphi + \varrho \boldsymbol{\omega} \times (\boldsymbol{\omega} \times \mathbf{r}). \tag{6}$$

But in the convective zone, $P = P(\rho)$ implying that $\boldsymbol{\omega} \times (\boldsymbol{\omega} \times \mathbf{r})$ is derivable from a potential. This is possible if and only if ω is a function of $r \sin \theta$.

The turbulent-viscous stress is only 5% of that associated with the rotational velocity but, nonetheless, it is enormous. The internal stress of isotropic turbulence is many orders of magnitude greater than the solar wind stress. The source of the torque necessary to drive this differential rotation seems to be the forces derived from anisotropic turbulence. (Wasiutynski, 1946; Kippenhahn, 1963; Cocke, 1967.)

Below the convective zone, P and ρ are separately variable. Furthermore, the distributions over spherical surfaces of both pressure and density, and hence temperature, are uniquely determined by the distribution of angular velocity. The proof of this is similar to the discussion in Dicke (1967).

It should be emphasized that only the dynamical problem is being considered here. Thermally driven currents (Eddington-Sweet) may occur, as well as thermally driven instabilities (Goldreich and Schubert, 1967; 1968).

The temperature distribution forced by a particular distribution in angular velocity may not be compatible with the requirement that the heatflow be solenoidal beyond the nuclear-reaction-core. If not, circulation currents are induced to maintain a heat balance (Eddington-Sweet).

These thermally driven currents have velocities so low $< 10^{-6}$ cm/sec that their dynamical effects are negligible. It should be remarked that in a zone of differential rotation, the velocity of the thermally driven currents can be one or two orders of magnitude greater than that in a uniformly rotating core.

There are several aspects to the problem of stability in stars with rapidly rotating cores. First, it should be noted that some 2×10^9 cm below the bottom of the convective zone the sun becomes strongly density-stratified, the temperature gradient being substantially below the adiabatic level. Under these conditions the density stratification permits differential flow without turbulence, the velocity gradient being normal to level surfaces. Instead of the Reynolds number criterion for turbulence, that of Richardson namely

$$- r \frac{d\omega}{dr} < 2 \sqrt{\frac{g}{\gamma} \left[(\gamma - 1) \frac{d \ln \varrho}{dr} - \frac{d \ln T}{dr} \right]^{1/2}} \tag{7}$$

must be satisfied. This criterion permits very large angular velocity gradients. Adopting the Weymann (1957) solar model Equation (7) yields the limiting derivative angular velocity gradients shown in Table II.

TABLE II

Maximum angular velocity
gradient under the Richardson criterion

r/r_0	$(r_0/\omega_0)\,(d\omega/dr)$
0.84	$-\ \ \ 90.$
0.80	$-\ \ 870.$
0.76	$-1230.$
0.72	$-1390.$
0.64	-1730
0.52	-2440

Dynamically driven turbulence is not the only possible source of instability in the sun. Howard *et al.* (1967) suggested that the existence of 'spin-down' would preclude the existence of a rapidly rotating core.

After a cup of tea is stirred to set it spinning, its rotation rapidly ceases, much more rapidly than would be expected from the diffusion of angular momentum to the walls of the cup. The slowing process, called 'spin-down' is due to the convection by Ekman currents of angular momentum to a thin layer at the bottom of the cup, the Ekman layer.

The reasons for these circulation currents are easily seen. The density of the tea is constant and, neglecting viscous forces, Equation (6) is applicable for purely rotational motion. But note that the implied rotation-on-cylinders is impossible if the boundary condition of zero rotation on the cup bottom is to be satisfied. Purely rotational motion cannot occur and furthermore the intrinsically weak viscous force becomes important because of a steep angular velocity of gradient near the bottom of the cup. The circulation current pumps the fluid into this thin layer (Ekman) at the bottom.

In a brief note (Dicke, 1967) it was remarked that the sun is 'no cup of tea', that the functional connection between pressure and density that forces Ekman currents does not occur deep in the sun, and that for each almost arbitrarily chosen angular velocity distribution there is a distribution of pressure and density compatible with purely rotational motion without dynamically driven circulation currents. Thus a density stratified interior which eliminates turbulence also permits separate variability of pressure and density and purely rotational motion without circulation (except for the thermally driven circulation currents discussed by Eddington and Sweet).

In a series of experiments, E. McDonald investigated the role of density stratification vis-à-vis the spin-down process (McDonald and Dicke, 1967). It was shown that a large angular velocity shear could be induced in a density stratified fluid without the fluid spinning down. For this to be true it was essential that the angular velocity gradient be induced gradually, by slowly changing the angular velocity of the cylindrical container of the fluid. A very small change in angular velocity of the container, if suddenly applied, would cause the fluid to spin-down.

The resulting spin-down was interesting to watch. The fluid would mix in two layers.

Circulation currents would separately mix the upper and lower halves of the contents of the container. This occurred by first establishing gravity waves of ever-increasing amplitude at mid-depths in the fluid. This wave behavior was followed by a period of turbulence in this thick layer and the establishment of circulation currents that separately mixed the upper and lower halves of the fluid. Meanwhile the layer of differential rotation became thinner. Finally there occurred the establishment of Ekman layers at the boundary of density change and at the container bottom causing the fluid to spin-down.

The reason for spin-down when the change in angular velocity is instantaneously established is easily seen. For purely rotational motion, a sudden change in the distribution of the angular velocity requires a sudden change in the density distribution. This is impossible to establish. Hence the actual density distribution is incompatible with such a sudden change in angular velocity, leading to the complicated set of motions described above.

If the angular velocity distribution is very slowly changed, very slow circulation currents (slow enough to be dynamically unimportant) permit the establishment of the new required density distribution.

Still another possible type of instability requires discussion. This is the thermally driven instability first discussed by Goldreich and Schubert (1967, 1968) and later by Fricke (1968). With this instability, angular momentum is transported toward the surface in outward and inward moving thin toruses, with a thickness in the θ direction of only a few km. The fluid must flow accurately parallel to these toruses if this instability is to develop. Thus circulation currents or oscillating motion in the θ direction would inhibit the instability and stabilize the rotation. Another means of stabilizing the rotation is provided by a gradient in mean molecular weight (Colgate, 1968; Goldreich and Schubert, 1968).

My overall impression of this instability is that it would occur if all the assumptions were satisfied, compositional gradients and magnetic fields being absent and the motion being accurately rotational. However, it is difficult to be certain that all these conditions would be satisfied in the deep solar interior.

One is reminded of the example of the thermo-haline instability discussed by Goldreich and Schubert (1967). Here, essentially the same analysis as that used by Goldreich, Schubert and Fricke can be employed to 'prove' that the salt concentration must be greater at the bottoms of all oceans than at the tops. If one's direct knowledge of the deep oceans were no better than that of the deep interior of the sun, he might be impressed by this 'proof'. Fortunately, we can observe the deep oceans, and the observations show quite the contrary, an increased salt concentration at the top in the tropics. The so-called 'proof' may be invalid because of oscillatory motions near the surface.

The viscosity of the solar medium below the convective zone is sufficiently low that a rapidly rotating core could exist for many times the solar age if diffusion controlled the flow of angular momentum. Over the depth range $r = 0.5$–0.85 the gaseous, or molecular, viscosity (Spitzer, 1962) is 4 to 10 times as great as that due to radiation (Thomas, 1930). These are tabulated in Table III, together with the kinematic viscosity η/ρ, of their sum $\nu = \eta/\rho$.

The effect of diffusion is one of widening the zone of differential rotation until it has a width of approximately $2.5\,(\nu T)^{\frac{1}{2}} \sim 3.8 \times 10^9$ cm.

TABLE III

The molecular and radiative viscosities in gm/cm sec and the kinematic viscosity $\nu = (\eta_m + \eta_r)/\rho$ in cm²/sec

r/r_0	η_m	η_r	ν
0.5	14.2	1.2	12.2
0.55	9.5	1.0	14.8
0.60	6.7	0.9	19.1
0.65	4.23	0.67	21.4
0.70	2.82	0.55	26.8
0.75	1.89	0.48	35.0
0.80	1.19	0.37	44.0
0.85	0.60	0.16	42.8

The diffusion equation can be written

$$\frac{\partial}{\partial r}\left(\nu \varrho r^4 \frac{\partial \omega}{\partial r}\right) = r^4 \varrho \frac{\partial \omega}{\partial t}. \tag{8}$$

The solution of this equation is simplified by treating ν as a constant and assuming that ϱr^4 varies with radial distance as $\exp(-kr)$. Based on Weyman's solar model, kr_0 has the values 2.7, 4.4, 5.7, 7.2 at $r/r_0 = 0.45, 0.55, 0.65, 0.75$ respectively. For these values $2.5\,(\nu T)^{\frac{1}{4}} k \ll 1$ and the variation of both ν and ϱr^4 can be neglected in a fairly good approximation.

In this approximation Equation (8) becomes

$$\nu \frac{\partial^2 \omega}{\partial r^2} = \frac{\partial \omega}{\partial t}. \tag{9}$$

Equation (8), or (9) if applicable, can be used to calculate the present distribution of angular velocity in the sun and the present solar torque density, assuming that the sun arrived on the main sequence uniformly rotating. It is assumed that the rotation of an outer mixed layer, was quickly slowed, and that below this outer layer angular momentum is transported outward by diffusion.

To simplify the diffusion problem, Equation (8) or (9) is solved with the initial condition $\omega = \omega_c$ for $r < r_m$, and with the boundary condition $\omega = \omega_0$ for $r = r_m$. Here r_m is the inner radius of the outer mixed zone which may be deeper than the convective zone of Weymann's model.

The solution to this equation is

$$\omega = \omega_0 + (\omega_c - \omega_0)\,\text{erf}\,[(r_m - r)/2\sqrt{\nu t}] \tag{10}$$

where $\mathrm{erf}(x)=(2/\sqrt{\pi})\int_0^x e^{-z^2}\,dz$. The equatorial torque density at the inner radius of the mixed zone is $v\varrho r_m^2\,(d\omega/dr)$. The equivalent equatorial torque density at the solar surface is

$$K = v\varrho r_m^2\,(r_m/r_0)^2\,(d\omega/dr)$$
$$= \varrho r_0^2\,(r_m/r_0)^4\,(v/\pi t)^{1/2}\,(\omega_c - \omega_0) \tag{11}$$

Values of the relevant quantities are given in Table IV for 3 different choices of r_m/r_0. It should be noted that the resulting values for the torque density are nearly equal to that obtained for the solar wind.

To test the assumption that the variation with r of $v\varrho r^4$ could be neglected in Equation (8), this equation was integrated numerically for $kr_0=4$. For several different values of $t\leqslant 4.5\times 10^9$ years the dependence of ω on r is closely approximated by Equation (10). See Figure 5. For the numerically analyzed example the computed surface torque agrees well with the approximate result given in Table IV.

TABLE IV

The equatorial torque density, K, at the solar surface calculated from a model for the diffusion of angular momentum (see Equation (11)). It is assumed that $\omega_c = 15\,\omega_0$, a value consistent with the observed solar oblateness. Note that the values given in the last column agree reasonably well with the torque density derived from the solar wind ($\sim 10^8$ dyne/cm)

r_m/r_0	v cm²/sec	ϱ gm/cm³	K dyne/cm
0.8	35	0.035	2.5×10^7
0.7	22	0.124	4.2×10^7
0.6	15	0.404	6.1×10^7

When the above calculation was first made (Dicke, 1964), the surface torque density obtained was $K=7\times 10^7$ dyne/cm, the difference from the results obtained here being due to the use of a cruder approximation.

The boundary condition assumed above at $r=r_m$ is equivalent to the assumption that the angular velocity of the outer mixed shell is instantaneously reduced from ω_c to ω_0. But this is inconsistent with the results given in Table I, particularly for $r/r_0 \leqslant 0.7$. Nonetheless the above calculation is applicable as a good approximation providing the actual mixing depth is slightly greater than the value assumed. Consider the example plotted in Figure 5. For a cut taken at $r/r_0=0.587$, ω is observed to vary with t, for $t<2$ b.y., in a manner similar to that expected for the whole mixed shell. Thus a correct calculation based on the proper boundary condition at this radius should give a solution very similar to that shown in Figure 5, and a torque nearly equal to that given by Equation (11).

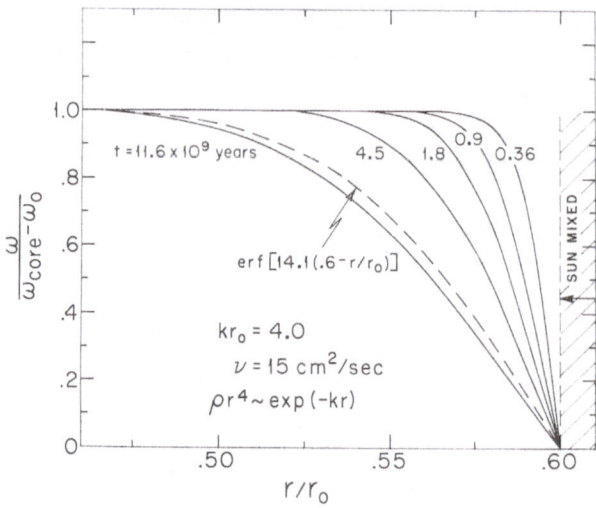

Fig. 5. Numerical integration of the diffusion of angular momentum from a rapidly rotating core, initially 0.6 in radius. The dashed curve shows an approximate analytic solution.

There is an additional way in which the above calculation may be an oversimplification. As mentioned earlier, the velocities of thermally driven circulation currents can be a couple of orders of magnitude greater in the zone of differential rotation than in the uniformly rotating core (Eddington-Sweet currents), and these velocities may be great enough to significantly transport angular momentum. The effect of nonuniform rotation on circulation has been previously considered by Baker and Kippenhahn (1959) and Mestel (1966).

In the absence of velocity and magnetic fields in the solar interior, surfaces of constant P, ϱ and φ (gravitational potential) coincide. If the mean molecular weight is a function of φ, temperature is also constant on these spherical surfaces and thermally driven circulation does not occur.

If the star is uniformly rotating, P, ϱ, and T are functions of the potential $\varphi - \frac{1}{2} \varrho r^2 \omega^2 \sin^2 \theta$, which includes the centrifugal term (Von Zeipel, 1924). In general for any set of magnetic and velocity fields leading to a quasi-steady state and such that the velocity and magnetic force density has the form $\varrho \nabla W$, Von Zeipel's relation is applicable to the effective potential $\varphi - W$. It must be emphasized that this relation, P, ϱ, and T being functions of $(\phi - W)$, is applicable also over any fraction of the star, or surface in the star, for which the magnetic and velocity force density is of the above form. One such example is provided by the solar surface. Only the gravitational quadrupole moment and the stresses in the 'seen layers' of the sun can affect the oblateness. If Von Zeipel's relations hold in the 'seen layers', their implications for the interpretation of the oblateness are valid independent of conditions below the surface.

In the presence of magnetic and velocity fields, thermally driven currents occur because the temperature distribution forced by the presence of magnetic and velocity fields is generally incompatible with the vanishing of the divergence of the heat flux outside the energy generating core. To preserve the temperature distribution demanded by the magnetic and velocity field distribution, matter must flow and this flow represents the circulation current. Two separate effects contribute to the circulation current. The contribution arising in a non-spherical gravitational potential φ is global in origin, the whole of the distorted mass distribution generating the distorted potential. The contribution from W is local in origin. Over any spherical surface W, and hence the contribution of W to the radial component of circulation velocity, is given by local values of the fields.

Except for rotation on cylinders, the centrifugal force density per unit mass of non-uniform rotation is not derivable from a potential, and Von Zeipel's functional relations are not satisfied. Instead, over spherical surfaces the variation of P and ϱ are separately determined (Dicke, 1967). But once again the distribution of temperature over such a surface is determined by the rotational distribution. Here, even though the distortion of surfaces of constant gravitational potential may be small, the variation of the temperature induced locally can be relatively large and the violation of the divergence condition correspondingly large.

In general the angular velocity distribution obtained from the diffusion Equation (8) is such that circulation could occur and this circulation would modify the distribution of rotation in the diffusion zone. For rotational distributions without circulation, angular momentum is transported by molecular diffusion but the transport rate would be affected by the modification in the rotational distribution. This question requires an analysis.

3. A History of the Sun's Rotation

In my *Nature* article (Dicke, 1964) a possible history of the sun's rotation was assembled with the realization that the sun's past is even more hidden than its interior. In the intervening half decade new data have cast some light on this void, but these data do not seem to require any very fundamental change in the story. The one significant change that I would make appeared already in 1965 and was not forced by any new data, but rather by the realization that my model was probably defective at one point. I shall discuss this change in the last section.

We are considering the history of the sun with the view to asking whether this history could reasonably lead to a physical understanding of the presence of a rapidly rotating core in the sun. But in this connection, it is essential that we consider the sun to be a normal star, not an exception! If we are correct, all normal, solar type stars, young or old, will have rapidly rotating interiors and by observing solar type stars of various ages some evidence concerning the effect of such a rotation might conceivably be unearthed.

We visualize the solar system as having formed through gravitational collapse of a condensation in a gas cloud, the condensation having more than enough angular momentum to supply a rapidly rotating star. The resulting rotation is limited by the maximum angular momentum possible for the collapsing sun, and it may also be reduced to a value far below this limit by the Schatzman (1962) torque.

The twisting of magnetic field in the solar nebula may not have been as effective in slowing the proto-sun's rotation as had been previously thought. The stresses induced by a toroidal magnetic field tend to make a rotating star prolate, also unstable. This instability, to be discussed elsewhere, occurs when roughly half of the star's kinetic energy is converted to magnetic energy. It results in a precession of the star relative to the rotation axis, putting the magnetic axis perpendicular to the rotation axis. In the perpendicular position the star's magnetic field becomes cut-off from the outside, greatly reducing the torque. If this picture is correct it may provide an explanation for stars which are magnetic variables as 'oblique rotators' (see Preston, 1967). Mestel (1968) has discussed this same problem showing that stellar wind torques may make the perpendicular orientation of the magnetic axis the preferred position.

Whatever the physical explanation for the rather low value for a star's surface rotation relative to the limiting value, after most of the star's mass has been accumulated in a central concentration, a large effective radius seems to be required for mass loss or gain.

If the angular momentum cut-off occurs during formation for all stars at the same fractional collapse (or average density), the angular momentum upon first arrival on the main sequence should vary as $M^{5/3}$. This can be best seen from a crude argument based on homology. For the radius at which centrifugal and gravitational forces balance,

$$GM = \omega^2 r^3$$

or $\omega \sim \varrho^{1/2}$. But then

$$J \sim Mr^2\omega \sim M^{5/3}. \tag{12}$$

It is found that, up to 10 solar masses, stars bluer than F0 show this dependence on mass. (See Figure 6.) But the reduction to lower values of J/M for $M/M_0 > 10$ suggests that this simple argument is inadequate. Also another significant change may occur for $M/M_0 < 1.6$. However, the situation is complicated for stars of lower mass, redder than F5. As discussed by Kraft (1968), a stellar wind torque, i.e. slowing on the main sequence, is expected for such stars. Thus the observations may not reflect the initial rotation on the main sequence.

The extension of the dotted curve to the left in Figure 6 may provide an appropriate value for the initial solar rotation on the main sequence. However, it will be shown below that a somewhat lower initial value is more likely. From this viewpoint all stars on the main sequence, 1 solar mass or greater, are rotators. They probably arrive on the main sequence uniformly rotating having passed through a state with a deep convective envelope. For stars redder then F5 the stellar wind driven by surface turbulence provides a torque to slow down the rotation of an outer shell (or the whole star). If our picture is correct only an outer shell is slowed. But the angular momentum content of such a shell is small and the angular momentum remains deeply buried in these stars.

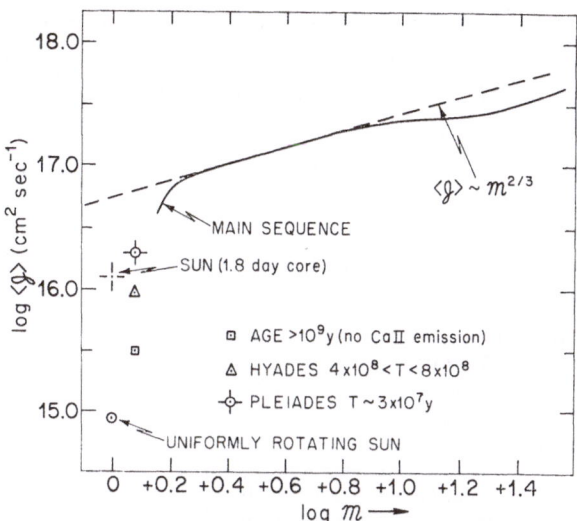

Fig. 6.   The rotation of stars of various masses (expressed as angular momentum per unit mass). Stars bluer than F5 are believed to be free of a stellar wind torque. The surface rotations of stars redder than F5 decrease with age. The three points for these stars and the 'uniformly rotating sun' assume uniform rotation with the observed surface velocity. The other point for the sun (1.8 day core) is consistent with the observed solar oblateness. (Based on Kraft (1968), Fig. 17.)

As the discussion in the next section will bring out, there is a non-trivial amount of support for this viewpoint. It will be shown that the sun's initial rotation on the main sequence was probably somewhat below the dotted extension of Figure 6, similar to that seen in the Pleiades. It will also be shown that the outer mixed shell, slowed by the solar wind, may go as deep as $r/r_0 = 0.55$.

I shall now briefly summarize our picture of the history of the sun's rotation. The sun is viewed as having approached the main sequence rotating at a rate of about 30 km/sec on the equator, similar to that seen in the Pleiades. Having passed through a convective state and containing magnetic fields, the sun was probably initially uniformly rotating. The solar wind torque slowed an outer mixed shell. The beginning solar wind is believed to have had a ratio of torque to angular velocity somewhat larger than today's value. How rapidly the outer shell was slowed depends upon both this torque and the shell thickness. In the next section it will be shown that the shell was probably fairly thick and that the e-folding time for slowing this shell may have been as great as 10^9 years.

Classical discussions of the formation of the solar system were always plagued with a persistent problem, how to remove the original angular momentum from the sun. If we are correct, this problem has vanished, for the angular momentum is still in the sun, but deeply buried.

4. The Sun as a Rotating Star

If the sun has a rapidly rotating core this is to be regarded as a normal condition for stars of about 1 solar mass, not as an exception. If the sun was originally rotating with a substantial equatorial velocity of 30 km/sec, this rotation should be seen in young stars. If the outer layers of the sun were slowed in 10^9 years, this slowing should be seen in stars as a function of age.

Figure 6 and Table V are based on R. Kraft's work, in particular on Figure 17 and

TABLE V[a]

	$\log (J/M)$
$M^{2/3}$ extrapolation	16.8
Pleiades $\sim 3 \times 10^7$ years	16.3
Hyades $\sim 6 \times 10^8$ years	16.0
Old $t > 10^9$ years	15.5
Initial main-sequence	16.32
Sun, uniformly rotating	14.9
Sun, 1.8 day core	16.1

[a] J/M (in cm²/sec) is the angular momentum per unit mass of rotating stars with $M/M_\odot = 1.2$ for rows 1–4. (Based on Kraft, 1968.) See Figure 6.

Table III of Kraft (1968). The solid curve is Kraft's, but the dotted line is mine. In Kraft's Figure 17 the point marked 'Dicke Sun' seems to have been incorrectly positioned. It has been relocated in Figure 6. The data for the three points representing stars of 1.2 solar masses were taken from Table III of Kraft's article.

There are several possible interpretations of the Pleiades point in Figure 6:

(a) These stars are so young that their surface angular velocities have not been appreciably affected by the stellar wind torque.

(b) The original angular velocity was consistent with the angular momentum given by the dotted curve, but the outer 10% of the star (by radius) was slowed to $\frac{1}{3}$ of its original rotation by a stellar wind like the present solar wind.

(c) The whole star was slowed by the factor of $\frac{1}{3}$ by a stellar wind torque 10^4 times as strong as the present solar wind, i.e. with a ratio of torque to angular velocity 500 times as great as the standard (solar wind) ratio.

(d) The outer 40% of the star, by radius, was slowed by a torque 10^3 times as great as that of the solar wind (50 times as great in ratio of torque to angular velocity).

The hypotheses (b) and (c) are unlikely for reasons, to be discussed below, connected with the depletion of lithium in young stars. These observations suggest that the outer 40% by radius, but only the outer 40%, is slowed by a stellar wind. If the argument given above for a standard ratio of stellar-wind torque to angular velocity is correct (for solar type stars), the hypothesis (d) is also unlikely.

That the interpretation (a) is most likely is seen by considering the Hyades point. If the Pleiades point represents approximately the initial main-sequence rotation at 1.2 solar masses, the e-folding time for slowing rotation in the Hyades is roughly 9×10^8 years. This would require a ratio of stellar wind torque to angular velocity consistent with the present solar wind torque if the outer 30% of the star, by radius, is being decelerated. If both the Pleiades and Hyades are assumed to be acted on by a stellar wind of the same torque ratio and have the same thickness shell decelerated, the point representing the initial value for the logarithm of angular momentum per unit mass would fall on Figure 6 only 0.015 above the Pleiades point. If the point marked 'sun, 1.8 day core' represents the solar rotation approaching the main sequence, this point is consistent with the interpretation of the Pleiades point as initial rotation on the main sequence.

Reasons were given earlier for believing that the initial solar wind torque, and stellar wind torque of other young solar-type stars, have a somewhat greater ratio of torque to angular velocity than that found associated with the present solar wind. In the case of the Hyades the whole star could be slowed by a torque 15 times as great. This possibility cannot be excluded by the above argument, but the argument to be discussed in the next section seems convincing.

5. The Rotation of the Sun and the Depletion of Lithium

Goldreich and Schubert (1967) have noted that the Howard *et al.* (1967) spin-down of the sun, initially rapidly rotating, would have transported lithium and berylium, from the outer parts of the sun below the radii of $r/r_0 = 0.6$ and 0.5, respectively, at which these elements would have been rapidly destroyed. They also note that diffusion associated with their thermally driven turbulence would have had a similar effect.

The argument that angular momentum transport and depletion of lithium and berylium should be connected seems convincing. It has been shown that angular momentum cannot be transported by molecular diffusion more than $0.05\, r_0$ during the lifetime of the sun. Thus, to remove angular momentum from the deep solar interior to the surface requires the transport to the surface of depleted solar material.

If the transport were to occur via the spin-down process, circulation currents outside the core would erode its surface, transporting its contents to the convective zone. Thermally driven currents inside the uniformly rotating core would be slow. Thus circulation currents would penetrate ever deeper, first destroying the lithium and then the berylium. After the depth of burning is reached for a given isotope, its destruction proceeds rapidly.

In similar fashion the Goldreich-Schubert instability, if it should occur, would transport angular momentum to the convective shell by a type of turbulent diffusion. The thin ring of high angular momentum material ejected from the core would not float all the way to the surface. It would be quickly destroyed by instabilities. The successive formation and destruction of these rings would lead to diffusion of angular momentum out from the surface of the steadily shrinking core and of lithium and berylium down to their zones of burning. Thus this process couples the loss of angular momentum from the interior to the depletion of lithium and berylium.

Goldreich and Schubert interpret the presence of lithium and berylium in the sun to mean that the sun was not rapidly rotating initially. But if this argument were valid, it should also be applicable to other stars. But solar type stars in very young clusters are observed to be rotating at about 400 km/sec whereas only 6 km/sec rotation is seen in old stars of the same mass. As in the case of the sun, lithium is observed to have suffered depletion, but not berylium.

We shall consider the following questions:

(a) What do the observations tell us about the depletion of lithium and berylium in stars?

(b) How much should lithium and berylium be decreased if the whole star is slowed by spin-down, or by turbulent diffusion of angular momentum?

(c) What seems to be the implications of the depletion of lithium for internal rotation in solar-type stars?

We consider first the observational material regarding the abundance of lithium and berylium.

Lithium is believed to be depleted in the sun by almost 3 orders of magnitude relative to that in chondritic meteorites. (See survey article by Wallerstein and Conti, 1969, for references.) The berylium abundance in the sun seems to be in good agreement with that seen in these meteorites. Herbig (1965) has noted a correlation between lithium abundance and stellar age of main-sequence stars of spectral type G. This correlation was studied by Wallerstein, Danziger, and Conti, and most recently by Danziger (1969) who found that the data for T Tauri stars, the Pleiades, the Coma Cluster, the Hyades, and the sun are consistent with the assumption that the lithium abundance was initially that found in chondritic meteorites but has since decreased exponentially with an e-folding time of 7×10^8 years.

It is questionable whether there is any appreciable amount of Li^6 in the sun or solar-type stars. Schmall and Schröter (1965) find that the profile of a lithium resonance line observed in sunspots agrees well with a Li^7 profile but that the fit is slightly improved by adding a small amount of the Li^6 profile. If the improved fit implies the presence of Li^6, the abundance ratio of Li^6 to Li^7 should be roughly 0.05 in reasonable agreement with the terrestrial value of 0.08.

The abundance of berylium in the sun appears to be essentially the same as that found in chondritic meteorites (Wallerstein and Conti, 1969). This also seems to be true for substantially all main-sequence field stars redder than F7. It seems clear from Danziger (1969), and Wallerstein and Conti (1969), that late F and early G stars arrive on the main sequence with the meteoritic abundance of both berylium and lithium. The lithium is depleted as the star ages, but not the berylium. Li^7 and Be^9 are burned at radii differing by only 10% of the solar radius.

A reasonable interpretation of the above observations is that the slowing on the main sequence of the rotation of the sun, and of other solar type stars, was accompanied by a mixing of the star down to a depth of about $r/r_0 = 0.6$. The depth of 0.6 lies well below the presumed bottom of the convective zone (0.85 for Weymann's model), but the angular momentum could have been transported outward to the convective zone by thermally driven circulation currents, lithium being carried inward.

TABLE VI

Fractional depth, r_b/r_0, at which burning takes place with indicated mean life (from Fowler et al., 1967)

Mean life	3×10^6 years	3×10^7 years	3×10^8 years
Li^6	0.57	0.6	0.63
Li^7	0.51	0.55	0.58
Be^9	0.42	0.45	0.47

TABLE VII

Minimum depletions of Li^6, Li^7, and Be^9. For the sun, the decrease in the logarithmic abundance relative to that in chondritic meteorites is given by $[X]_m - [X]_s = (1/2.3)(\Delta M/(1 - M_b))$ and is tabulated in the last two columns. ΔM_1 and ΔM_2 are the mass fractions of the sun outside the rapidly spinning core of adopted radii $r_1 = 0.55$ and $r_2 = 0.40$ respectively but inside the radii of burning, r_b. Relative to calcium the observed depletion for Li^7 is $[Li/Ca]_m - [Li/Ca]_s \cong +2.8$. No depletion of Be^9 is observed.

Isotope	r_b/r_0	$M_b(r)$	ΔM_1	ΔM_2	$(\Delta M/2.3)(1 - M_b)$ #1	#2
Li^6	0.63	0.976	0.028	0.158	0.51	2.86
Li^7	0.58	0.961	0.013	0.143	0.01	1.59
Be^9	0.47	0.896	—	0.078	—	0.33

Another possibility is that turbulent diffusion driven by the Goldreich-Schubert-Fricke effect could result in the diffusion of angular momentum outward from a zone of molecular diffusion within which this effect is for some reason inoperative. The turbulent diffusion of angular momentum should be accompanied by a diffusion of lithium inward to the zone of burning. Both of these possibilities for transporting lithium inward and angular momentum outward will be discussed.

To simplify the discussion it will be assumed that a sharp boundary marks the zone of rapid burning of each of the isotopes Li^6, Li^7 and Be^9. The values adopted for the radii of these boundaries are given in Table VII. As there is no observable depletion of beryllium, the boundary of the rapidly rotating core (or minimum radius from which angular momentum is convected to the surface) is assumed to be greater than 0.47.

It is possible to calculate a minimum value for lithium depletion using a simple argument. Define two concentric shells for the sun, one ranging from the core radius r_c to the outer boundary for burning a lithium isotope, the second lying between this boundary and the solar surface. The material containing angular momentum but without lithium originally in the inner shell is assumed to be transported to the convective zone, where it is mixed with the original material. This results in a dilution of the original lithium content. A lower value for the dilution factor is obtained if the zone of mixing is assumed to be the whole of the outer zone instead of just the convective layer. For the latter case, the depletion of lithium relative to calcium, in comparison with meteorites is given by,

$$[Li^7/Ca]_s - [Li^7/Ca]_m = -\Delta M/(1 - M_b) \ln 10 \qquad (13)$$

where the brackets represent logarithms of abundance ratios, ΔM is the mass transferred from the lithium burning shell, and $1 - M_b$ is the mass fraction lying outside
r_b, (i.e. the mass of the mixed zone). The relevant numbers are given in Table VII. It
is evident that this lower bound for the depletion factor is much smaller than the
observed depletion.

If the sun were originally rotating as rapidly as the G stars in the Pleiades, $\omega_p \sim 5\omega_0$
and were slowed down to its presently observed rotation, ω_0, by the Goldreich-
Schubert-Fricke instability, or by the spin-down process of Howard *et al.* (1967), the
decrease in the logarithmic abundance of berylium would be expected to be at least

$$-[Be/Ca]_s + [Be/Ca]_m =$$

$$+\left(1 - \frac{\omega_0}{\omega_p}\right) M_b/2.3(1 - M_b) \qquad (14)$$

$$= +0.8 \times 0.896/2.3 \times 0.104 = +3.00.$$

By contrast, no appreciable depletion is observed.

Instead of computing a lower bound for lithium depletion one might attempt to
make a calculation from a model based on an assumed transport means for angular
momentum. It will be assumed that outside the core of radius r_c, rotating uniformly
with an angular velocity ω_c, there is a differentially rotating shell whose outer surface
has a radius r_m and angular velocity ω_m. It is assumed that through some unspecified
means the Goldreich-Schubert-Fricke instability is inoperative in this shell of differential rotation and the flow of angular momentum is limited by molecular diffusion.
Thus the angular velocity distribution is of the type shown in Figure 5. (See the last
section for a discussion of one of several possible means for stabilizing the flow in this
shell.) It is assumed that outside of this shell to the bottom of the convective zone at
r_v angular momentum is transported by the Goldreich-Schubert-Fricke thermally-
driven turbulent diffusion. Because of the effectiveness of this diffusion process it must
operate near threshold conditions to provide the low angular momentum flow rate
given by molecular diffusion in the inner shell. Thus the angular velocity in the
intermediate shell varies inversely as the square of the radius and

$$\omega_m/\omega_0 = (r_v/r_m)^2.$$

(See Goldreich and Schubert, 1967). Again ω_0 designates the angular velocity of the
outer convective zone.

In the intermediate shell the turbulent diffusion of angular momentum is governed
by Equation (8) with the kinematic viscosity v automatically adjusted to provide the
correct angular momentum flux and $\omega \sim r^{-2}$. These conditions require $2\varrho v = kr$, where
k is the angular momentum flux density.

The diffusion of lithium from the bottom of the convective zone to r_b, the radius of burning is governed by the diffusion equation

$$\frac{\partial}{\partial r}\left(v\varrho r^2 \frac{\partial}{\partial r} F\right) = r^2 \cdot \varrho \frac{\partial F}{\partial t} \tag{15}$$

where $F\varrho$ is the mass density of the isotope in question. Because the diffusion is turbulent, the diffusivity v is the same for both angular momentum and matter. Integrating Equation (15) outward from a spherical surface just inside the convective zone gives

$$\left(v\varrho r^2 \frac{\partial F}{\partial r}\right)_v = \frac{1}{4\pi} \frac{M_v F}{\tau_F} \tag{16}$$

where the quantities on the left side are to be evaluated below but near r_v. M_v is the mass of the convective zone. τ_F is the mean life for the decay of the isotope. $(\partial F/\partial r)$ is evaluated from a numerical integration of the lowest normal decay mode of Equation (15). In similar fashion, Equation (8) is integrated to yield the analogue of (16) where τ_ω would represent the e-folding time for increasing ω_0 if the solar wind torque were to be suddenly switched off. Combining these equations yields

$$\tau_F = \frac{2\tau_\omega F_v}{r_v (\partial F/\partial r)_v}. \tag{17}$$

There is considerable uncertainty about the depth of the convective layer in the sun and Weymann's (1957) value for $r_v = 0.85$ may be too great. It has been suggested by Sears and Weymann (1965) that convection may exist down to $r_v = 0.7$. Equation

TABLE VIII

The mean decay times τ_6 and τ_7 for the destruction of Li^6 and Li^7 by turbulent diffusion. Three different radii are chosen for r_v, the bottom of the convective zone. τ_ω, from Table I, is the decay time for slowing of the convective zone based on the torque derived from solar wind measurements. The radii adopted for rapid burning of Li^6 and Li^7 are $r_b/r_0 = 0.63$ and 0.58 respectively. (See Table VI.) In order to avoid burning berylium while permitting the burning of Li^7, r_m should satisfy $0.5 < r_m/r_0 < 0.58$. The radius of the rapidly spinning core is approximately $r_c/r_0 \sim 0.5$. For the sun, the 'observed' decay time of Li^7 is $\tau_7 = 7 \times 10^8$ years (Danziger, 1969). The calculated values are based on Equation (17).

r_v/r_0	τ_ω (years)	τ_6 (years)	τ_7 (years)
0.86	1.84×10^8	5.3×10^8	10.2×10^8
0.78	5.1×10^8	4.5×10^8	8.8×10^8
0.70	11.2×10^8	3.1×10^8	7.1×10^8

(17) has been used to determine τ_F for both isotopes of lithium with three choices for r_v; see Table VIII. For each of these choices τ_ω is derived from the values given in Table I. For Li^7 to be depleted but not berylium, $0.5 < r_m/r_0 < 0.58$. Thus the radius of the rapidly rotating core should be approximately $r_c/r_0 \sim 0.5$.

The dimensionless ratio $(r_v/F_v)(\partial F/\partial r)_v$ is determined from a numerical integration of the eigen value equation

$$\frac{\partial}{\partial r}\left(v\varrho r^2 \frac{\partial}{\partial r} F\right) + \Lambda r^2 \varrho F = 0, \tag{18}$$

obtained from (15). The lowest decay mode is found for the appropriate boundary condition at r_v. The higher decay modes disappear in a few hundred million years. In (18), $r\varrho v = \frac{1}{2}r^2 k$ is assumed to be constant.

From Table VIII, the mean decay time of Li^7 is in good agreement with the value observed by Danziger (1969). The calculated decay time of Li^6 is less by a factor of $\frac{1}{2}$. This implies a greater depletion in the sun of Li^6 relative to Li^7 by 2 or 3 orders of magnitude. As was noted above there is not yet conclusive evidence for the presence of Li^6 in the sun.

Another possible model yielding a depletion of lithium employs thermally driven circulation currents outside the zone of molecular diffusion of angular momentum. These currents would transport angular momentum upward and lithium isotopes downward into their burning zones. It is assumed that the currents do not extend deep enough to cause a depletion of berylium.

It is assumed that the currents carry both of the isotopes so deep as to destroy them completely. Upward currents are then devoid of these isotopes.

The calculation leading to Equation (13) is now applicable to this problem providing the zone of mixing is assumed to be the convective zone. Angular momentum is conserved along a stream line of the circulation and many cycles of circulation are required to transport the angular momentum. ΔM of Equation (13) must contain the number of cycles as a multiplier. For reasonable values of this number, ΔM, and the mass fraction in the convective zone, the depletion is far too great. It is concluded that of these two models, the turbulent diffusion is best.

6. The Solar Magnetic Field

One aspect of the formation of the solar system requires some additional discussion. This concerns the magnetic field trapped in the gas when the proto-sun is first formed. In 1964 I pictured this field as stirred by convection during the Hayashi phase. The long lived magnetic modes were assumed to be completely converted into short lived ones by the convection. The short lived modes would rapidly decay after the magnetic field became frozen into the radiative core upon termination of the Hayashi phase.

There can be little doubt that this mechanism would work if the sun was properly stirred, but, I now doubt that the stirring would be sufficiently thorough and of the right type to completely destroy the long lived modes. The destruction of long wavelength modes must be extremely thorough for a poloidal field with axial symmetry characterized by a magnetic mode with long wavelengths would generate toroidal magnetic modes of ever increasing strength under the influence of differential rotation.

I now consider it unlikely that the destruction of such modes (long wavelength) could be so thorough. This is the point mentioned earlier where I have had second thoughts about my 1964 model.

Two other possibilities now seem more likely. The first involves magnetic flux exclusion from the radiative core by the 'beer foam process'. The sun is pictured as initially completely convective, and the magnetic field is assumed to be thoroughly twisted. This is similar to the previous picture. However the low magnetic modes are not destroyed at this stage. Instead the twisted field configuration is assumed to contain null-field surfaces, where regions of oppositely oriented magnetic fields meet. The fluid in these null-field regions is free of magnetic pressure and hence has a greater density. It is conceivable, but by no means certain, that a radiative core would slowly grow by developing a central spherical cavity filled with this dense fluid. The field-free fluid is pictured as filling the cavity by flowing along null-field channels. As these channels close, the diffusion of magnetic field would generate more field-free fluid. It must be emphasized that this picture is highly conjectural.

The second view of the magnetic field is also conjectural, but it may be correct as there is a non-trivial amount of observational support for the picture. According to this viewpoint, the sun arrived on the main sequence rotating with its magnetic field of a few thousand gauss trapped in a perpendicular axis configuration, this configuration having been induced by the instability mentioned earlier. This field is assumed to be deeply buried at this stage, cut off by differential rotation. Thus the interior of the sun is pictured to be something like a perpendicular rotator model of a magnetic A star.

If this picture is valid for the sun, it should also be applicable to other stars arriving on the main-sequence to the right of F5. From this viewpoint solar type stars are both magnetic and rapid rotators but most of the angular momentum and magnetic field is deeply buried.

The work on this magnetic model is being published elsewhere. Here I shall only summarize the central conclusions. This model is too complicated to permit a complete theoretical study and all that has been accomplished is to analyze parts of the problem using approximation methods.

The most interesting feature of the model is the torsional oscillation of the rapidly rotating core. The completely embedded approximately perpendicular dipolar-field gives the core an elasticity for torsional modes. The lowest frequency mode is a torsional mode for which the north and south magnetic poles move on ellipses, oscillating back and forth between the northern and southern hemispheres. If the core magnetic field strength is of the order of magnitude of 10^5 gauss the oscillation period can be made to agree with the 22 year sunspot cycle.

Another feature of the torsional oscillation is the generation of toroidal magnetic fields of alternating polarity. This toroidal field is generated by differential rotation acting on the (oscillating) axial component of the dipolar field. The toroidal field appears in the differentially rotating shell in the vicinity of $\pm 45°$ latitudes with opposite polarities in the two hemispheres. Owing to magnetic buoyancy, strands of this toroidal field slowly float up to the solar surface where magnetic field breaking through the surface could be the source of the sunspot phenomena.

The various observational tests of this hypothesis are being published elsewhere, and only a single example will be considered here.

One obvious implication of this magnetic model is that, owing to the extremely high Q of the torsional oscillation, the sunspot cycle should be timed by a very precise central clock. Owing to the delayed arrival of the torsional magnetic field at the surface there could occur random phase errors, i.e. delays in the time of arrival of the sunspot maximum. But such a phase error would not be cumulative. It should later disappear.

Under the sunspot theory of Babcock (1961) and Leighton (1969), the magnetic field of a new sunspot cycle is regenerated from the old. This model does not contain a tuned oscillator, and phase shifts are cumulative. If for a couple of cycles the time of sunspot maximum were to be delayed a year or two, this delayed phase would be expected to continue indefinitely into the future.

A least squares fit of a straight line has been made to the epoch of sunspot maximum plotted against sunspot cycle number N for the period of 1615 to 1958, the following relation is obtained for the years of sunspot maximum (see Allen (1963) for a table of sunspot maxima):

$$y = 1749.41 + 11.0814\,N. \tag{19}$$

With few exceptions the times of observed sunspot maxima agree with this formula to within a year. One pronounced anomaly occurs for two successive maxima, in 1778 and 1788, when the time of maximum sunspot eruption appeared $4\frac{1}{4}$ and $5\frac{2}{3}$

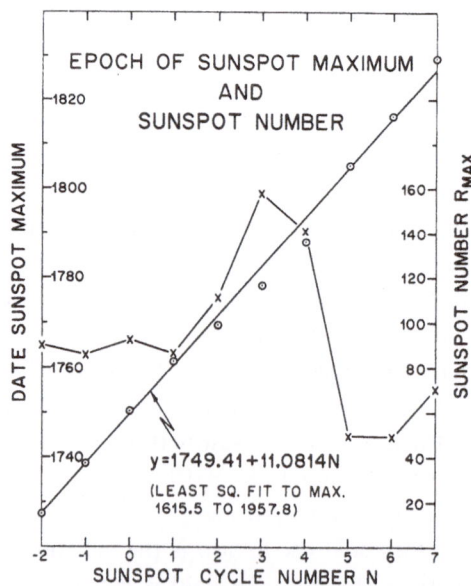

Fig. 7.   The phase shift in sunspot maximum observed for cycle numbers 2–4. Note the recovery of the correct phase by #5 contrary to expectation under the Babcock-Leighton theory. The correction of the phase error by 1805 suggests that the sunspot cycle is controlled by a tuned oscillator, perhaps the torsional oscillation of the core suggested in the text.

years too early respectively. This anomaly was accompanied by unusually intense sunspot activity, the average number of sunspots being approximately twice as great as normal. (See Figure 7.) It should be noted that this large phase shift disappeared permanently at the next sunspot maximum. Apparently some internal clock remembered the old phase throughout this massive disturbance. Under the Babcock-Leighton theory this phase shift would be expected to continue indefinitely into the future.

The presence of an oscillation in a rapidly rotating core could be significant for the Goldreich-Schubert-Fricke instability. Such an oscillation would induce a meridional component of velocity in the differentially rotating shell. This could be as great as 1 m/sec, eliminating the instability (see Goldreich and Schubert, 1967).

7. Summary and Conclusions

In my 1964 article in *Nature* it was suggested that the sun might have a rapidly rotating core and consequently an appreciable quadrupole moment that could advance the perihelion of Mercury's orbit about $4''$ arc/century to bring the observations in agreement with the scalar-tensor theory of gravitation. The solar wind torque density was estimated to be 1×10^8 dyne/cm. Whereas the torque density necessary to keep the outside of the sun rotating slowly while angular momentum diffused out of the core was evaluated from a solution of the diffusion equation and found to be comparable with this value.

In 1966 and 1967 the oblateness of the sun was measured and found to be consistent with the presence of such a gravitational quadrupole moment, and a value of $\omega = 5$ for the coupling constant of the scalar-tensor theory. The Mariner II space probe measurements of the solar wind flux and magnetic field strengths in the solar wind give a solar wind torque density consistent with the earlier estimate.

Other information not available in 1964 tends to support the original conjecture providing the sun is believed to be a typical star and not a special case. Kraft has shown that surfaces of very young solar type stars are rotating with substantially the same angular velocity postulated for the rapidly rotating solar core. Hyades solar-type stars show a loss of angular momentum of the amount expected if the stellar wind torques were equal to the solar wind torque and the stars were mixed down to a radius of 0.7. Reasons are given supporting the contention that stellar-wind torques should be roughly equal for main-sequence stars of the same mass redder than F5.

For angular momentum to be transported from the deep stellar interior in the absence of magnetic stresses requires mass transport and the destruction of berylium carried below $r = 0.5$. Thus to slow the rotation of the deep interior of a star by convection or the Goldreich-Schubert-Fricke effect implies a loss of almost all of its berylium. For solar-type stars of all ages, including the sun, berylium appears to be present with the meteoritic abundance. However, for such stars, lithium which burns at $r \sim 0.6$ appears to be depleted relative to the meteoritic value, the abundance of lithium decreasing exponentially with time with an e-folding time of 7×10^8 years.

These observations suggest that in the sun an outer mixed slowly rotating shell has an inner radius $0.5 < r < 0.6$. The moment of inertia of the shell implied by this value implies an e-folding time for slowing reasonably consistent with the present solar wind torque and the observed rotation in the Hyades.

The radius assumed in 1964 for the rapidly rotating core was about $r = 0.75$ whereas the lithium data seem to force a value of approximately 0.5. The observations seem to support a picture of turbulent mixing outside a molecular diffusion shell about 0.05 in thickness. Such turbulence could be driven by the Goldreich-Schubert-Fricke instability which for some reason is inoperative below this. A torsional oscillation of the rapidly rotating core, if it contains a crossed dipolar magnetic field, provides at least one means for stabilizing rotation against this instability in the molecular diffusion zone. Reasons are given for eliminating 'spin-down' as a source of instability.

In conclusion, if the sun is a normal main-sequence star, the notion that it has a rapidly rotating core receives support from the observations of rotation in young solar-type stars, the observations of the depletion of lithium in such stars, and the lack of berylium depletion. All of these observations are consistent with the core rotation needed to account for the solar oblateness.

References

Alfonso-Faus, A.: 1967, *J. Geophys. Res.* **72**, 5576.
Allen, C. W.: 1963, *Astrophysical Quantities*, Athlone Press, London.
Babcock, H. W.: 1961, *Astrophys. J.* **133**, 572.
Baker, N. and Kippenhahn, R.: 1959, *Z. Astrophys.* **48**, 140.
Bergmann, P.: 1948, *Ann. Math.* **49**, 255.
Brans, C. and Dicke, R. H.: 1961, *Phys. Rev.* **124**, 925.
Chazy, J.: 1928, *La théorie de la relativité et de la mécanique céleste,* Gauthier-Villars, Paris.
Cocke, W. J.: 1967, *Astrophys. J.* **150**, 1041.
Coleman, P. J., Jr.: 1966, *J. Geophys. Res.* **71**, 5509.
Colgate, S. A.: 1968, *Astrophys. J.* **153**, L81.
Danziger, I. J.: 1969, *Astrophys. Letters* **3**, 115.
Deutsch, A. J.: 1967, *Science* **156**, 236.
Dicke, R. H.: 1962, *Phys. Rev.* **125**, 2163.
Dicke, R. H.: 1964, *Nature* **202**, 432.
Dicke, R. H.: 1966, *Stellar Evolution* (ed. by R. F. Stein and A. G. W. Cameron), Plenum Press., New York, p. 319.
Dicke, R. H.: 1967, *Astrophys. J.* **149**, L121.
Dicke, R. H.: 1970a, *Astrophys. J.* **159**, 1.
Dicke, R. H.: 1970b, *Astrophys. J.* **159**, 25.
Dicke, R. H. and Goldenberg, H. M.: 1967, *Phys. Rev. Letters,* **18**, 313.
Dicke, R. H. and Peebles, P. J. E.: 1965, *Space Sci. Rev.* **4**, 419.
Fowler, W. A., Caughlan, G. R., and Zimmerman, B. A.: 1967, *Ann. Rev. Astron. Astrophys.* **5**, 525.
Fricke, K.: 1968, *Z. Astrophys.* **68**, 317.
Gilvarry, J. J. and Sturrock, P. A.: 1967, *Nature* **216**, 1283.
Goldreich, P. and Schubert, G.: 1967, *Astrophys. J.* **150**, 571.
Goldreich, P. and Schubert, G.: 1968, *Astrophys. J.* **154**, 1005.
Herbig, G. H.: 1965, *Astrophys. J.* **141**, 588.
Howard, L. N., Moore, D. W., and Spiegel, E. A.: 1967, *Nature* **214**, 1297.
Jordan, P.: 1948, *Astron. Nach.* **276**, 1955.
Jordan, P.: 1959, *Schwerkraft und Weltall*, Vieweg, Braunschweig, Germany.
Kippenhahn, R.: 1963, *Astrophys. J.* **137**, 664.

Kraft, R.: 1967, *Astrophys. J.* **150**, 551.

Kraft, R.: 1968, 'Stellar Rotation' in *Stellar Astronomy*, Vol. 2 (ed. by H. Y. Chiu, R. Warasila and J. Remo), Gordon and Breach, New York.

Leighton, R. B.: 1969, *Astrophys. J.* **156**, 1.

McDonald, B. E. and Dicke, R. H.: 1967, *Science* **158**, 1562.

Mestel, L.: 1966, *Z. Astrophys.* **63**, 196.

Mestel, L.: 1968, *Monthly Notices Roy. Astron. Soc.* **140**, 177.

Modiesette, J. L.: 1967, *J. Geophys. Res.* **72**, 1521.

Neugebauer, M. and Snyder, C. W.: 1966, *J. Geophys. Res* **71**, 4469.

Newcomb, S.: 1897, *Suppl. Am. Ephemeris and Nautical Almanac*.

O'Connell, R. F.: 1968, *Astrophys. J. (Letters)* **152**, L11.

Plaskett, H. H.: 1965, *The Observatory* **85**, 178.

Preston, G. W.: 1967, *Astrophys. J.* **150**, 547.

Roxburgh, I. W.: 1964, *Icarus* **3**, 92.

Schatzman, E.: 1962, *Ann. Astrophys.* **25**, 18.

Schmal, G. and Schröter, E. H.: 1965, *Z. Astrophys.* **62**, 143.

Sears, R. L. and Weymann, R.: 1965, *Astrophys. J.* **142**, 174.

Shapiro, I. I.: 1965, *Icarus* **4**, 549.

Spitzer, L.: 1962, *Physics of Fully Ionized Gas* (2nd ed.), Interscience, New York, p. 146.

Sturrock, P. A. and Gilvarry, J. J.: 1967, *Nature* **216**, 1280.

Thirry, Y. R.: 1948, *Comptes-Rendus* **226**, 216.

Thomas, L. H.: 1930, *Quart. J. Math.* **1**, 239.

Von Zeipel, H.: 1924, *Monthly Notices Roy. Astron. Soc.* **84**, 665.

Wallerstein, G. and Conti, P. S.: 1969, *Ann. Rev. Astron. Astrophys.* **7**, 99.

Wasiutynski, J.: 1946, *Astrophys. Norv.* **4**.

Weber, E. J. and Davis, L., Jr.: 1967, *Astrophys. J.* **148**, 217.

Weymann, R.: 1957, *Astrophys. J.* **126**, 208.

Discussion

Roxburgh: You already stated that a small non-conservative distortion produces a large temperature variation. This shows that only a very small distorting effect produces a large oblateness on temperature constant surface. If it is such surfaces that are effectively being measured then the oblateness is to be expected from the magnitude of the disturbing forces.

Of course one should construct model photospheres and integrate the energy flux coming to an observer; have you done this?

Dicke: The shapes of isophotes are not measured. Rather the integrated flux passing the occulting disk is used to determine both the oblateness of the extreme limb and the equatorial polar difference in surface brightness. By making measurements with several different amounts of exposed limb, the two effects are separated. The answer to your question is "yes". The significance of the atmosphere on the above interpretation has been considered, also the effect on the model atmosphere of the observed latitude independence of surface brightness. In the absence of surface magnetic and velocity fields (other than rotation), the position of the limb is determined by a surface of constant potential to within 3 meters and the limb brightness is remarkably latitude independent. Surface stresses induce both an oblateness and a latitude dependence in surface brightness. But the observations show that the sun's brightness is latitude independent and this implies a latitude independence of the atmospheric model for the normal photosphere.

Deutsch: Consider a solar-type star of the same age as the sun, which formed with the angular velocity in the core double that in the sun. Can you say how you would expect the rotational velocity of the hydrogen convection zone to compare with that we see in the sun?

Dicke: The surface angular velocity would be expected to be doubled if the stellar wind torque per unit angular velocity were the same as that of the sun and convective transport of angular momentum is unimportant, both of which are doubtful.

Abt: How did you allow for differential refraction?

Dicke: The refractive correction was computed from temperature and barometric pressure at the observatory. This was subtracted from the data. The residual showed no significant variation with time of the form characteristic of the computed refractive term.

ON THE EQUATORIAL ACCELERATION OF THE SUN

IAN W. ROXBURGH

Queen Mary College, University of London, England

Abstract. The interaction of rotation and turbulent convection gives rise to a latitude dependent turbulent energy transport. Energy conservation demands a slow meridional circulation in the solar outer convective zone. The transport of angular momentum by this circulation is balanced in a steady state by the turbulent viscous transport across an angular velocity gradient. Models are constructed which give equatorial acceleration as observed on the sun.

Fundamental Equations

In this paper we shall consider as simple a model as possible that still retains the fundamental points of the theory. We are, after all, interested primarily in understanding the phenomenon of equatorial acceleration; the full quantitative treatment will be very complex and is postponed to a later date. We shall, for example, ignore the energy carried by radiation within the convective zone. This is satisfactory over most of it but obviously breaks down at the boundary where convective energy transport ceases and all the energy is carried by radiation. With this approximation we write the turbulent energy flux as

$$\mathbf{F} = k\nabla S \tag{1}$$

where

$$S = c_v \ln(P/\varrho^\gamma) \tag{2}$$

is the entropy, P the pressure and ϱ the density. We shall throughout ignore the effects of ionisation and take $\gamma = \frac{5}{3}$. The 'turbulent conduction' k varies from place to place, being a function both of depth and latitude.

In the energy balance equation we neglect viscous dissipation; if we thought it worthwhile this effect could be included, but it would only alter the quantitative not the qualitative result, so we shall neglect it here. With the 'conductivity' k varying with latitude the divergence of the turbulent energy flux will not in general be zero and there will be a general circulation also carrying energy. Energy balance is then expressed by

$$\nabla \cdot F + \tfrac{3}{2} P \mathbf{v} \cdot \nabla S = 0. \tag{3}$$

The convective zone is assumed to be a viscous layer so that in a steady state it satisfies the equation

$$\nabla \frac{V^2}{2} - \mathbf{v}\,\text{curl}\,\mathbf{v} + \frac{\nabla P}{\varrho} + \nabla \Phi + \frac{1}{\varrho}\mathbf{R} = 0$$

where $\mathbf{v}$ is the total velocity, the sum of the circulation velocity $\mathbf{v}_m$ and the rotation. Φ is the gravitational potential and $\mathbf{R}$ the viscous force.

A. Slettebak (ed.), Stellar Rotation, 318–320. All Rights Reserved

The turbulent viscous force in a convective layer is a notoriously difficult thing to estimate. We shall here assume that it has the same form as kinematic viscosity with the coefficient of viscosity $\eta = \varrho v_t l$ where ϱ is the density, v_t the mean turbulent velocity and l the mixing length. The azimuthal (ϕ) component of the viscous force is then:

$$R_\phi = -\frac{1}{r^3}\sin\theta\,\frac{\partial}{\partial r}\left(r^4\eta\,\frac{\partial\Omega}{\partial r}\right) - \frac{1}{r\sin^2\theta}\frac{\partial}{\partial\theta}\left(\sin^3\theta\eta\,\frac{\partial\Omega}{\partial\theta}\right).$$

In an axially symmetric situation this viscous force balances the convection of angular momentum by the thermally driven circulation **v**, hence

$$-\frac{\varrho}{r\sin\theta}\left[v_r\,\frac{\partial}{\partial r}\,(\Omega r^2\sin^2\theta) + \frac{v_\theta}{r}\frac{\partial}{\partial\theta}\,(\Omega r^2\sin^2\theta)\right] = R_\phi.$$

If v_θ is known this equation determines $\Omega(r,\theta)$.

EXAMPLE

As an example we shall consider the case where

$$k = k(r,\theta) = K_0(r)\,(1 + \varepsilon(r)\,P_2(\cos\theta))$$

and $\varepsilon \ll 1$. Using a series expansion in ε, gives

$$\frac{k_0}{r^2}\frac{d}{dr}\left(r^2\,\frac{dS_0}{dr}\right) + \frac{dk_0}{dr}\frac{dS_0}{dr} = 0$$

for the zero-order state, and

$$\nabla P_0/\varrho = -\nabla\phi_0.$$

If we further assume that v is always so small as to be negligible in the hydrostatic balance the terms of order ε yield

$$v_r = \frac{d\varepsilon}{dr}\left(\frac{2k_0}{3P}\right)P_2 = \frac{\eta}{\varrho}\frac{d\varepsilon}{dr}P_2,$$

since $2k/3\varrho = \eta$ in the simple mixing length theory. Hence the stream lines of v are given by $S(r,\theta) = \psi(r)\sin\theta\,P_2'$ so that

$$v_r = \frac{6}{r^2\varrho}\,\psi P_2,\quad v_\theta = -\frac{1}{r\varrho}\frac{d\psi}{dr}P_2'.$$

If we now turn to the equation that determines Ω we note that for η large the solution is uniform rotation $\Omega = \Omega_0 = $ constant. For large but finite η we write $\Omega = \Omega_0 + \omega_1(r) + \omega_2(r)\,P_2$ and find

$$\frac{1}{r^3}\frac{d}{dr}\left(r^4\eta\,\frac{d\omega_2}{dr}\right) - \frac{10\omega_2\eta}{r} = -4\Omega_0 r^2\,\frac{d}{dr}\left(\frac{\eta}{6r}\frac{d\varepsilon}{dr}\right).$$

The solution is obviously independent of the magnitude of η, but does depend on ts variation in space. The solution depends on the interaction of rotation and convection through $\varepsilon(r)$. To simplify the numerical work we shall introduce variables.

$$s = \eta/\eta_0, \quad y = \omega_2/\varepsilon_0\Omega_0, \quad x = r/R, \quad \varepsilon = \varepsilon_0 f$$

so that

$$\frac{1}{x^3}\frac{d}{dx}\left(x^4 s\frac{dy}{dx}\right) - \frac{10ys}{x} = -\tfrac{2}{3}x^2\frac{d}{dx}\left(\frac{s\,df/dx}{x}\right)$$

subject to the free surface boundary conditions $dy/dx = 0$ at $x = 1$, x_i, where the convective zone extends from $r = R$ to $r = r_i = x_i R$.

PARTICULAR CASE $s = 1$, $d\varepsilon/dx = 1$

This is a highly idealised case but is sufficient to give some insight into the nature of the solution. y satisfies the equation

$$\frac{d^2y}{dx^2} + \frac{4}{x}\frac{dy}{dx} - \frac{10y}{x^2} = \frac{2}{3}x,$$

which has the solution

$$y = \frac{1}{18}\frac{(1 - x_i^6)x^2}{(1 - x_i^7)} - \frac{x_i^6(1 - x_i)}{45(1 - x_i^7)x^5} - \frac{x}{9}.$$

For $x_i = 0.8$, an approximate solar model, this gives $y_s = -0.06$. So that applying our model to the sun gives

$$\frac{\Omega_{eq}}{\Omega_{pole}} = \frac{(1 - \omega_{2/2}/\Omega_0)}{1 + \omega_2/\Omega_0} = 1 - 3\omega_2/2 = 1 + \varepsilon_0\,0.09,$$

that is an equatorial acceleration.

FURTHER WORK

The problem and solution given above is in very simplified form. Before any definite conclusion may be reached a detailed model including both radiative and convective energy transport must be calculated. This is being done by Dr. Durney and myself. Already one detailed solution has been obtained and we expect to publish details of this work in the future.

SPECTROSCOPIC OBSERVATIONS OF DIFFERENTIAL ROTATION WITH HEIGHT IN THE SOLAR ENVELOPE*

W. C. LIVINGSTON

Kitt Peak National Observatory†, Tucson, Ariz., U.S.A.

Abstract. The problem of determining the spectroscopic rotation of the sun is not one of instrumental sensitivity, which can be about 10 m/sec, but rather of abstracting the rotational component from the hierarchy of long-lived surface currents. Required are continued daily observations over long periods of time and this task is not presently feasible at most observatories. Past observations have indicated a height gradient of rotational angular velocities. We have determined the photosphere and chromosphere angular velocity gradient from observations across the disk of C 5380 Å ($h \simeq -250$ km), CN 3882 Å ($h \simeq 0$ km), Ca^+K_3 ($h \simeq +5000$ km), together with other Fraunhofer lines of intermediate origin while K-line spectra of prominences are used to sense the chromosphere-corona interface.

Preliminary results for the disk measurements indicate a constant rotation through the photosphere, but an *increase* through the chromosphere up to 20–30% for the Ca^+K_3 layers.

1. The Observational Problem

The task of measuring the rotation of the sun, as compared with a star, is simplified by the enormous available light and the fact that we deal with a resolved disk. High light intensity permits a spectroscopic Doppler sensitivity of about 10 m/sec for a Fraunhofer line with an equivalent width of about 100 mÅ. The resolved disk allows us, in principle, to directly measure the motion of rotation. Unfortunately, this

TABLE I

Approximate sizes, lifetimes and velocities of convective disturbances at the solar surface

	s (km)	τ	V (km/sec p. to p.)	
			Photo	Chrom.
('White-light') granulation	$\simeq 10^3$	$\simeq 7^m$	1–2	
Oscillatory elements	2×10^3	5^m	0.5	4
				$(Ca^+ K_3)$
Supergranulation	30×10^3	1^d	0.5	
'Giant granules'	300×10^3 ?	$> 1^d$?	0.2 ?	

rotational aspect is somewhat masked by the hierarchy of velocity fields found on the solar surface. Some properties of these currents – in order of increasing spatial extent – are summarized in Table I.

The confusion of the fine granules and photospheric oscillatory fields can be made

---

* Kitt Peak National Observatory Contribution No. 531.
† Operated by The Association of Universities for Research in Astronomy, Inc., under contract with the National Science Foundation.

negligible by spatial averaging at the telescope – using an out-of-focus image or an integrating lens. However, a reduction of the chromospheric oscillations (because of their magnitude) and the supergranular and hypothetical 'giant granules' (because of their long life and spatial extent) requires the averaging of observations over many days and even weeks. So far, limitations in telescope time have prevented us from making such an average and thus determining fully the rotation of the sun.

What one can do with limited observational material is to study how the angular velocity of rotation varies with height through the solar envelope. Past observations have indicated a surprising increase in the rotational velocity of the sun for the higher layers. Certain Fraunhofer lines, suitably chosen because of their height of formation, permit a sensing of velocities at different layers in the photosphere and chromosphere. Further out, velocities in the chromosphere-corona interface can be obtained from a statistical analysis of prominence radial velocities. A program of these measurements is being pursued – although many observations are needed because of the large random motions of the individual prominences.

2. The Disk Observations

All disk observations are made using the 'Doppler compensator' of the solar magnetograph. This magnetograph is an accessory to the 13.7 m vertical spectrograph which is normally fed by the main 80-cm image. At the focus of the spectrograph the position of the Fraunhofer line is sensed by a double slit arrangement, the light from either wing of the line being directed to photomultipliers (Figure 1). The difference signal from these photomultipliers turns a torque motor, connected to the parallel glass plate, until the line is centered and the signal becomes zero. The position angle of this plate is sensed by a coaxial microsyn generator whose output is recorded on a strip chart. A knowledge of the spectrograph dispersion, the index of refraction and thickness of the tipping plate, together with the angle-display sensitivity of the microsyn. suffice to reduce any line shift to change of wavelength and hence relative velocity change. The r.m.s. system noise is about 10 m/sec with a 10 Hz bandwidth. This noise is completely negligible compared with the above mentioned solar effects. (There are actually two separate systems so that two lines can be studied simultaneously.)

The solar image is scanned across the spectrograph entrance slit at a constant rate and at a constant heliographic latitude. Figure 2 indicates a typical photospheric velocity and brightness recording. A straight line fit by eye is made to the velocity trace giving most weight to undisturbed regions, and ignoring certain limb effects. The slope of this line, reduced to angular rotation (sidereal, deg/day) becomes one data point for subsequent averaging. The equatorial rates reported here consist of a straight average of eight double scans, i.e. across and return, in two degree latitude increments. Thus a mean over $\pm 8°$ latitude is obtained.

A list of the lines used, how the exit slits are positioned with respect to the line profile and the estimated height of formation of that part of the line being analysed. is given in Table II. Note that, except for the Ca^+K_3 line, the core of the line is not

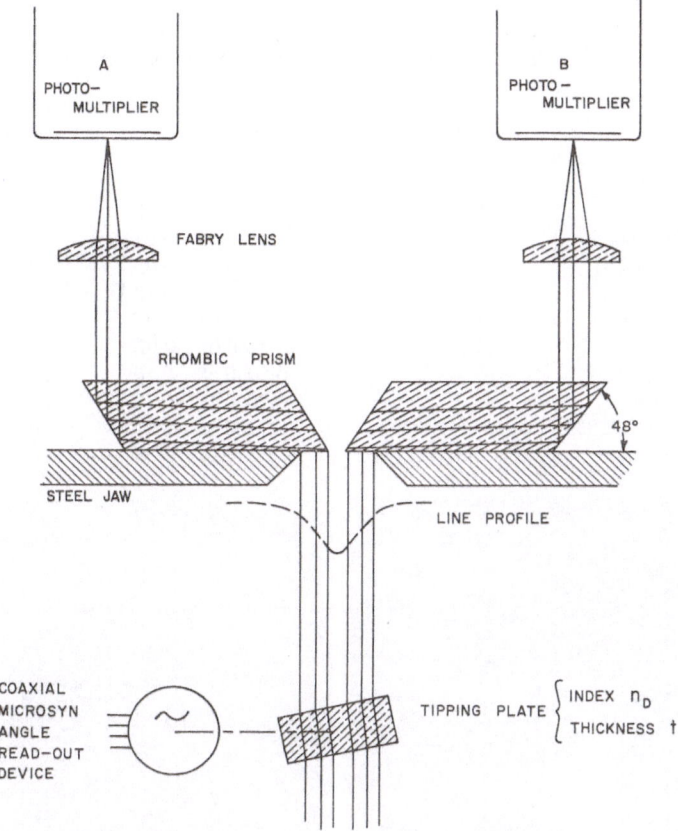

Fig. 1.   Arrangement at the focus of the spectrograph for measurement of relative Fraunhofer line displacement. The system exists in duplicate, permitting the simultaneous recording of two lines.

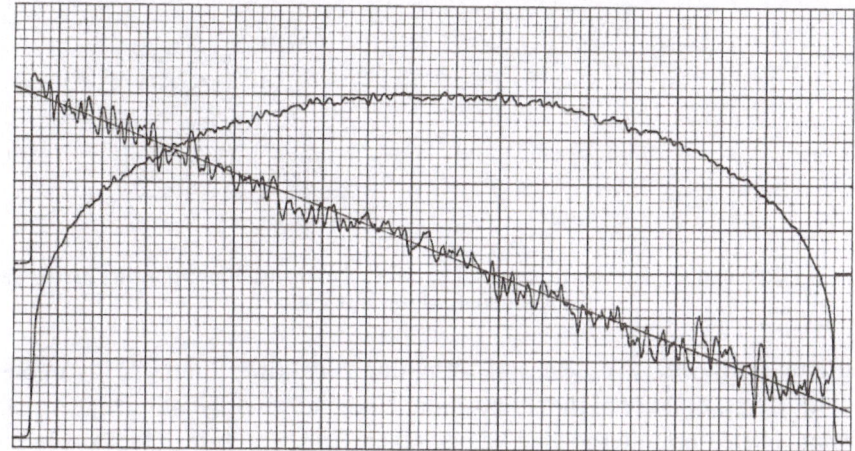

Fig. 2.   Typical superposed line displacement (velocity) and brightness records for a single scan across the heliographic equator. A linear fit is made to the velocity record.

TABLE II

Equivalent width, excitation potential, double slit widths and separations, and estimated height of formation for lines used in this study

	w (mÅ)	e.p. (volts)	Slits (Å)	h (km)
C 5380.3	26	7.68	0.03–0.08 –0.03	−250
Fe 3891.9	88	3.41	0.02–0.05 –0.02	
Fe 3944.9	72	2.99	0.02–0.05 –0.02	
Fe 4842.7	44	4.22	0.03–0.09 –0.03	−150
Fe 5233.0	346	2.94	0.18–0.15 –0.18	
Fe 5250.2	62	0.12	0.03–0.08 –0.03	
CN 3882.5	127	–	0.02–0.05 –0.02	0
Hα 6562.8	–	–	0.50–0.50 –0.50	2000
Hβ 4861.3	–	–	0.20–0.165 –0.20	2500
Ca⁺K₃ 3933.7	–	–	0.12–0.10 –0.12	5000

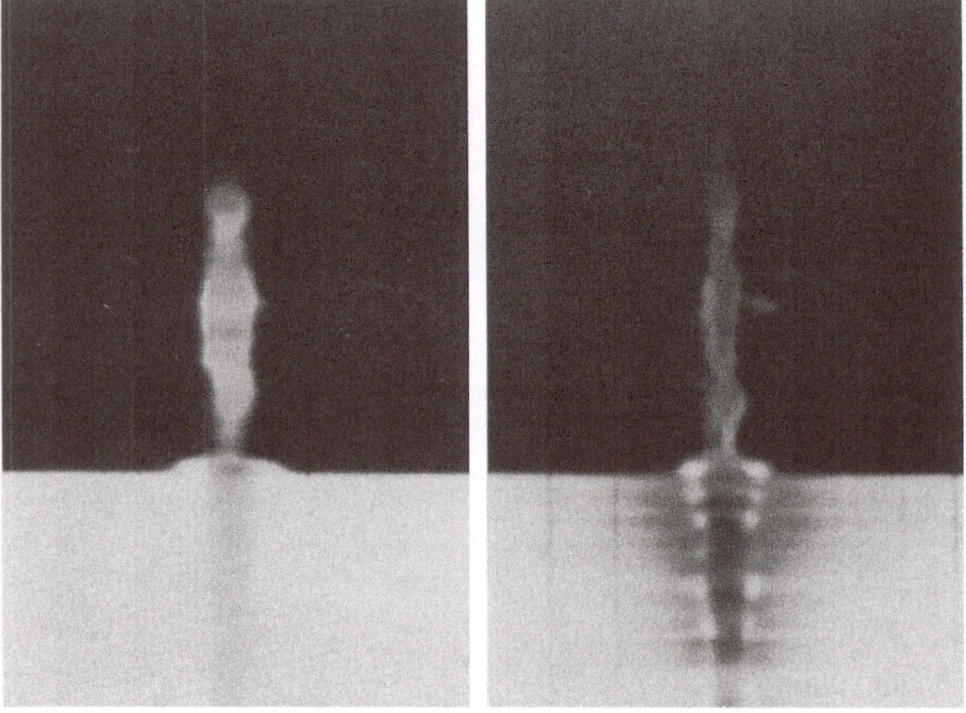

Fig. 3.   Hα (left) and Ca⁺K (right) spectra taken with slit radial to limb, showing narrow features that are only visible at Ca⁺K.

sensed and so the effective height of formation runs lower than usually quoted for these lines. Thus, because of the slit position, Hβ is shown originating higher than Hα. The heights listed are interpolated values taken from De Jager (1959), while other spectroscopic data are based on Moore *et al.* (1966).

Results for observations in 1968–69 are given in Table III and discussed in Section 4.

TABLE III

Equatorial angular velocity deg/day, sidereal, derived for height sensitive lines 1968–69

	h (km)	1968 Dec. 30	1969 May 8	Jul. 2	Jul. 3	Jul. 4	Aug. 10	Sep. 4	Sep. 5
C 5380.3	− 250	13.65							13.72
Fe 3891.9							13.70		
Fe 3944.9			13.58	13.70		13.09			13.60
Fe 4842.7	− 150					14.19			
Fe 5233.0		13.66			13.81	13.75			
Fe 5250.2		13.70			13.82	13.88			
CN 3882.5	0						13.65		
Hα 6562.8	2000	14.90			13.57	14.13		14.57	14.75
Hβ 4861.3	2500					14.52			
Ca⁺K₃ 3933.7	5000		17.09	17.25		17.25			18.10

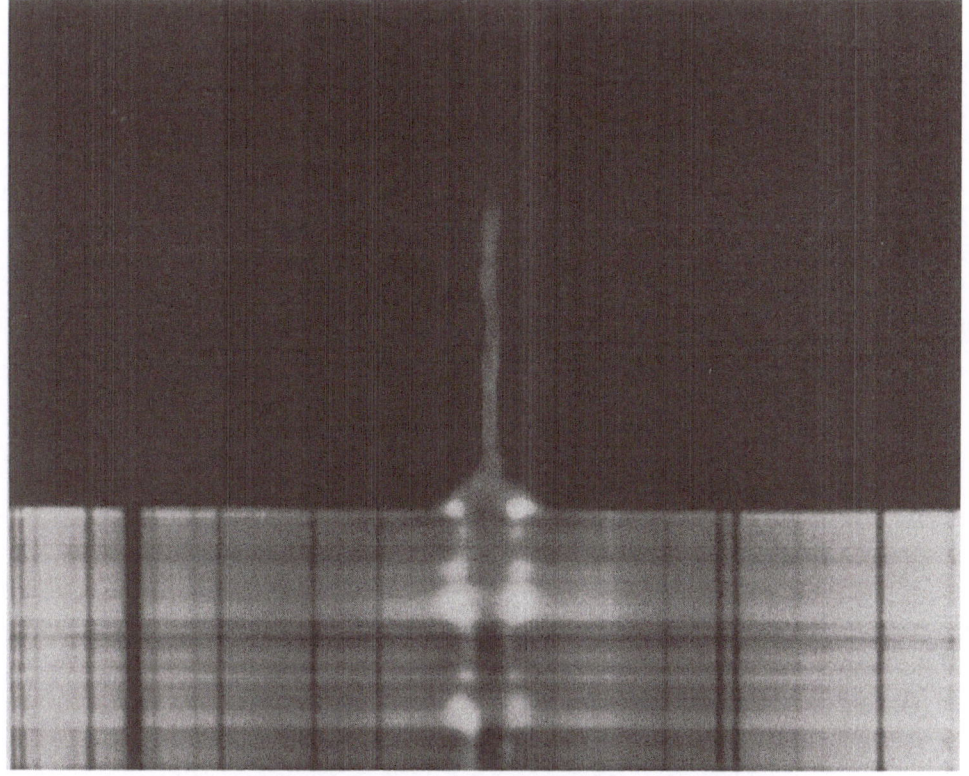

Fig. 4.   Quiescent prominence in Ca⁺K, radial slit.

3. The Prominence Observations

In an interesting series of papers Evershed (1927, 1929, 1935, 1945) has reported a striking increase in angular velocity with height based on displacements of the H and K lines in prominences. Although disputed by others (cf. Perepelkin, 1932) the conflicting observations have generally been made at comparatively low dispersion or using Hα (cf. Liszka, 1969). O. Engvold pointed out to the writer that the K-line in certain prominences is very narrow. High dispersion (0.09 Å/mm) material has shown that frequently a prominence has a broad turbulent part co-mingling with a very narrow central core (Figure 3).

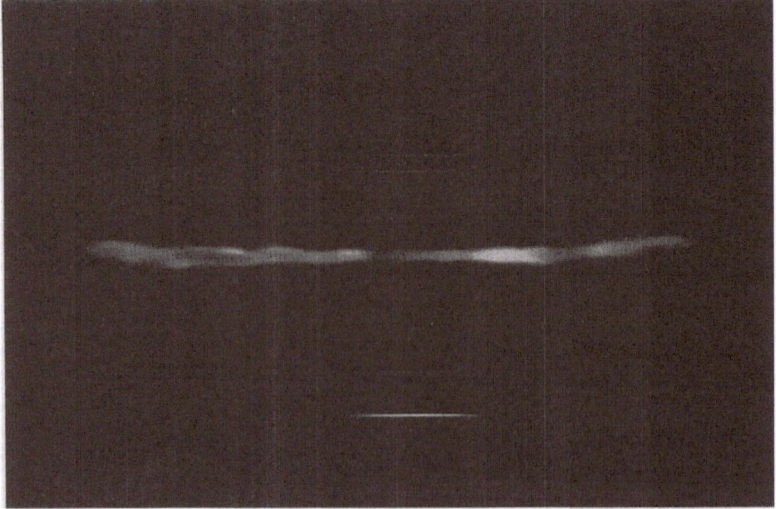

Fig. 5.   Hα (top) and Ca⁺K (bottom) spectra of quiescent prominence taken with slit 27 000 km above and tangent to the limb. Comparison spectrum is thorium.

Often quiescent prominences show extensive regions having the narrow K-line characteristic (Figures 4 and 5). We have recently begun to systematically obtain K-line spectra of prominences and in the reduction of these plates for solar rotation special weight is given to the narrow line measurements. At the time of writing (Sept. 1969) 60 narrow line objects have been measured by Mr. Daniel Gezari, but more material is needed before the results of Evershed can be confirmed.

4. Discussion of Results for the Disk

An inspection of Table III discloses two trends. First the rotation rates vary with date. This variable characteristic is further amplified by our previous results extending three years back (Livingston, 1969a). We presume this scatter is mainly a consequence of our inability to average-out the long-lived and large scale convective patterns discussed in the first section.

Second, an increasing angular velocity with height is noticed. The chromospheric lines taken as a whole show almost a deg/day, or a 7%, increase over the low lying photosphere. No certain variation is detected within the photosphere (see also Livingston, 1969b). The velocity obtained for Ca^+K_3 is particularly large, but then this line may be formed appreciably higher than $H\alpha$ and $H\beta$.

A summary of the observed rotation with height within the solar envelope is given in Figure 6. Aslanov (1963) found a wide range of velocity within the photosphere, but this is yet to be confirmed. Our own findings are not new but rather parallel the discovery of Adams (1911) who first showed the increase of velocity with height as so indicated. Finally the 1967 K-corona tracer (coronal condensations) measurements of Hansen *et al.* (1969) indicate a possible retardation of velocity in the intermediate

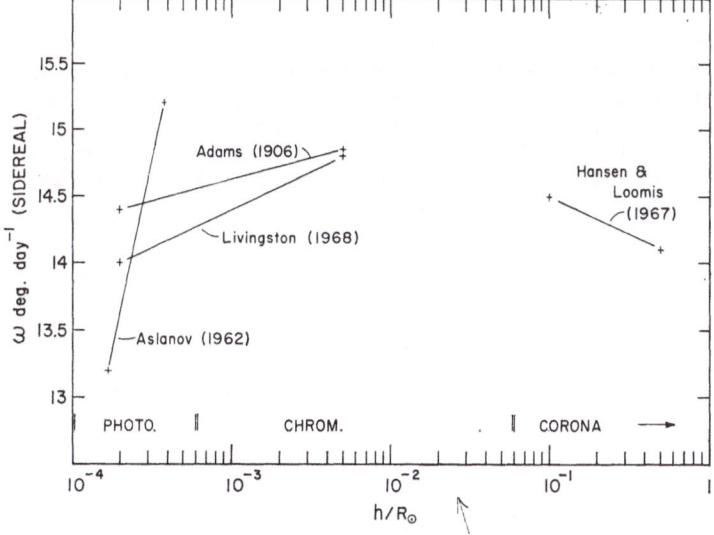

Fig. 6.   Summary of the observational data of variation of angular rotation with height. See text .

corona. One concludes from these data that there is an increase of velocity with height through the chromosphere of up to about 7% (30% if the K_3 data prove correct) and then a subsequent fall off through the outer corona.

What are the stellar consequences of these observations? At this Colloquium we have heard a number of theories discussed wherein momentum is transferred outward from the star's surface to a stellar wind, with a resulting braking action being exerted on the star's rotation. Brandt (1966) has similarly considered this effect on the sun. If, however, the angular velocity increases outward the reverse would seem to be the case, i.e. momentum would flow downward. The detailed mechanism for creating these conditions has, of course, yet to be given. The latitude dependence of this height gradient, not given here, may prove crucial.

References

Adams, W. S.: 1911, *Carnegie Inst. of Washington Publ.*, No. 138.
Aslanov, I. A.: 1963, *Astron. Zh.* **40**, 1036 (*Soviet Astron. – AJ* **7**, 794).
Brandt, J. C.: 1966, *Astrophys. J.* **144**, 1221.
De Jager, C.: 1959, *Handbuch der Physik* (ed. by S. Flügge), Springer-Verlag, Berlin.
Evershed, J.: 1927, *Monthly Notices Roy. Astron. Soc.* **88**, 126.
Evershed, J.: 1929, *Monthly Notices Roy. Astron. Soc.* **89**, 250.
Evershed, J.: 1935, *Monthly Notices Roy. Astron. Soc.* **95**, 503.
Evershed, J.: 1945, *Monthly Notices Roy. Astron. Soc.* **105**, 204.
Hansen, R. T., Hansen, S. F., and Loomis, H. G.: 1969, *Solar Phys.*, in press.
Liszka, L.: 1969, *Kiruna Geophys. Obs. Report*, No. 691.
Livingston, W. C.: 1969a, *Solar Phys.* **7**, 144.
Livingston, W. C.: 1969b, *Solar Phys.* (in press).
Moore, C. E., Minnaert, M. G. J., and Houtgast, J.: 1966, *N.B.S. Monograph 61*.
Perepelkin, E. J.: 1932, *Pulkovo Obs. Circ.* No. 1.

Discussion

Dicke: One possible means of increasing the rotational velocity of the chromosphere requires back bombardment by the corona. Angular momentum is transferred from the sun to the corona by magnetic stresses but gas flow along magnetic flux strands to the chromosphere from the corona would transfer angular momentum to the upper chromosphere.

Roxburgh: It need not be as difficult to transport angular momentum from a region of low momentum to one of the high momentum. Non-linear waves could do this at the same time as heat is transported up to the corona.

SPIN DOWN OF BOUSSINESQ FLUID IN THE CIRCULAR CYLINDER, AS A SIMULATION OF THE SOLAR SPIN-DOWN PROCEDURE

TAKEO SAKURAI

Dept. of Aeronautical Engineering, Kyoto University, Kyoto, Japan

Abstract. As the simplest model to simulate the solar spin down procedure from the hydrodynamical viewpoint, a spin down problem with a simple geometrical configuration within the frame work of the Boussinesq approximation is investigated. A new proposal of the origin of the solar differential rotation is made on the basis of the calculated asymptotic distribution of the angular velocity. A comment on the present day solar spin down controversy is also made.

1. Introduction

In Hoyle's theory of the origin of the solar system and several other studies of stellar rotation (Hoyle, 1960; Lüst and Schlüter, 1955; Dicke, 1964; Brandt, 1966), it is shown that the solar angular momentum is lost by the magnetohydrodynamic interaction with the interplanetary medium, and that the primeval solar angular momentum rotating once a day is lost in the time duration of the solar age or of an order of magnitude larger.

What effect does this have on the rotation in the sun? This question is fundamental not only in itself but also as the mediator of the solar spin down controversy. Dicke (1964) made an important suggestion that the solar interior rotates once a day in spite of the observational evidence that the solar surface rotates once in 27 days. Howard *et al.* (1967) opposed Dicke on the basis of the spin down procedure in the solar interior; that is the angular momentum rearrangement by the meridional circulation induced by the pumping mechanism of the Ekman's boundary layer. Thereafter, several authors discussed the validity of Dicke's hypothesis (Fricke and Kippenhahn; Dicke 1967; Bretherton and Spiegel, 1968). However, no conclusive result has yet been obtained.

As the simplest model to simulate the unsteady rotational motion of the sun under the influence of the angular momentum loss, the following is investigated in this paper: Viscous heat conducting compressible fluid rotates rigidly with the containing circular cylinder rotating uniformly around the vertical axis of symmetry under the influence of the constant gravitational force and the stable distribution of the temperature. The angular velocity of the cylinder is changed abruptly at a certain instant of time while the horizontal wall of the cylinder is made to keep to the original temperature and the vertical wall is made to keep to the original temperature or to be held thermally insulated. Our problem is to study the response of the fluid to the above abrupt change within the time scale of the incompressible spin down time and within the frame work of the Boussinesq approximation. The asymptotic distribution of the angular velocity and of the temperature especially are studied in detail.

A. Slettebak (ed.), Stellar Rotation, 329–339. All Rights Reserved
Copyright © 1970 by D. Reidel Publishing Company, Dordrecht-Holland

The overall character of the response is divided into three steps according to the respective time scales: Within the time scale of the rotational period (the reciprocal of the angular velocity), Ekman's boundary layers are established on the horizontal walls and the meridional circulation is generated by the pumping mechanism of the Ekman's boundary layer. Within the time scale of the spin down time (the geometrical mean of the rotational period and the following diffusion time), the angular momentum is transferred by the meridional current and the main bulk of the inviscid region is spun down. On the other hand, the above spin down procedure is partially quenched by the effect of the stable stratification, and there appears a non-uniformity in the angular velocity distribution. Finally, within the time scale of the diffusion time (the square of the representative length divided by the kinematic viscosity), the above non-uniformity is relaxed by the effect of the viscous diffusion to realize the final state of the rigid rotation corresponding to the disturbed angular velocity of the containing cylinder.

The rotational period of the sun is about a day or 27 days (according to Dicke's hypothesis or to the observational evidence), while the spin down time is estimated to be of the order of 10^{5-7} years and the diffusion time to be of the order of 10^{11} years. The solar age, on the other hand, is estimated to be of the order of 10^9 years. Thus, the present status of the solar rotation is understood as the asymptotic state within the time scale of the spin down time.

The above problem is a source of controversy also in the field of hydrodynamics. Holton (1965) investigated the spin down of rotating stratified fluid in the circular cylinder by a heuristic treatment and showed the existence of the spin down procedure similar but not identical to that in the incompressible fluid. Pedlosky (1967) re-examined the same problem from the viewpoint of complementing Holton's treatment and gave a clearer formulation. Pedlosky's conclusion for the case with a thermally insulated vertical wall is that the inviscid region is spun down by a strictly diffusive process within the time scale of the diffusion time, which conclusion is completely at variance with that of Holton. Holton and Stone (1968) examined Pedlosky's solution, pointed out the inconsistency in the solution, and suggested a further study of the same problem.

Recently, Sakurai (1969a, b) investigated the above problem in detail and resolved the Holton-Pedlosky controversy analytically. He did not, however, make a detailed numerical study of the asymptotic distribution of the angular velocity and the temperature. Thus, the present study is to be taken as the extension of his previous investigations. Since the analytical treatment is completely the same as that in his previous investigations, the mathematical detail is omitted except that which is essential to understand the underlying mathematics.

2. Basic Equations

The basic equations governing the axisymmetric motion of viscous heat conducting compressible fluid within the frame work of the Boussinesq approximation written in cylindrical coordinates rotating with the angular velocity Ω about the vertical axis of symmetry are:

$$\frac{\partial q_r}{\partial t} - 2q_\theta = -\frac{\partial p}{\partial r} + E\mathscr{L}q_r, \tag{1}$$

$$\frac{\partial q_\theta}{\partial t} + 2q_r = E\mathscr{L}q_\theta, \tag{2}$$

$$\frac{\partial q_z}{\partial t} = -\frac{\partial p}{\partial z} + T + E\,\varDelta q_z, \tag{3}$$

$$\sigma\left(\frac{\partial T}{\partial t} + Sq_z\right) = E\,\varDelta T, \tag{4}$$

$$0 = \frac{1}{r}\frac{\partial}{\partial r}(rq_r) + \frac{\partial q_z}{\partial z}, \tag{5}$$

$$E = \frac{\nu}{L^2\Omega}, \quad S = \frac{\alpha g\,(\bar{T}_1 - \bar{T}_0)}{L\Omega^2}, \quad \sigma = \frac{\nu}{\kappa}, \tag{6}$$

$$\mathscr{L} = \varDelta - \frac{1}{r^2}, \quad \varDelta = \frac{\partial^2}{\partial r^2} + \frac{1}{r}\frac{\partial}{\partial r} + \frac{\partial^2}{\partial z^2}, \tag{7}$$

where

$$(\bar{r}, \theta, \bar{z}) = (Lr, \theta, Lz), \quad \bar{t} = \frac{t}{\Omega}, \tag{8}$$

$$\bar{\mathbf{q}} = \varepsilon\Omega L\mathbf{q}, \quad \bar{p} = \bar{p}_S + \varepsilon\Omega^2 L^2\bar{\varrho}_S p, \quad \bar{T} = \bar{T}_S + \varepsilon\frac{\Omega^2 L}{\alpha g}T, \tag{9}$$

$$\bar{\varrho}_S = \bar{\varrho}_0\left\{1 - \frac{\alpha(\bar{T}_1 - \bar{T}_0)}{L}z\right\}, \quad \bar{p}_S = \bar{p}_0 - \int_0^{\bar{z}} g\bar{\varrho}_S\,d\bar{z}, \quad \bar{T}_S = \bar{T}_0 + \frac{\bar{T}_1 - \bar{T}_0}{L}z, \tag{10}$$

and (q_r, q_θ, q_z), p, ϱ, T, (r, θ, z), t, ν, κ, α, g, L and ε are the velocity, the pressure, the density, the temperature, the position vector, the time, the viscosity, the thermal conductivity, the coefficient of thermal expansion, the gravitational acceleration, the height of the cylinder and the parameter corresponding to the small deviation from the state of rotating equilibrium, and suffixes S, 1 and 0 and bars on letters refer to the state of rotating equilibrium, the upper and lower bottom of the cylinder and to the original physical quantity with dimension, respectively. In the derivation of the above linear equations from the full basic equations, terms of the order of ε^2 and $\Omega^2 L/g$ are neglected as is exemplified in the steady flow (Barcilon and Pedlosky, 1967). E is assumed to be very small while S and σ are taken to be of the order of 1. Hence, our investigation falls within the parameter range of compressibility dominant cases, in which the Ekman's pumping mechanism does not work, according to Barcilon and Pedlosky's classification based on the steady solution.

The introduction of the stream function of the meridional current leads to the following:

$$E\tilde{\mathscr{L}}\left\{E^{1/2}\tilde{\mathscr{L}} - \frac{\partial}{\partial \tilde{t}}\right\}\tilde{\psi} + 2\frac{\partial \tilde{q}_\theta}{\partial \tilde{z}} = \frac{\partial \tilde{T}}{\partial \tilde{r}}, \tag{11}$$

$$\left\{E^{1/2}\tilde{\mathscr{L}} - \frac{\partial}{\partial \tilde{t}}\right\}\tilde{q}_\theta - 2\frac{\partial \tilde{\psi}}{\partial \tilde{z}} = 0, \tag{12}$$

$$\left\{E^{1/2}\tilde{\Delta} - \sigma\frac{\partial}{\partial \tilde{t}}\right\}\tilde{T} + \frac{\sigma S}{\tilde{r}}\frac{\partial}{\partial \tilde{r}}(\tilde{r}\tilde{\psi}) = 0, \tag{13}$$

where

$$q_r = \frac{\partial \psi}{\partial z}, \quad q_z = -\frac{1}{r}\frac{\partial}{\partial r}(r\psi), \tag{14}$$

$$r = \tilde{r}, \quad z = \tilde{z}, \quad t = E^{-1/2}\tilde{t},$$
$$\psi = E^{1/2}\tilde{\psi}, \quad q_\theta = \tilde{q}_\theta, \quad T = \tilde{T}. \tag{15}$$

The transformation (15) corresponds to the time duration in which our investigation is aimed. The quantities with tildes are assumed to be of the order of 1. Hence, (15) implies the existence of the meridional current with the same order of magnitude as that in the incompressible flow, showing the essential difference between the steady and the unsteady motion.

The initial conditions are obtained by the expression that the fluid is in rotating equilibrium until a certain instant:

$$\tilde{\psi} = \tilde{q}_\theta = \tilde{T} = 0, \quad \text{for} \quad 0 \leqslant \tilde{z} \leqslant 1, \quad 0 \leqslant \tilde{r} \leqslant r_0, \quad \tilde{t} < 0 \tag{16}$$

where r_0 is the ratio of the radius to the height of the cylinder.

The boundary conditions are obtained by the expression that the angular velocity of the cylinder is abruptly changed after the above instant while the horizontal wall of the cylinder is made to keep to the original temperature and the vertical wall is (A) to keep to the original temperature or (B) to be held thermally insulated:

$$\tilde{\psi} = \partial\tilde{\psi}/\partial\tilde{z} = \tilde{T} = 0, \quad \tilde{q}_\theta = \omega\tilde{r}, \quad \text{for}$$
$$\tilde{z} = 0 \quad \text{or} \quad 1, \quad 0 \leqslant \tilde{r} \leqslant r_0, \quad \tilde{t} > 0 \tag{17}$$

(A) $$\tilde{\psi} = \partial\tilde{\psi}/\partial\tilde{r} = \tilde{T} = 0, \quad \tilde{q}_\theta = \omega r_0, \tag{18}$$

(B) $$\tilde{\psi} = \partial\tilde{\psi}/\partial\tilde{r} = \partial\tilde{T}/\partial\tilde{r} = 0, \quad \tilde{q}_\theta = \omega r_0. \tag{19}$$

for

$$0 \leqslant \tilde{z} \leqslant 1, \quad \tilde{r} = r_0, \quad \tilde{t} > 0.$$

3. Similar Solution

Before going ahead directly into the examination of the numerical results of the asymptotic distributions calculated on the basis of the solution of (11) to (13) under the initial and boundary conditions (16) to (19), it is interesting to note that there is the following similar solution for the limiting case with the infinitely large radius of the cylinder:

$$\psi = r\phi(z, t), \quad q_\theta = rv(z, t), \quad T = T(z, t) \tag{20}$$

where tildes over letters are omitted for the sake of simplicity.

The substitution of (20) into (11) to (13) leads to the following:

$$E \frac{\partial^2}{\partial z^2} \left\{ E^{1/2} \frac{\partial^2}{\partial z^2} - \frac{\partial}{\partial t} \right\} \phi + 2 \frac{\partial v}{\partial z} = 0, \tag{21}$$

$$\left\{ E^{1/2} \frac{\partial^2}{\partial z^2} - \frac{\partial}{\partial t} \right\} v - 2 \frac{\partial \phi}{\partial z} = 0, \tag{22}$$

$$\left\{ E^{1/2} \frac{\partial^2}{\partial z^2} - \sigma \frac{\partial}{\partial t} \right\} T + 2\sigma S\phi = 0. \tag{23}$$

Equations (21), (22) and the corresponding initial and boundary conditions are nothing but those treated in the incompressible case (Greenspan and Howard, 1963). It is concluded, therefore, that the state of rigid rotation is established within the time duration of the incompressible spin down time without regard to the stratification in the present limiting case. The important point to be stressed here is that this limiting case is not singly isolated from other cases but proceeds continuously as r_0 is made larger and larger. Another important phenomenon which the above similar solution implies is that the thickness of the thermal boundary layer is of the order of $E^{1/4}$, unlike the velocity boundary layer, as is seen by (23). The meridional current driven by the Ekman's boundary layer modifies not only the velocity but also the temperature outside the boundary layer. The matching of this convectively modified temperature with the wall temperature is achieved via the thermal boundary layer of the above thickness where thermal conduction is predominant. The appearance of this thermal boundary layer on the horizontal surface is characteristic of the transient variation within the time duration of the order of the incompressible spin down time and has no analogue in the steady flow problem. It is naturally expected that, for the case with finite r_0, the intermediate layer of the thickness of the order of $E^{1/4}$ coexists together with the inner layer of the thickness of the order of $E^{1/2}$ also along the vertical wall.

Therefore, the flow field in the general case is divided into five regions: In the inviscid region, the Coriolis force, the buoyancy force and the pressure dominate in the flow field while the temperature is governed by the convection. In the horizontal inner layer, that is in the Ekman's boundary layer, the Coriolis force is balanced by the shearing stress in the flow field, and the peripheral velocity is coupled with the meridional velocity. In the horizontal intermediate layer, the non-diffusive temperature in the inviscid region is relaxed into the wall temperature by thermal conduction. In the vertical inner layer, the buoyancy is balanced by the shearing stress of the meridional current while thermal convection is balanced by thermal conduction, and the meridional current is coupled with the temperature. In the vertical intermediate layer, the inviscid peripheral velocity is relaxed into the wall velocity by the viscous diffusion.

The mathematical formulation of the above features was made by Sakurai (1969a, b) who gave a detailed analytical discussion of the flow field. Therefore, we do not get into trouble with the mathematical complexity except noting the following: Since the boundary layers are very thin within our time scale, the inviscid region consists of the main bulk of the flow field, and the mathematical function of the former is to provide us with the boundary conditions to be satisfied by the latter. Our problem reduces, thus, to the solution of the initial boundary value problem with respect to the inviscid flow. The solution is obtained by the expansion with respect to the fundamental solutions which, in turn, are obtained by the method of the separation of variables. The convergence of the asymptotic form of this expansion becomes poor for the temperature in the neighborhood of $r/r_0 = 0$ and 1. This is the reason why we get some extraordinary behavior in Figures 3, 4, 7 and 8 in the same region.

Finally, the meridional circulation dies away in the asymptotic state, and the asymptotic distribution of the angular velocity and the temperature is governed by a single parameter $\sqrt{S}/r_0$. This leads to the equivalence of the case with infinite radius of the cylinder to that with vanishing stratification. Thus, our solution certainly includes the former as a special case.

4. Results and Discussion

The numerical calculation of the asymptotic distribution of the angular velocity and the temperature based on Sakurai's analytical formulation was performed on TOSBAC-3400 electronic computer of the Research Institute of Mathematical Science of Kyoto University, and the results are shown in Figures 1 to 8. These figures clearly show the situations where the effect of the meridional current is quenched, and thus,

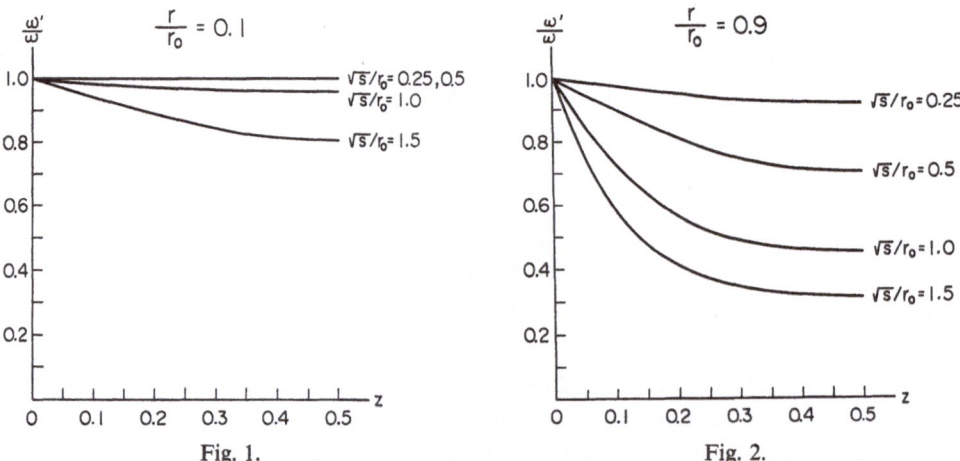

Figs. 1 and 2.   Asymptotic distributions of the perturbation angular velocity as functions of the height with a radius and stratification numbers as parameters. Case A.

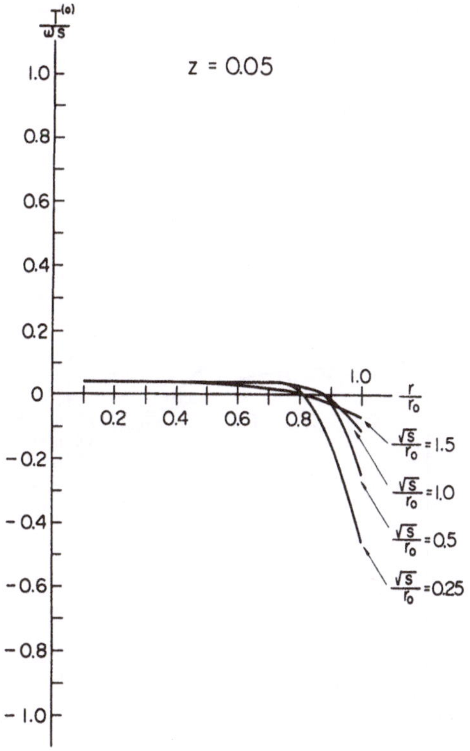

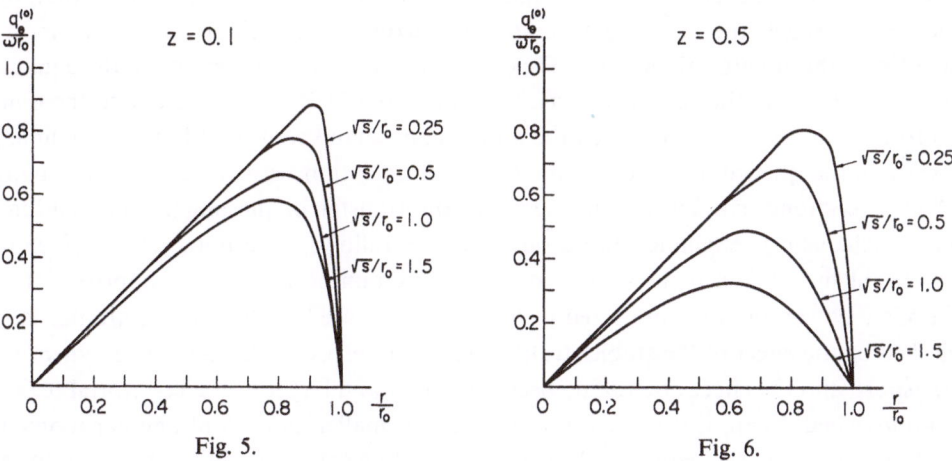

Fig. 3. 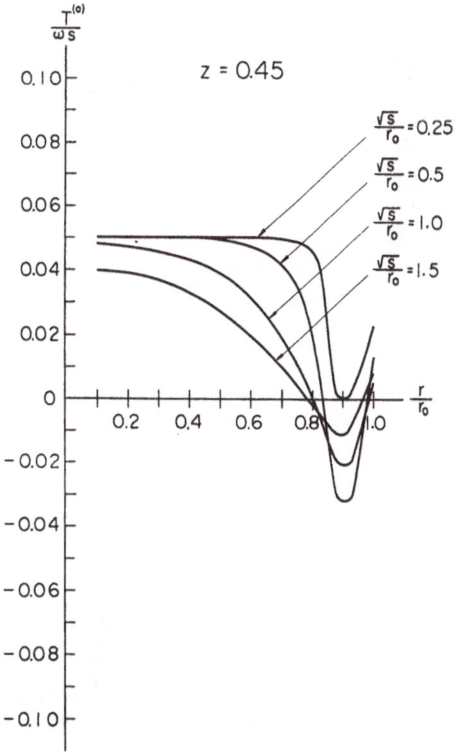   Fig. 4.

Figs. 3 and 4.   Asymptotic distributions of the perturbation temperature as functions of the radius with a height and stratification numbers as parameters. Case A.

Fig. 5.   Fig. 6.

Figs. 5 and 6.   Asymptotic distributions of the perturbation peripheral velocity as functions of the radius with a height and stratification numbers as parameters. Case B.

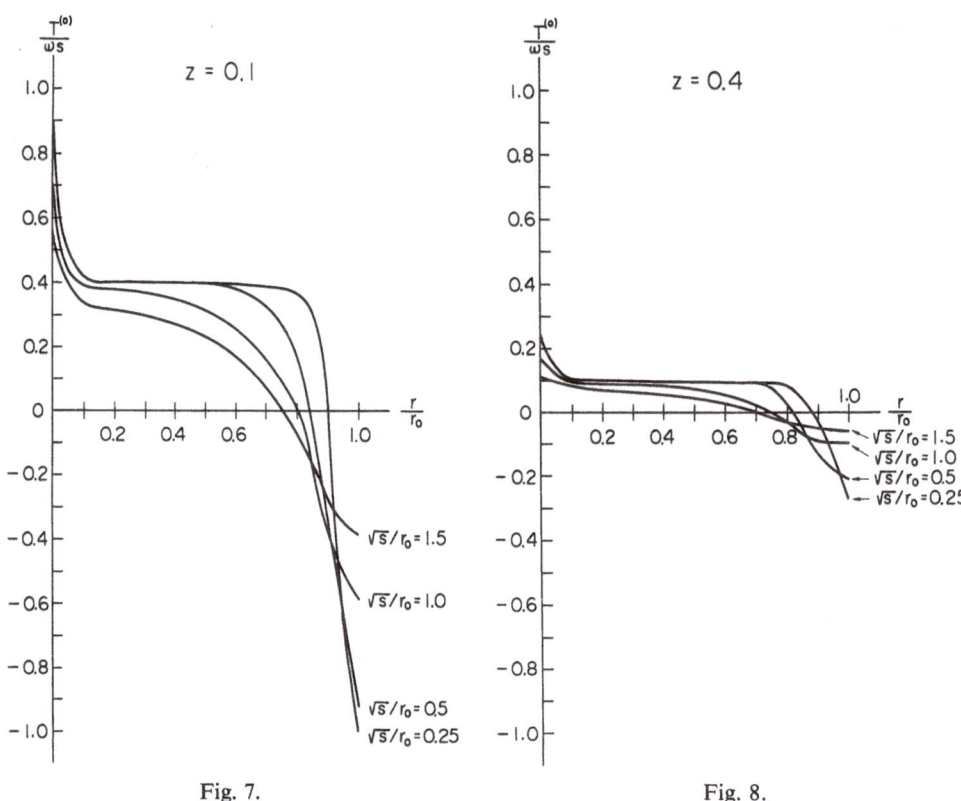

Fig. 7.   Fig. 8.

Figs. 7 and 8.   Asymptotic distributions of the perturbation temperature as functions of the radius
with a height and stratification numbers as parameters. Case B.

the non-uniformity in the flow field is augmented by the stable stratification.
The most interesting feature, however, is that our asymptotic angular velocity
distribution in case A shows the equatorial acceleration to be similar to that on
the solar surface: the absolute value of the perturbation peripheral velocity is the
smallest in the equatorial region on the constant r-plane, corresponding to the equatorial acceleration in the spin down case (Figures 1 and 2). This may be due to the non-
uniformity in the conflict between the pumping mechanism of the Ekman's boundary
layer and the quenching action of the stable stratification with respect to the driving
of the meridional circulation and hence to the transfer of the angular momentum.
The mechanism is explained more intuitively as follows: The horizontal surface of
our problem is taken to be equipped with many pumps which have approximately
the same power and are connected to bad hoses (Figure 9). These hoses are partially
choked by the effect of the stable stratification with respect to the angular momentum
transport, and the effect of choking becomes larger and larger as the equatorial region
is approached. Then, it is evident that we have a smaller amount of angular momentum transport in the equatorial region than elsewhere, to obtain the equatorial
acceleration in the spin down case. It is noted that the qualitative nature of this procedure is quite general without regard to the geometrical configuration and the origin

of the pumping mechanism. In effect, the above mechanism does work in the solar interior where we have a stronger effect of the stratification and a more powerful pumping mechanism by Bretherton and Spiegel (1968). Thus, our result is taken to suggest a possible explanation of the origin of the solar differential rotation.

The details of the procedure are, of course, influenced by (1) the nature of the mechanical interaction between the hydrogen convection layer and the inner stable region, (2) the latitudinal distribution of the magnetohydrodynamical breaking torque

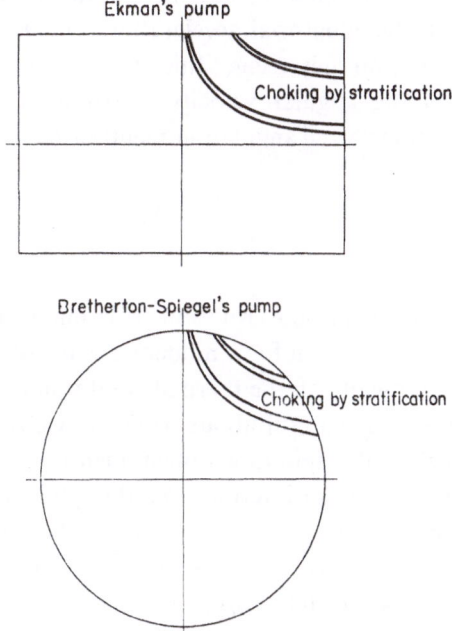

Fig. 9.   Intuitive explanation of the origin of equatorial acceleration.

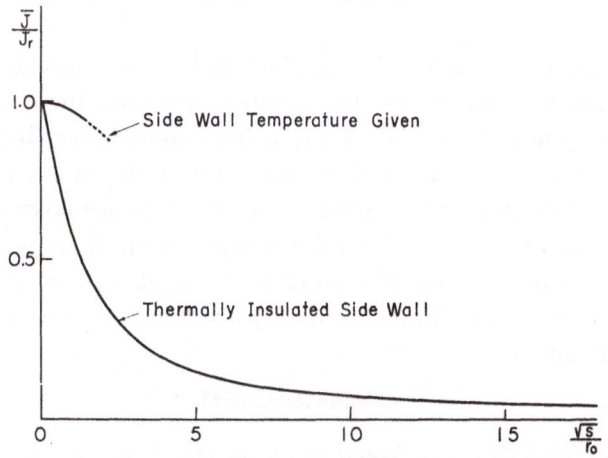

Fig. 10.   Asymptotic amounts of the transferred angular momentum as functions of the stratification number.

on the surface of the sun, (3) the solar magnetic field, and so on. However, the general features may remain the same.

The explanation based on the non-isotropy of the eddy viscosity in the hydrogen convection layer (Sakurai, 1966) has the fatal defect that the estimated order of magnitude of the meridional current on the solar surface is much larger than that of the observed one. The present explanation has no such defect, since the meridional circulation dies away in the asymptotic state. Recently, Kato and Nakagawa (1968, 1969) showed that the Rossby wave is excited in the differentially rotating solar photosphere and that the thus excited Rossby wave induces such an angular momentum transfer as to sustain the equatorial acceleration. Our mechanism of the origin of the solar differential rotation may trigger this effect of the Rossby wave.

The non-uniformity of the angular velocity distribution is also shown by the asymptotic amount of the transferred angular momentum J/J_r, as is shown in Figure 10, where

$$J = 2\pi \int_0^L \int_0^{Lr_0} \bar{r}^2 \bar{\varrho} \bar{q}_\theta \, \mathrm{d}\bar{r} \, \mathrm{d}\bar{z},$$ (24)

and the suffix r refers to the rigid body rotation. It is interesting that the amount is quite remarkable in the case (A) even for a moderate value of S. It is also interesting that the amount depends critically on the thermal condition on the vertical wall. This is a very important point to be careful about from the viewpoint of constructing a more concrete model of the solar spin down phenomenon.

Let us apply our result to the solar interior, even though it is not permitted because of the limitations on the Boussinesq approximation. The term $\alpha(\bar{T}_1 - \bar{T}_0)$ in (6) is taken to be of the order of 100 since it represents the fractional order of the density variation. Then, the main parameter $\sqrt{S}/r_0$ is estimated to be of the order of 100 corresponding to the rate of 1 rotation per day. The asymptotic amount of the transferred angular momentum is very small as is seen in Figure 10, and we cannot expect the effective spin down of the solar interior, which result seems to ascertain Dicke's hypothesis.

However, we are more interested in the fact that our results clearly show the occurrence of the spin down within our time duration. We want, therefore, to be prudent not to give any judgement about the solar spin down controversy. In effect, Bretherton and Spiegel (1968) proposed that a slow circulation in the convection zone induced by the gradual reduction in the rotation of the solar surface layer pumps fluid into the solar interior and gives rise to spin down currents which are much more intense than those resulting from a solar Ekman layer. Thus, all we can say is that the only way to resolve the solar spin down controversy is to investigate the unsteady rotation of the model solar interior.

Acknowledgement

The author wishes to express his cordial thanks to Mr. S. Yamamoto for his assistance in performing numerical calculations.

References

Barcilon, V. and Pedlosky, J.: 1967, *J. Fluid Mech.* **29**, 1.
Brandt, J. C.: 1966, *Astrophys. J.* **144**, 1221.
Bretherton, F. P. and Spiegel, E. A.: 1968, *Astrophys. J.* **153**, L77.
Dicke, R. H.: 1964, *Nature* **202**, 432.
Dicke, R. H.: 1967, *Astrophys. J.* **149**, L121.
Fricke, K. and Kippenhahn, R.: *Publ. Univ. Sternwarte Göttingen.*
Greenspan, H. P. and Howard, L. N.: 1963, *J. Fluid Mech.* **17**, 385.
Holton, J. R.: 1965, *J. Atmos. Sci.* **22**, 402.
Holton, J. R. and Stone, P. H.: 1968, *J. Fluid Mech.* **33**, 127.
Howard, L. N., Moore, D. W., and Spiegel, E. A.: 1967, *Nature* **214**, 1297.
Hoyle, F.: 1960, *Quart. J. Roy. Astron. Soc.* **1**, 28.
Kato, S. and Nakagawa, Y.: 1968, Res. Note of H.A.O.
Kato, S. and Nakagawa, Y.: 1969, Res. Note of H.A.O.
Lüst, V. R. and Schlüter, A.: 1955, *Z. Astrophys.* **38**, 190.
Pedlosky, J.: 1967, *J. Fluid Mech.* **28**, 463.
Sakurai, T.: 1966, *Publ. Astron. Soc. Japan* **8**, 174.
Sakurai, T.: 1969a, *J. Phys. Soc. Japan* **26**, 840.
Sakurai, T.: 1969b, *J. Fluid Mech.* **37**, 689.

Discussion

Roxburgh: This summer in Boulder I heard two colloquia, one by Clark and the other by someone else – both of whom claimed that spin down would not work in a strongly stratified fluid. Ekman pumping worked in a thin layer where the stratification was weak but did not penetrate the stable layers.

Sakurai: I have not read the papers you referenced. Therefore, I can say nothing about them. I ask you, however, to give due care to the possible existence of the solar spin down current driven by the Bretherton-Spiegel type pumping mechanism.

Dicke: If I understand Dr. Sakurai's calculation, it is based on the assumption that the angular velocity of the container is changed discontinuously. In this connection, the results of an experiment performed by my student E. McDonald are significant. In his investigations of rotation of a density stratified fluid it was found that the density stratification stabilized the fluid against the formation of circulation currents and Ekman layers, but *only* if the rotation of the container was changed slowly and continuously. The physical reason for this is easily found. The state of rotation of the fluid determines the distribution of density, for rotation without circulation. Thus a sudden change in the rotation of the boundary will result in the wrong density distribution in the vicinity of the boundary. Thus purely rotational motion is impossible. McDonald found that a very small change in the angular velocity of the fluid container *if suddenly applied* would induce spin down of the fluid. Circulation currents would divide the fluid into two zones of uniform but different density. An Ekman layer would form at the container bottom and at the boundary between the two fluids. He also found that large differential rotations could be induced without spin down if the rotation of the fluid container was changed slowly.

THERMAL STABILITY OF NONSPHERICAL STARS
AND THE SOLAR OBLATENESS PROBLEM

KLAUS J. FRICKE

Universitätssternwarte Göttingen, Germany

Abstract. A brief account of the stability of distorted stars is given. From this appropriate criteria emerge, which are subsequently applied to rotational models of the sun in connection with the solar oblateness problem.

1. Stability of Distorted Stars

If we are not interested in perturbation time scales longer than the thermal diffusion time scale, we may neglect the weaker dissipative effects as viscous or material diffusion in the linear stability analysis. Then we get a polynomial of the third degree for a dimensionless growth rate:

$$\sigma^3 + \frac{\tau_h}{\tau_r}\sigma^2 + A_1\sigma + A_0\frac{\tau_h}{\tau_r} = 0. \tag{1}$$

Here $\tau_h = |r/g_{eff}|^{1/2}$ is the local hydrodynamic time scale with $\mathbf{g}_{eff}$ the acceleration of gravity modified by the distorting force and r the distance from the centre; τ_r is the local thermal relaxation time scale, which is of the order of $\tau_{KH} L^2/R^2$, where τ_{KH} is the Kelvin-Helmholtz time scale of the entire star of radius R, and L is the size of the disturbance. $1/\tau_h$ is used as scaling factor of the growth rate. The coefficients A_1 and A_0 are of the order of 1 and λ respectively, λ being the local fraction of the distortion of the star. The third degree in σ of Equation (1) is obtained from the second order time derivatives in the momentum equation and from the entropy change rate in the energy balance for nonadiabatic processes. From Equation (1), the Hurwitz criteria provide the following three conditions

$$A_1 \geqslant 0, \tag{2a}$$

$$A_1 \geqslant A_0, \tag{2b}$$

$$A_0 \geqslant 0. \tag{2c}$$

Inequalities (2) must be satisfied simultaneously in order to insure stability. They are the necessary conditions for dynamical, vibrational and secular stability respectively. This may be verified intuitively by interchanging material elements along an arbitrary path l in the meridional plane (axially symmetric perturbations) in an adiabatic, quasiadiabatic or diabatic manner. If the translation energy of the elements is increased by one of these processes, the system is unstable with respect to one of these three modes of instability.

In vibrationally stable layers, Equation (2b) prevents overstable convection. Equation (2) then tells us, that in vibrationally stable layers the secular stability condition is always stronger than the dynamic one. The explanation is simply that the stabilizing effect of buoyancy forces caused by temperature differences is wiped out if the motions are slow enough to permit an appreciable amount of heat exchange. Each of the conditions (2) has in general the form

$$a_i q^2 + b_i q + c_i \geqslant 0, \quad i = 1, 2, 3,$$ (3)

where $q = l_\theta / l_r$ is the ratio of the tangential and the radial component of the displacement vector of the two elements mentioned above. Equation (3) then splits into two conditions

$$c_i \geqslant 0 \quad \text{and} \quad 4a_i c_i - b_i^2 \geqslant 0, \quad i = 1, 2, 3,$$ (4)

because q may vary arbitrarily. The first set of criteria corresponds to radial perturbations ($q = 0$) and the second one to non-radial perturbations ($q \neq 0$).

We now consider the special case of a rotating star having a spherical distribution of angular velocity $\Omega(r)$ and mean molecular weight $\mu(r)$. This case is relevant for a subsequent application to the sun. The rotation rate shall be slow, i.e. $\lambda = \Omega^2 r / |g_{\text{eff}}| \ll 1$. We obtain the following scheme for the six stability criteria valid in the equatorial plane:

mode	vibrational	dynamical	secular
radial	$\nabla_{\text{ad}} - \nabla_T \geqslant 0$	$\nabla_{\text{ad}} - \nabla_T$ $+ \nabla_\mu - 2\lambda \nabla_j \geqslant 0$	$\nabla_\mu - 2\lambda \nabla_j \geqslant 0$
non-radial	$-\lambda^2 K^2 \geqslant 0$	$\nabla_{\text{ad}} - \nabla_T$ $+ \nabla_\mu - 2\lambda \nabla_j - \lambda^2 C^2 \geqslant 0$	$\nabla_\mu - 2\lambda \nabla_j - \lambda^2 C^2 \geqslant 0$

(5)

where $j = \Omega r^2$, $\nabla_x = \text{d} \ln x / \text{d} \ln p$, $C = \frac{1}{2} (r/h)^{1/2} \nabla_J$, $J = \Omega r^4$, $\lambda K = |g_{\text{eff}} \times \textbf{grad } T|$; p, T, h, ∇_{ad} denote pressure, temperature, pressure scale height and adiabatic temperature gradient respectively.

According to the scheme (5) the transition from dynamical to secular modes shows an increase in strength of the respective stability criteria if $\nabla_T \leqslant \nabla_{\text{ad}}$. The same statement holds for the transition from radial to non-radial modes. This is not surprising as the non-radial modes represent more general disturbances than the radial ones. Such relations concerning the strength of stability criteria hold also if effects of magnetic fields are included. Incidently, for non-radial vibrational modes of rotationally distorted stars a general local instability is obtained. This result will not be discussed here. The important conclusion from the above synopsis of the stability conditions is that the 'non-radial secular' criterion provides the strongest restraint for a stellar rotation law. It can be written as

$$\tfrac{1}{4} r^2 \Omega'^2 \leqslant g_{\text{eff}} (\ln \mu)',$$ (6)

the prime denotes $\text{d}/\text{d}r$.

For 'normal' composition gradients $\mu' < 0$, the r.h.s. of (6) is positive. If in addition $\Omega' < 0$, the steepest stable Ω-gradient at a certain radius is given by

$$\Omega' = -\frac{2}{r} |g_{\mathrm{eff}} (\ln \mu)'|^{1/2}. \tag{7}$$

In a chemically homogeneous layer with no magnetic field, only uniform rotation can be stable.

2. Discussion of Dicke's Model

According to Dicke and Goldenberg (1964) (hereafter referred to as D.G.) the sun shows an oblateness of $\Delta r / r = (5 \pm 0.7) \times 10^{-5}$, a value which is larger by a factor of 5 than what one should expect if the sun were to rotate rigidly with the observed surface angular velocity $\Omega_s = 2.8 \times 10^{-6}$ rad/sec. In the present Colloquium Professor Dicke has provided evidence against a possible distortion of the solar atmosphere by stresses of velocity or magnetic surface fields and he has argued in favour of the D.G. model of the angular velocity distribution in the sun (this volume, p. 289). In the D.G. model the radiative core of the sun rotates at the high primordial rate of 4×10^{-5} rad/sec, while the outer convection zone and the atmosphere have been spun down to the present surface rotation rate by the solar wind torque. Thus, below the bottom of the convection zone at 0.8 solar radius a thin transition layer of 0.05 solar radius thickness is postulated in which Ω decreases by a factor larger than 10. This model encountered two main objections:

(1) The transition zone cannot be in hydrostatic equilibrium (Howard et al., 1967),

(2) the transition zone must be thermally unstable (Goldreich and Schubert, 1967; Fricke, 1969a).

I shall not consider the onset of an Ekman flow, which transports angular momentum through the transition zone, to be imperative. As Fricke and Kippenhahn (1967) have pointed out, the stellar gas can adjust its temperature distribution to a quasi-hydrostatic state as long as the Eddington-Vogt circulations are small compared to the velocity of sound; the latter still holds in the D.G. model. I consider the instability objection to be the important one. Goldreich and Schubert used only the 'radial secular' condition of the restraints (5) to show the instability of the Ω-transition. Obviously, it is also unstable against secular non-radial motions as Equation (6) (with $\mu = $ const.) shows. Dicke (1967) has proposed the following stabilizing effects: (i) magnetic fields, (ii) velocity fields.

I have investigated secular stability in the presence of toroidal or poloidal magnetic fields (Fricke, 1969a). I found toroidal fields and differential rotation to be separately unstable with respect to non-radial secular perturbations. On the other hand, a mutual stabilization is possible in principle but requires in the case of the D.G. model an unlikely physical situation. The weaker condition for radial motions already demands a field strength of the order of 10^5 Gauss with a field gradient of the order of 1 Gauss/km pointing outward. In addition, Schubert and Fricke (to be published) have shown that rotating stars with toroidal fields are secularly unstable with respect to non-axisymmetric modes. Poloidal fields cannot suppress the instability of differential rotation against non-radial disturbances, if the rotation is steady. Bearing these results in mind the situation does not appear promising for stability even if a field of a more sophisticated structure is present, since the instabilities are of a local nature.

Dicke argues furthermore that the shear produced by meridional motions may prevent the instability motions from growing, in a similar manner as ocean currents prevent the growth of salt fingers. In stars the Eddington-Vogt circulations may take over this role, although we know very little about this mechanism. These currents have velocities of the order of 10^{-4} λ cm/sec in the solar core and are probably much less than the turbulent motions caused by the instability. Within the D.G. transition zone the currents are faster by a factor of 10^4 because of the Baker-Kippenhahn correction and a correction due to the steep Ω-gradient. There the velocities become of the order of λ cm/sec, where $\lambda \approx 10^{-3}$. However, the latter estimate for the velocity of the currents is not relevant for Dicke's argument. The instability according to Equation (6) operated from the beginning of the spin down process, when no Ω-transition has been present. Thus, I presume that such a transition layer could not have been formed.

3. Stable Models

The question arises whether stable rotation laws can be constructed, which can also account for the solar oblateness. On the basis of the condition (6) for non-radial secular stability I have shown that this is not possible (Fricke, 1969b). The use of the criterion for radial motions led Goldreich and Schubert (1968) to the opposite result.

The maximum possible angular momentum content which can be stably distributed in the sun can be deduced by use of Equation (7). In the absence of a μ-gradient in the sun's interior the solar oblateness would be produced by rigid body rotation with the present value of the surface velocity of 2.8×10^{-6} rad/sec. This gives $\Delta r / r = 1.0 \times 10^{-5}$. If we assume a stabilizing μ-gradient obtained from evolution theory (Schwarzschild, 1958) for the present age of the sun, 5×10^9 yr, and use the present surface velocity, Equation (7) can be integrated to give a unique Ω-distribution, which provides a first upper bound to the solar oblateness from stability arguments. The value thus obtained is 8.5×10^{-5} and is only slightly higher than the observed value of D.G. Actually, the μ-gradient has been evolved from zero to the present value by nuclear burning in the core. In order to take into account this evolution effect, two earlier Ω-distributions for the ages 3×10^7 yr and 4×10^8 yr have been integrated using the corresponding μ-distributions and surface velocities. The latter have been taken from Kraft (1967). The minimum curve resulting from the three calculated Ω-distributions is given in Figure 1. This curve represents an Ω-distribution which could have evolved from an initially constant rotation (taken as the Keplerian angular velocity at the solar equator) by three simultaneous effects: (i) evolution by nuclear burning, (ii) spin-down by the solar wind torque and (iii) angular momentum transport as an effect of the instability. This distribution produces an oblateness of 1.4×10^{-5} only, which is about a factor 4 smaller than the measured value. If the latter is correct, we must conclude that the large measured oblateness indicates an unstable Ω-distribution to be present in the solar core.

This last possibility cannot be excluded completely as long as accurate time scales for the secular instabilities are unknown. The time scale for an unstable Ω-distribution to be maintained must not be smaller than the braking time scale, which is presumably of the order of 10^8 yr. Up to now, no reliable theory is known which could support such a high value for the time scale of the instability.

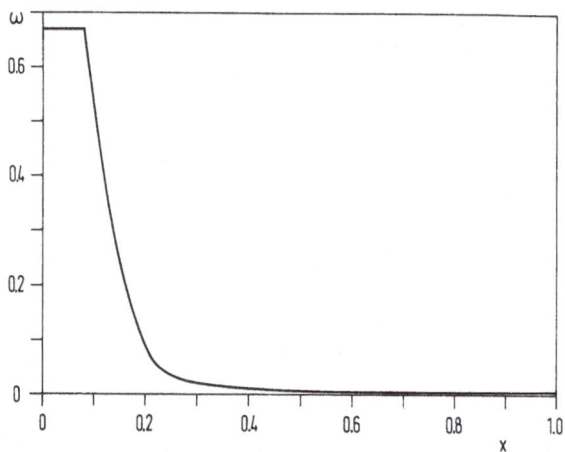

Fig. 1.   The limiting angular velocity distribution of the present sun obtained from Equation (7) with regard to the history of the μ-gradient and the surface velocity. $\omega^2 = \Omega^2 R^3/GM$, where M and R are the solar mass and radius and G is the constant of gravitation; x is the relative radius r/R.

References

Dicke, R. H. and Goldenberg, H. M.: 1964, *Phys. Rev. Letters* **18**, 313.
Dicke, R. H.: 1967, *Science* **157**, 960.
Fricke, K.: 1969a, *Astron. Astrophys.* **1**, 388.
Fricke, K.: 1969b, *Astrophys. Letters* **3**, 219.
Fricke, K. and Kippenhahn, R.: 1967, *unpublished.*
Goldreich, P. and Schubert, G.: 1967, *Astrophys. J.* **150**, 571.
Goldreich, P. and Schubert, G.: 1968, *Astrophys. J.* **154**, 1005.
Howard, L. N., Moore, D. W., and Spiegel, E. A.: 1967, *Nature* **214**, 1297.
Kraft, R. P.: 1967, *Astrophys. J.* **150**, 551.
Schwarzschild, M.: 1958, *Structure and Evolution of the Stars*, Princeton.

Discussion

Roxburgh: Could you explain why your angular velocity distribution was constant in the central regions of the sun?

Fricke: I simply assumed that the sun reaches the main sequence with the Keplerian velocity and is initially in a state of rigid body rotation. During the subsequent spin down process the angular velocity is lowered in the outer parts, whereas the central distribution remains unaffected for a long time. Thus, I truncated the increasing distribution inwards at the Keplerian velocity.

Ruben: Dr. F. Krause, of the Central Institute for Astrophysics in Potsdam, asked me to make the following comment:

Using a theory of mean fields in a turbulent conducting medium one can prove the existence of an alternating field dynamo. To get such an alternating field dynamo, it is necessary that the gradient of the angular velocity at the bottom of the hydrogen convection zone be great. At the moment, it is difficult to say exactly how much the angular velocity should change. Probably for the sun the change will be of the order of the difference $\omega_{eq} - \omega_{pole}$. But it cannot be smaller than this. It is to be hoped, that in the future the analysis of the butterfly diagram and related phenomena will give some limits on the structure of the outer solar layers. The theoretical butterfly diagrams have been published in *Astronomische Nachrichten* **291** (1969), 49.

Dicke: I should like to make two comments:

(1) In connection with Dr. Fricke's paper I should like to return to my comment on Monday. A linear stability analysis cannot be carried out without first assuming a model for the stellar interior. Thus one is faced with the disagreeable task of evaluating the stability of all possible models before a general proof of instability is obtained.

(2) The Goldreich-Schubert-Fricke instability leads to the transport of angular momentum, along with the material containing the angular momentum, to the surface of the sun. Kraft's observations show that young G-type stars on the main sequence possess a large amount of angular momentum. Assuming that this angular momentum is transported to the surface implies a strong depletion of beryllium at the surface. Most of the angular momentum of a uniformly rotating solar-type star lies below $r = 0.5$, the radius at which beryllium rapidly burns. Apparently depletion of beryllium is not observed implying that the original angular momentum has not been transported to the surface from below $r = 0.5$ by this instability.

Fricke: First, I should like to mention that the local stability analysis does not require any specification of the equilibrium model. Conversely, the linear stability criteria prescribe the local properties of thermally stable models. Using these constraints I found (i) none of the proposed models for an oblate sun is compatible with the stability requirements, and (ii) stable models which yield a sufficiently high oblateness cannot be of a simple structure. Thus, the existence of a solar quadrupole moment of the required amount is not presently understood.

Concerning the second point, my opinion is that the depletion of lithium and the sensible deficiency of beryllium in the solar atmosphere (cf. a paper by N. Grevesse in *Solar Physics*, 1968) are favourable to the idea of mixing between core and envelope, although I have not considered the problem quantitatively.

DIFFERENTIAL ROTATION IN THE SOLAR INTERIOR

MAURICE J. CLEMENT

David Dunlap Observatory, University of Toronto, Canada

1. Introduction

One of the big problems in stellar rotation which has been the object of much debate recently concerns the magnitude of the angular velocity in the central regions of the sun. It is a good example of our general ignorance of the distribution of angular momentum in the interiors of stars. There is good reason, of course, for this ignorance. One can't make any direct observations and from a theoretical point of view there are many real problems such as the lack of a good theory of convection and meridian circulation, and our ignorance of the structure and magnitude of magnetic fields in the deep stellar interior. These problems among others make it very difficult, for example, to specify a surface condition on the angular velocity. It was pointed out recently (Clement, 1969; this paper is referred to hereinafter as Paper I) that such a condition might enable us to estimate the magnitude of the interior stellar rotation.

In this paper, an equilibrium model for the distribution of angular velocity in the sun is presented. The model has a surface distribution of velocity which is the same as that of the sun (by assumption) and also differs significantly in its interior properties from the model of Roxburgh (1964) for which the angular velocity is constant on spheres. In view of Roxburgh's review paper in this volume, one might ask: Why look for equilibrium distributions of velocity when we know they are likely to be unstable? It is possible that an equilibrium distribution is related in some way with the actual, time-dependent distribution. Also, there still appears to be some doubt as to the stability of distributions which differ from those with cylindrical symmetry by only a small amount. The equilibrium distributions for upper-main-sequence stars which are presented in Paper I do have approximate cylindrical symmetry with no meridian motions and so their stability or instability is, in fact, open to some question.

2. Formulation

To illustrate how this problem is solved it is necessary to show some of the basic equations. The mathematical formulation and notation is essentially that of Schwarzschild (1947) and Roxburgh (1964). Slow rotation is assumed and the dimensionless physical quantities are written as the sum of a spherically symmetric quantity with subscript u and an axisymmetric quantity with subscript d and coefficient λ which is proportional to the square of the angular velocity at the pole. Thus,

$$x = \frac{r}{R_u}, \quad \lambda = \frac{\Omega_p^2 R_u^3}{GM}, \tag{1}$$

A. Slettebak (ed.), Stellar Rotation, 346–351. All Rights Reserved
Copyright © 1970 by D. Reidel Publishing Company, Dordrecht-Holland

$$p = \frac{4\pi R_u^4}{GM^2} P = p_u + \lambda \frac{p_u}{t_u} p_d, \tag{2}$$

$$t = \frac{k}{\mu H} \frac{R_u}{GM} T = t_u + \lambda t_d, \tag{3}$$

$$\phi = \frac{R_u}{GM} \Phi = \phi_u + \lambda \phi_d, \tag{4}$$

$$l = \frac{4\pi r^2}{L_u} F_r = l_u + \lambda \frac{x t_u^b}{C p_u^e} l_d, \tag{5}$$

where P, T, M, etc. have their usual meaning. The parameters b and e are associated with the assumed opacity law:

$$\kappa = \kappa_0 \varrho^{e-1} T^{-s} \quad \text{with} \quad b = s + e + 3. \tag{6}$$

The axisymmetric quantities are expanded in a series of Legendre polynomials in the standard way; e.g.

$$p_d = \sum_j p_j(x) P_{2j}(\cos\theta). \tag{7}$$

Substitution of the foregoing expressions into the standard equations of stellar structure and the elimination of the angular velocity from the equation of hydrostatic equilibrium (cf. Paper I) yields the following equations which are to be applied to the radiative core:

$$x \frac{dp_j}{dx} = Q_j + q_j - (2_j + 1)(p_j - \phi_j) + U_u p_j - V_u t_j, \tag{8}$$

$$x \frac{dt_j}{dx} = U_u(e p_j - b t_j) - l_j, \tag{9}$$

$$x \frac{d\phi_j}{dx} = q_j, \tag{10}$$

$$x \frac{dq_j}{dx} = - q_j + 2j(2j + 1) \phi_j + W_u(t_j - p_j), \tag{11}$$

$$x \frac{dl_j}{dx} = - X_u l_j - 2j(2j + 1) t_j - Y_u[2p_j + (v - 2) t_j], \tag{12}$$

where

$$Q_j = \frac{(2j - 1)(4j + 1)}{2j(4j - 3)} [(4j - 3)(p_{j-1} - \phi_{j-1}) - Q_{j-1}], \tag{13}$$

U_u, V_u, W_u, and X_u are as given in Paper I, and

$$Y_u = \left(\frac{d \ln l_u}{d \ln x}\right)\left(\frac{d \ln t_u}{d \ln x}\right) = \left(\frac{d \ln l_u}{d \ln x}\right) U_u. \tag{14}$$

This last quantity makes allowance for the nuclear energy generation in the radiative core. In Equation (12), v is the temperature exponent of the energy generation law.

The foregoing equations point out the source of difficulty inherent in this problem of differential rotation. Equations (8)–(12) are homogeneous except for the quantity $Q_j(x)$ which clearly couples equations of different Legendre orders. As the problem now stands, Q_0 (say) is undefined. Given a function $Q_0(x)$, all the equations (including the relatively simple ones for the convective envelope) could be integrated in a straightforward way. And if there were no conditions to satisfy other than the standard boundary conditions, then Q_0 would be quite arbitrary. Thus, an unlimited number of velocity distributions could be generated by choosing different Q_0. For example, Q_0 can be chosen in quite a simple way such that the angular velocity is constant on spheres. This is essentially how Roxburgh obtained his model. But solutions found in this way are mathematical ones and have little or no physical basis. One should choose a Q_0 which gives a velocity distribution satisfying some justifiable boundary condition. This has been done for the upper-main-sequence stars (cf. Paper I). The particular condition applied to these stars will not be discussed here because it has no relevance for lower-main-sequence stars which have convective envelopes. However, in the case of the sun, there is an obvious boundary condition which is just the observed distribution of angular velocity on the surface; viz. (cf. Allen, 1963)

$$\Omega/\Omega_p = 1.0000 + 0.2312 \sin^2 \theta. \tag{15}$$

What has been done is to determine that unique $Q_0(x)$ which gives this surface distribution.

3. Assumptions

A number of assumptions have been made; some of these are obvious from the foregoing equations.

(i) The only effective forces are those of pressure, gravitation, and rotation: This means that there are no acceleration forces due to a solar-wind torque or to meridian circulation; i.e., there is steady state and no motions other than pure rotation. This assumption also means that there are no viscous or magnetic stresses. It is a purely hydrodynamic model and if it turns out that the magnetic field in the deep interior of the sun is very strong then the model presented here may have to be modified.

(ii) Uniform chemical composition: A gradient in the mean molecular weight may affect the equilibrium distribution of angular velocity; it has already been shown to be important for stability. This problem will be examined at another time.

(iii) Adiabatic convective envelope: The effect of this assumption is to make the angular velocity in the envelope constant on cylinders and so the velocity distribution at the bottom of the convective region is directly related in an obvious way to the observed distribution at the surface. We know that the temperature gradient near the surface is actually superadiabatic and what effect this will have on the velocity distribution at the bottom of the envelope is not clear. However, numerical experiments show that the angular velocity at the center of the sun depends only weakly on the actual distribution at the bottom of the convective envelope. So unless viscous stresses, meridian circulation, and/or a superadiabatic temperature gradient drastically affect the rotation law in the envelope, the assumption that these effects are negligible will not affect the model's central rotation rate in a significant way.

4. Results

As already indicated, the object of the present problem is to find that particular $Q_0(x)$ which gives the observed solar velocity distribution. This is essentially a matter of curve-fitting and to this end Q_0 is written as a polynomial in x^2 (there can be no linear term; otherwise, the pressure gradient at the origin would differ from zero as can be easily verified). The polynomial coefficients are determined by matching the various physical quantities in the radiative core with those in the convective envelope. The number of terms $J+1$ (say) in the expansion for Q_0 must equal the number of Legendre orders to be included in the analysis. So far, terms up to and including $P_{10}(\cos\theta)$ have been considered and this means terms up to x^{10} for $Q_0(x)$ (i.e., up to $J=5$). By considering solutions with a smaller number of terms, one can, of course, see how the model converges. In fact, $Q_0(x)$ converges quite well because the curves corresponding to $J=3$, 4, and 5 are almost indistinguishable on the scale shown in Figure 1 (for clarity, only one curve has been drawn). That is, the model described by terms up to only order $P_6(\cos\theta)$ is almost as accurate as one including terms up to $P_{10}(\cos\theta)$ as far as Q_0 is concerned. Unfortunately, the central angular velocity depends not on

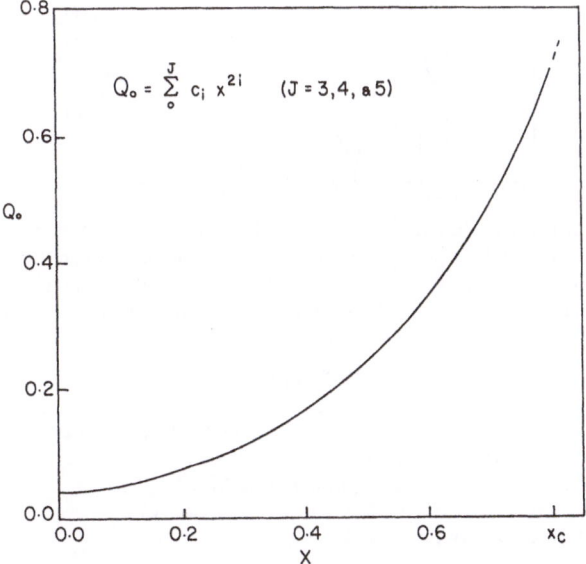

Fig. 1. The function $Q_0(x)$ for a solar model in which the radiative core extends out to $x_c = 0.80$. The same quantity can be found in the convective envelope with the aid of Equation (8) but there it is determined mainly by the assumed velocity distribution on the surface.

Q_0 but on the *second derivative* of Q_0 at the origin. As with all curve-fitting problems, a polynomial with a finite number of terms can't represent the higher derivatives of a function as accurately as the function itself. Consequently, the central angular velocity is not as well determined as the convergence of Q_0 would indicate.

Figure 2 illustrates the velocity distribution corresponding to the $Q_0(x)$ for $J = 5$. The central angular velocity is 40% higher than that at the pole and less than 15% higher than the equatorial value. This relatively low angular velocity does not change significantly if the envelope is made to rotate rigidly or if the opacity and nuclear-energy-generation laws are changed. It is evident from the figure that the distribution of velocity has no cylindrical symmetry in the radiative core and is therefore likely to be unstable in view of the analysis of Goldreich and Schubert (1967). If there really are no stable equilibrium velocity distributions then one might wonder how the time-average of the actual velocity compares with the equilibrium distribution. The velocity at different points may oscillate about the equilibrium values in which case the latter would be representative of the actual distribution. The resolution of this problem must await future investigations.

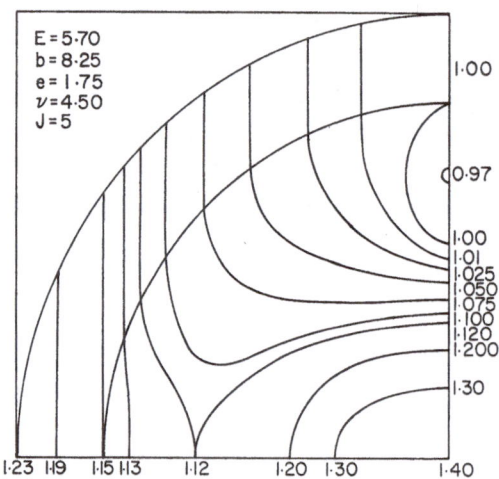

Fig. 2.   The curves of constant angular velocity (Ω/Ω_p) in a quadrant of a meridian plane. The equator is at the bottom left and the pole at the top right. The region between the two quarter-circles is the convective envelope which is governed by the Schwarzschild parameter $E = 5.70$. The other parameters characterizing this model are shown at the upper left of the figure (cf. text for definitions).

References

Allen, C. W.: 1963, *Astrophysical Quantities*, 2nd ed., University of London, Athlone Press, p. 179.
Clement, M. J.: 1969, *Astrophys. J.* **156**, 1051.
Goldreich, P. and Schubert, G.: 1967, *Astrophys. J.* **150**, 571.
Roxburgh, I. W.: 1964, *Monthly Notices Roy. Astron. Soc.* **128**, 157.
Schwarzschild, M.: 1947, *Astrophys. J.* **106**, 427.

Discussion

Livingston: The models indicate a definite change of angular velocity (differential rotation) right to the solar equator. I will just comment that, observationally, there is no differential motion seen through the sunspot zone. At least this seems to be the case spectroscopically, and at this part of the solar cycle.

Roxburgh: I would be surprised if you got approximately cylindrical rotation. Since we know there are no equilibrium solutions in which this is the case and one would expect that the departures from cylindrical symmetry are as big as the rotation itself.

Clement: This did not turn out to be the case for the upper main-sequence models for which the distribution of angular velocity *is* approximately constant on cylinders. In fact, in the limit of slow rotation, the departures from cylindrical symmetry are *independent* of the magnitude of the angular velocity and depend only on the mass, chemical composition, and boundary conditions.

SOLAR OBLATENESS AND DIFFERENTIALLY
ROTATING POLYTROPES

G. G. FAHLMAN, M. D. T. NAYLOR and S. P. S. ANAND

David Dunlap Observatory, University of Toronto, Canada

Dicke and Goldenberg (1967a) measured the solar oblateness to be $\sigma = (5 \pm 0.7) \times 10^{-5}$ and subsequently interpreted this measurement as evidence that the solar interior rotates with a period of $1\overset{d}{.}8$. With this interpretation, they then showed that the observed oblateness causes an 8% discrepancy in the Einstein prediction of the perihelion advance of Mercury. The stability analysis of Goldreich and Schubert (1967) seems to preclude such a fast rotation of the solar interior although magnetic field effects could alter their conclusions (Dicke, 1967). More recently Goldreich and Schubert (1968) and Fricke (1969) have calculated upper bounds to the solar oblateness essentially by finding the steepest distribution of angular velocity that is consistent with secular stability at each point in the equatorial plane of the sun. Fricke's result of $\sigma_{\max} = 1.4 \times 10^{-5}$ is based on a stronger stability criterion than that of Goldreich and Schubert who found $\sigma_{\max} = 1.4 \times 10^{-4}$; Fricke, however, suggests that this may be in error and should actually be $\sigma_{\max} = 3.4 \times 10^{-5}$. In their calculations of $\sigma_{\max}$ the above authors assumed that the outer convective layers of the sun are rotating uniformly and that the angular velocity in the interior is a function of the radial distance from the center of the sun only. We note that while these assumptions are reasonable, neither of them is supported by the observed solar rotation.

Detailed calculations based on uniformly rotating polytropic models led Anand and Fahlman (1968) to suggest that a modest differential rotation extending throughout the entire sun may explain the observed oblateness without implying a significant change in the calculated perihelion advance of Mercury. We report here the results of some calculations based on applying the differential rotation law of Stoeckly (1965) to solar polytropes.

The Stoeckly differential rotation formula is

$$\omega = \omega_0 \exp\left(-cs^2/R^2\right), \tag{1}$$

where ω_0 is the central angular velocity, s is the cylindrical radial coordinate measured from the axis of rotation, R is the solar equatorial radius and the constant parameter c determines the amount of differential rotation. If the Rayleigh stability criterion, $\partial/\partial s\,(\omega s^2) \geqslant 0$, is to be everywhere satisfied then we must have $c \leqslant 1$. Rayleigh's criterion can be stabilized by a suitable gradient of mean molecular weight (Goldreich and Schubert, 1967) so that we may expect $c > 1$ in the evolved solar interior.

In our calculations we have assumed that the Stoeckly law can be applied throughout the entire sun, including the convection zone. This assumption will limit the usefulness of our results because the Stoeckly law does not give the observed latitude variation in angular velocity on the solar surface. The equator of the sun is observed to be

A. Slettebak (ed.), Stellar Rotation, 352–355. All Rights Reserved
Copyright © 1970 by D. Reidel Publishing Company, Dordrecht-Holland

rotating faster than the layers at higher latitudes whereas the Stoeckly law implies that the equator is rotating slower than the pole by a factor of e^{-c}. We may however assume differential rotation throughout the convection zone since it appears that a modest differential rotation can be supported by an anisotropic convective viscosity (Cocke, 1967). If this theory is applicable we would generally require that the aniso-tropy be a function of both radius and latitude in order to give the angular velocity both a radial latitude dependence. In view of the large differences in mixing length and convective viscosity the convection zone such a possibility can not be excluded.

The perihelion advance of Mercury, $\Delta\bar{\omega}$ radians per orbital period, caused by the mass quadrupole moment of the rotationally distorted sun can be written (Anand and Fahlman, 1968)

$$\Delta\bar{\omega} = \frac{3(C-A)}{M} D.$$

(2)

Here C and A are the principal moments of inertia of the sun in the equatorial plane and along the rotation axes, M is the solar mass and D is a quantity which depends only on the elements of Mercury's orbit. If the outer layers of the sun are uniformly rotating then it can be shown that σ is analytically related to $(C-A)$ and hence to the perihelion advance. For non-uniform rotation, this is not the case and we must cal-culate $(C-A)$ independently of σ to obtain $\Delta\bar{\omega}$.

The models have been obtained by employing the technique devised by Stoeckly (1965) which gives accurate results for both fast uniform rotation and the restricted type of non-uniform rotation given by Equation (1). The Stoeckly scheme replaces the differential equations by difference equations which are then linearized and the solution is obtained by iteration. The actual dependent variable is the difference between two successive approximations to the physical variable. Critical uniformly rotating models were obtained (Naylor, 1968) for the polytropic indices 1.5, 2.0 and 3.0. The results agree with those of James (1964) to better than 1%.

In the present work we have required that the dimensionless equatorial angular velocity of the models be equivalent to the solar value (Allen, 1963). The first model in each series was static and provided a reference for the differentially rotating models which were obtained by increasing the parameter c from $c=0$ in steps of 0.05. The double integration that is necessary to obtain the value of $C-A$ was accomplished by using an eleven point Gauss-Legendre quadrature for the angular dimension and a repeated application (at 40–50 points) of Simpson's rule for the radial dimension. The final results should be accurate to a few percent or better.

In Table I we give values of the oblateness, σ and the perihelion advance, $\Delta\bar{\omega}$, now measured in sec of arc per century, for various values of c and for three different polytropic indices, n, which should bracket the solar structure. In general for the same value of c, polytropes with less mass concentration are more oblate and have higher quadrupole moments than those which are more centrally condensed. For the case $n=3$, the maximum oblateness consistent with the Rayleigh stability criterion is 3.54×10^{-5}. Further calculations for this case show that we require $c=1.25$ to obtain

TABLE I
Oblateness and perihelion advance

	$n = 2.5$		$n = 3.0$		$n = 3.25$	
c	$\sigma \times 10^{-5}$	$\Delta\bar{\omega}$	$\sigma \times 10^{-5}$	$\Delta\bar{\omega}$	$\sigma \times 10^{-5}$	$\Delta\bar{\omega}$
0.85	3.11	0.23	2.92	0.10	2.84	0.06
0.90	3.31	0.24	3.15	0.11	3.02	0.06
0.95	3.50	0.26	3.31	0.11	3.33	0.07
1.00	3.81	0.28	3.54	0.12	3.46	0.08

$\sigma = 5 \times 10^{-5}$. This value of c corresponds to the centre rotating 3.5 times faster than the equator. In all cases the perihelion advance of Mercury due to the rotational distortion is entirely negligible.

Thus we have at least shown that it is possible for a stellar surface to be rotating slowly at the equator and still have a fairly large oblateness but a small mass quadrupole moment. These results are not directly applicable to the sun because of our use of the Stoeckly law as discussed earlier. Nevertheless, they do not contradict the results of Dicke and Goldenberg (1967b) who used a simple perturbation technique to show that the effect of differential rotation on the solar surface is unimportant. In their perturbation analysis it is assumed that the angular velocity function has approximate spherical symmetry in the outer layers and can be represented as a two-term Legendre expansion. If the angular velocity law deviates from spherical symmetry so that the higher order terms in the Legendre expansion are not negligible, as is the case for the Stoeckly law, their analysis is no longer applicable. However, in view of the observed solar rotation law, it is difficult to see how the case discussed here can be relevant to the sun. We can only conclude that if the solar oblateness is to be explained by differential rotation then we must have a non-spherical rotation law but to find such a law that is also compatible with the observed solar rotation appears to be very difficult.

Acknowledgements

This research was supported in part by the National Research Council of Canada and the Department of University Affairs of the Province of Ontario.

References

Allen, C. W.: 1963, *Astrophysical Quantities*, Athlone Press, London.
Anand, S. P. S.: 1968, *Astrophys. J.* **153**, 135.
Anand, S. P. S. and Fahlman, G. G.: 1968, *Icarus* **8**, 492.
Chandrasekhar, S.: 1933, *Monthly Notices Roy. Astron. Soc.* **93**, 390.
Cocke, W. J.: 1967, *Astrophys. J.* **150**, 1041.
Dicke, R. H.: 1967, *Science* **157**, 960.
Dicke, R. H. and Goldenberg, H. Mark: 1967a, *Phys. Rev. Letters* **18**, 313.
Dicke, R. H. and Goldenberg, H. Mark: 1967b, *Nature* **214**, 1294.
Fricke, K.: 1969, *Astrophys. Letters* **3**, 219.

Goldreich, P. and Schubert, G.: 1967, *Astrophys. J.* **150**, 571.
Goldreich, P. and Schubert, G.: 1968, *Astrophys. J.* **154**, 1005.
James, R. A.: 1964, *Astrophys. J.* **140**, 552.
Naylor, M. D. T.: 1968, unpublished.
Stoeckly, R.: 1965, *Astrophys. J.* **142**, 208.

Discussion

Dicke: The rotating model considered by Mr. Fahlman has a negligibly small gravitational quadrupole moment, and the oblateness is determined wholly by rotation at the surface. Two different models with the same surface rotation but different internal rotations will have the same oblateness if their quadrupole moments are negligibly small. Mr. Fahlman's models give more rapid rotation at the pole and are not applicable to the sun. For the observed rotation the corresponding oblateness is 0.81×10^{-5}.